Biochemistry Explained

Biochemistry Explained

A Practical Guide to Learning Biochemistry

Thomas Millar
University of Western Sydney, Nepean
Kingswood, Australia

Taylor & Francis
Taylor & Francis Group
Boca Raton London New York

CRC is an imprint of the Taylor & Francis Group,
an informa business

Published in 2002 by
CRC Press
Taylor & Francis Group
6000 Broken Sound Parkway NW, Suite 300
Boca Raton, FL 33487-2742

International Standard Book Number-10: 0-415-29942-X (Softcover)
International Standard Book Number-13: 978-0-415-29942-8 (Softcover)

Library of Congress Cataloging-in-Publication Data
Catalog record is available from the Library of Congress

Taylor & Francis Group is the Academic Division of Informa plc.

**Visit the Taylor & Francis Web site at
http://www.taylorandfrancis.com**

**and the CRC Press Web site at
http://www.crcpress.com**

Contents

To the authors Mary White and Densey Clyne, and the media personality Clive Robertson. All critically examine themselves and the world around them, remove facades, and reveal the fundamentals whether they be pleasing or distasteful. They express themselves with a passion that engages their audiences and readers, and a simplicity that appears to mask the real depth of their subjects. If my readers judge me to have emulated these qualities in but a small way, I will be pleased.

Preface

This book approaches the learning of biochemistry by establishing its fundamentals and from this standpoint complex reactions can thereby be fully appreciated and understood. The book is designed to be interactive (and hopefully not put you to sleep!); questions are embedded in the text and answers lie immediately below.

What promoted this approach to the learning of biochemistry was an observation that as an undergraduate dutifully learning the structures of ATP, GTP, DNA and sugars, I never realised their common ground. This led me to question at a later stage "Why didn't someone tell me?". Often the obvious is not stated, nor is a pathway for learning mapped. Without establishing clear priorities, students often treat all biochemical concepts with equal importance and as having no beginning or end.

Parts of the book contain information which is rote learning material, for example, the structure of an alcohol. Such information needs to be learnt and practised until it becomes part of your vocabulary. It is only with these basics that a logic can be built. Other parts of the book contain more conceptual material. Many of you will already know these concepts, but not realise their significance.

Most importantly, enjoy learning this subject: biochemistry is fun. Learn from and laugh at your mistakes; after all, you are a student and no one expects you to be perfect — not until the exams at least! If you do this, you will learn the most important lesson of all: to have confidence in your knowledge.

Using the techniques outlined in this book, my most recalcitrant student ever, who was absolutely determined to make learning biochemistry an unendurable torture (for me mainly), towards the end of the course commented: "I really hate to admit this, but biochemistry is really enjoyable and it has made understanding all of my other subjects so much easier". I hope you feel the same way.

When exploring this text you will note a variety of different structural features have been used to help learning:

There are two types of questions. The first type of question actually forms part of the text and as such the answer is provided immediately below it. This allows you to think about the answer and then immediately have it available without the inconvenience of looking elsewhere. To make best use of this type of question, it is recommended that a piece of paper is used to cover the answers while reading the text. This type of question is also useful during exam preparation. Once again, the answer can be covered with a blank sheet of paper and your attempt at the answer can be done on this piece of paper. The answer can then be checked — 'look, cover, write, check'. The second type of question is numbered and the answers are given at the end of the chapter. It is important to note that many of the answers give additional information and explanations, and therefore doing these questions and referring to the answers as you progress through the text is essential in order to gain maximum benefit from the text.

Rather than a glossary, particular terminology is explained as it appears in the text under the heading nomenclature. There are sections in the book which are called logic process. These sections outline a strategy for thinking in a particular manner so that learning is enhanced. In addition to these there are boxed sections called major points. For the most part these major points are rather simple, but they have been selected as key pieces of knowledge to enhance the understanding of biochemistry.

More complex examples of biochemical processes or structures are boxed under the heading of higher level. These should be read by all students, and if you are right on top of the subject, then you should have no difficulty learning these higher level elements. If you are running short of study-time before an exam, then detailed study of these areas should be left till last.

For the Lecturer

I recommended that this text be used in conjunction with the classic biochemistry texts, not to replace them. I have found that students actually prefer to use such texts as references only, because as a learning tool they can be difficult to understand. As lecturers, we often fail to appreciate the time constraints placed on students that limit their time to devote to studying one subject. This book is designed so that over the semester or term a student builds up knowledge, not just programs their minds.

In order to keep the book focussed on the foundations of biochemistry, some generalisations have been made at the expense of complete accuracy, and the myriad of exceptions or alternate pathways have been omitted. This can be used to advantage because it provides scope for additional information to be delivered in lectures.

Several elements which you may have expected to see in such a text have been omitted. These include:

1.   Reference to various notables in the field of biochemistry. This allows the book to retain its focus, which is to stimulate interest on biochemical fundamentals. Clearly, the famous researchers in biochemistry can be introduced to the students in lectures and by way of investigative assignment.

2.   The paucity of genetics and molecular biology that now comprises almost half of the current biochemistry texts. In most institutions, these areas have separated from biochemistry and grown into subjects in their own right, e.g. molecular biology, genetics and molecular biology. These subjects consequently do not use a biochemistry text but a dedicated text.

3.   The presentation of special examples in call-out boxes. Students often overlook such examples, with a belief that the crucial learning material is contained in the body of text. By omitting such 'boxes', the lecturer has the opportunity to introduce such information to the students as the specific needs dictate.

4.   Detailed discussions of enthalpy and entropy and its relationship to energy transfer in biochemical reactions. Without doubt, this information is important, but for the most part, it is enough to know if a reaction is favoured energetically or not. When specific quantities are given to these energy transfers in a text, students seem to feel that learning these quantities is the most important aspect of this topic rather than the principles of energy transfer.

Higher level sections have been included to assist students in determining the material and learning required to place them in this realm. For some students this will form an encouragement as they will perceive that they are coping with higher level material and will continue to strive to attain this level. Such knowledge also helps with time management. To a degree determining what is 'higher' level information is subjective. Some of the areas I have listed as higher level are used as a pedagogic strategy. Strictly speaking, I would regard some of them as only slightly harder.

Above all, I have tried in my approach to make the learning of biochemistry interesting and fun, and hence the chatty style. It is designed for students to develop confidence in applying their knowledge and to learn biochemistry for life, not just for the end of the semester exam which is usually followed by rapid biochemical amnesia.

1 Fundamental Concepts and an Introduction to the Cell

In this chapter you will:

- be introduced to the idea that lipid and water solubility is a major foundation stone of biochemistry
- learn which chemical structures determine water and lipid solubility
- be introduced to the terms hydrophobic and hydrophilic
- revise subcellular compartments and their functions
- learn the function of the cytoskeleton.

Lipid and water solubility

It will not be a surprise to hear that the major component of our bodies is water. For example, 75% of muscle tissue and 60% of red blood cells is water. What most of you have probably not considered is how do we keep the water inside us and why doesn't it just run away? Clearly, it is not contained in something that is water soluble, but equally, it is not contained in glass or plastic either. The solution to this problem is found by knowing what happens if you mix oil and water together – they do not mix and oil, which is less dense than water, goes to the top. This means that it would be possible to contain water in a bag of oil and fundamentally, this is what occurs in our bodies. Our cells are lipid bags containing water.

Nomenclature: Another name for oil or fat is lipid.

> **Major point 1:** Lipids and water do not mix.

Major point 1 is the basis for 90% of your biochemistry. We will refer to it time and time again.

> **Major point 2:** Lipids are the basis of cellular membranes.

If we now combine major point 1 with major point 2, we come up with another major point.

> **Major point 3:** For something to readily cross a cell membrane it must be lipid soluble.

The significance of these points lies in the concept that the body uses an aqueous environment, the blood, to transport fuels, nutrients, and chemical signals such as hormones and neurotransmitters. For the most part, these substances are water soluble. Cell membranes, being composed mainly of lipids, provide a barrier to these chemicals, and therefore there must be special (bio)chemical mechanisms for the nutrients or signals to pass across the cell membrane. By contrast, the blood also needs to be able to transport lipids or lipid-soluble substances, and again there is a (bio)chemical process that enables this to happen. It is also clear that a cell will be signaled from time to time to carry out a different function and for this to occur, something must change within a cell. Most often, the something is a small part of a molecule becoming more or less water soluble. This small change in water solubility causes the whole molecule to change its shape and consequently its activity. Therefore, it is essential to understand which chemical structures are likely to be water soluble and which are lipid soluble.

> ✱ **Logic process:** Whenever you look at the structure of biochemical molecules determine if they are lipid or water soluble. If they are lipid soluble, they will readily cross a cell membrane and if they are water soluble, they will not.

Which of the following are lipid or water soluble?
(a) sodium chloride
(b) cholesterol, oestrogen, testosterone
(c) vitamin A, vitamin D, vitamin E, vitamin K
(d) vitamin B, vitamin C
(e) ointment

Answer: *Sodium chloride and the vitamins B and C are water soluble, whereas cholesterol, oestrogen, testosterone and the other vitamins are lipid soluble. Ointments are lipid soluble so that the medication they contain passes across cell membranes (i.e. is absorbed through the skin).*

From all of these, you may have only known with certainty that sodium chloride (table salt) is water soluble. Other substances which are water soluble are other salts such as potassium chloride, acids such as acetic acid and bases such as sodium hydroxide. Therefore, in terms of their chemical structure, what do these substances have in common that makes them water soluble?

> **Q&A 1:** Draw the chemical structures of sodium chloride, acetic acid, and sodium hydroxide.

You will be aware from your previous chemistry that all of these substances prefer to exist in their ionic or charged form.

Now let us examine the structures of some lipid-soluble substances. These often have a long chain of carbon atoms as their backbone, others are ring structures. Therefore, the molecule carries no charge or the vast proportion of the molecule carries no charge.

By comparing the structures of the water-soluble salts and acids with the fat-soluble substances, there is a major difference that determines water or lipid solubility.

Fatty acid (stearic acid)

Fatty acid (oleic acid)

Phenanthrene

Steroid
(perhydrocylopentanophenanthrene)

Major point 4: A substance that readily dissolves in water is charged and a substance that readily dissolves in lipids is uncharged.

Some molecules are particularly interesting because part of their structure is charged and part is uncharged. Examine the following molecules and determine which part would be water soluble and which part would be lipid soluble.

A soap

A detergent
sodium dodecyl sulphate (SDS)

Answer: *The parts that are circled are water soluble because they are charged and the rest of the molecule is non-charged and would dissolve easily in lipid (hydrophobic or lipophilic). As indicated, one is a soap and the other a detergent. This property of both water and lipid solubility enables soaps and detergents to be used in water to remove grease from surfaces.*

Nomenclature: hydro = water; phobia = fear; lipo = fat; philus = beloved.

Q&A 2: Use the Latin terms above to form a word that means: fear of water; water loving; fear of fats; fat loving.

* Logic process: You would already know that glucose is our main energy source and is needed by cells in high concentrations. To get glucose inside the cell, it must be helped from its aqueous environment where it is in low concentration, across the lipid membrane to the inside of the cell where it is in high concentration. The structure of glucose will be dealt with in a later chapter, but it is enough here to realise that it is strongly water soluble and weakly lipid soluble. Being partially lipid soluble, and with a high concentration of glucose inside a cell, there is a strong tendency for glucose to leak back out of the cell towards extracellular fluid where it is in low concentration. Having spent a large amount of effort getting the glucose to the inside of the cell, the cell does not want the glucose to leak out again. How can it be prevented? The cell makes the glucose less lipid soluble by <u>adding negative charges</u> to it. The negative charges come from a phosphate group that is transferred from adenosine triphosphate (ATP) which is found inside cells.

Add phosphate here

Puts a negative charge here

Glucose

Glucose-6-phosphate
very water soluble

This example of a biochemical process illustrates how water and lipid solubility can be exploited by a cell to its advantage, i.e. to retain glucose inside the cell. In the next section, as the structure of a cell is reviewed, we will see how this property of lipid and water solubility is exploited even further.

The structure of a cell

A cell is more than a lipid bag containing water as was described earlier. It is really an outer lipid bag or cell membrane containing several smaller lipid bags or compartments. One such bag, or <u>subcellular compartment,</u> is the nucleus.

Q&A 3. What is contained within the nucleus?

Q&A 4. List 4 other subcellular compartments

Note that each of the subcellular compartments is surrounded by a lipid membrane and sometimes two. These membranes, like the cell membrane, are neither freely permeable to positive nor negative ions because these are water soluble.

> ✱ **Logic process:** This simple fact means that you can package different substances into different cellular compartments and therefore different functions can be carried out in the different subcellular compartments. One advantage is that a chemical can be synthesised in one compartment at the same time it is being broken down in another.

> **MAJOR POINT 5:** Different subcellular compartments carry out different functions.

Major point 5 means that (1) a cell can carry out a variety of processes at the same time, and (2) substances are transported from one subcellular compartment to another. A large amount of biochemistry is examining how and why molecules are transported across the membranes of the subcellular compartments and what controls this process.

> **Q&A 5:** Test your knowledge of subcellular compartments. What is the function of:
> a. the nucleus?
> b. mitochondria?
> c. smooth endoplasmic reticulum?
> d. rough endoplasmic reticulum?
> e. Golgi apparatus?
> f. peroxisomes?

In addition to the functions carried out by the subcellular compartments, cells have a particular shape. They can also change their shape and move when required. How can a cell have a particular shape, yet change it? You know this occurs with epithelial cells. When you remove epithelial cells by cutting your skin, the cells surrounding the wound eventually replace the damaged ones. The only way they can replace the missing cells is to migrate (move) into the wound. Muscle cells also move. They become shorter during contraction.

The shape of a cell and its ability to move comes from a group of structural chemicals called cytoskeletal (cell skeleton) proteins (see Chapter 5 for details on protein structure). Some very important examples are actin and myosin. These interact to cause movement in cells, particularly muscle cells. Others, fodrin, filamin and actin act together to give a cell its shape. There is an area of study in biochemistry that centres on how and why cells change their shape and the controlling mechanisms for these processes.

References and further reading

Alberts, B., Bray, D., Lewis, J., Raff, M., Roberts, K. and Watson, J.D. (1994) *Molecular Biology of the Cell*, 3rd edn. New York: Garland Publishing.

Brennen, P.J. and Nikaido, H (1995) The envelope of mycobacteria. *Annu. Rev. Biochem.* **64**:29-63.

Campbell, N.A. (1996) *Biology*, 4th edn. Redwood City, CA: Benjamin/Cummings.

Kellogg, P.J., Moriz, M. and Alberts, B.M. (1994) The centrosome and cellular organization. *Annu. Rev. Biochem.*, **63**:639–674.

Goodsell, D.S. (1991) Inside a living cell. *Trends Biochem. Sci.* **11**:437-485.

Kleinsmith, J.L. and Kish, V.M. (1995) *Principles of Cell and Molecular Biology*, 2nd edn. New York: Harper Collins College Publishers.

Luna, E.J. and Hitt, A.L. (1992) Cytoskeletal-plasma membrane interactions. *Science* **258**:955-964.

Rothman, J.E. (1992) The compartmental organization of the Golgi apparatus. *Sci. Amer.*, **253(3)**:74–89.

Sedava, D.E. (1993) *Cell Biology: Organelle Structure and Function*. Boston: Jones and Bartlett Publishers.

Weber, K. and Osborn, M. (1985) The molecules of the cell matrix. *Sci. Amer.*, **244(6)**:100–120.

Q&A Answers

1

Na^+ Cl^-	CH_3COO^- H^+	Na^+ OH^-
sodium chloride	acetic acid	sodium hydroxide

2 Hydrophobic: hydro = water, phobic = fearing, i.e. water insoluble.
- Hydrophilic: philic = loving, i.e. water soluble.
- Lipophobic: lipo = fats or oils, phobic fearing, i.e. fat insoluble.
- Lipophilic: fat loving, i.e. fat soluble.

3 DNA (deoxyribonucleic acid) is contained within a cell nucleus. Note that it is an acid and hence very water soluble.

4 Subcellular compartments are:
- Mitochondria
- Smooth and rough endoplasmic reticulum
- Golgi apparatus
- Lysosomes

5 a. The nucleus contains the genetic code for the animal, the information for producing proteins.

b. Mitochondria are the subcellular compartments where fuel (carbohydrates) is oxidised (burnt) and the energy released from this is converted to chemical bonds (usually ATP).

c. Smooth endoplasmic reticulum is where proteins are modified by having lipids or sugars added to them. This is called <u>post-translational modification</u> of proteins. An

other important function of smooth endoplasmic reticulum is that there are very high concentrations of
calcium (Ca^{2+}) within the smooth endoplasmic reticulum compartment and hence isolated from the cytoplasm of the cell. When some cells are activated, this
calcium can be released into the cytoplasm of the cell
where it binds to specific proteins, giving them a positive charge, and a resultant change of shape and function. This is dealt with in detail in Chapter 15.

d. Rough endoplasmic reticulum comprises membranes
(like the smooth endoplasmic reticulum) studded with
ribosomes. This is where proteins are made.

e. The Golgi apparatus is where many newly made proteins are packaged and further modified for shipping
to the cellular membrane for excretion or for incorporation into intracellular vesicles or lysosomes.

f. Peroxisomes are vesicles containing proteins for breaking down oxygen free radicals. These are highly reactive species of oxygen, which are formed in chemical
reactions in cells and would cause considerable damage
to the cell if they were not broken down. It is much
safer for the cell to separate these from the rest of the
cell by putting them in a separate compartment.

2 Basic Biochemical Structures

In this chapter you will:

- learn to look for functional groups in molecules rather than looking at molecules in their entirety
- learn to identify the most important functional groups:
  - the hydroxyl group;
  - the carboxyl (carboxylate) group;
  - the phosphyryl (phosphate) group;
  - the amino group.
- predict possible linkages between molecules based on the functional groups present. Possible linkages are:
  - an ester bond;
  - an amide bond;
  - an ether bond;
  - a phosphoester bond;
  - a phosphodiester bond;
  - glycosidic bonds: O-glycosidic and N-glycosidic.
- learn the similarity between these bonds
- recognise that phosphorylation of a molecule (placing a phosphate on it) places a large negative charge on it and makes it particularly water soluble.

Esters and their cousins

Esters

The most important structure you can ever learn in biochemistry, I believe, is the ester bond. Learning this structure and how it is formed will enable you to understand 95% of all biochemical interactions and structures. To learn the structure of an ester you have to know the structure of 2 other molecules: an acid; and an alcohol.

Test your knowledge by drawing the general structure of a carboxylic acid, and acetic acid.

Answer:

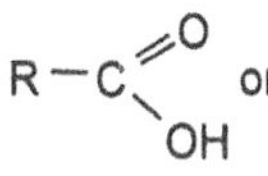

$R-C \overset{O}{\underset{OH}{\lessgtr}}$ or R-COOH $H_3C-C\overset{O}{\underset{OH}{\lessgtr}}$ or CH_3COOH

| general carboxylic acid | | acetic acid |

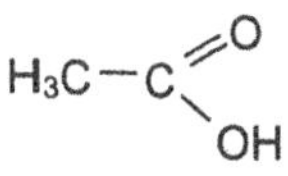

Nomenclature: The -COOH is a carboxyl group – <u>learn this.</u> Carboxylic or organic acids contain this functional group.

Draw the general structure of an alcohol and ethanol.

Answer:

HO – R or ROH   HO – CH_2CH_3 or CH_2CH_3OH

| general alcohol | | ethanol |

Nomenclature: The -OH is called an hydroxyl group – <u>learn this.</u>

Q&A 1: For practice, which of the following are alcohols?

A
H_2C-OH
|
$HC-OH$
|
H_2C-OH

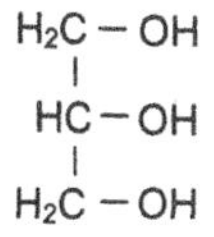

B
$HO-CH_2-CH_2-\overset{\overset{+}{CH_3}}{\underset{CH_3}{N}}-CH_3$

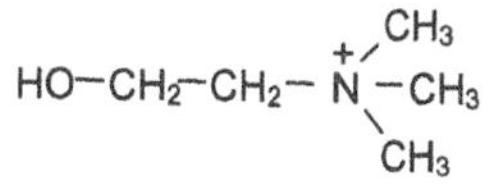

C
$NH_2-\overset{\overset{H}{|}}{\underset{\underset{OH}{|}}{\underset{C=O}{C}}}-CH_2-CH_2-C\overset{O}{\lessgtr}-OH$

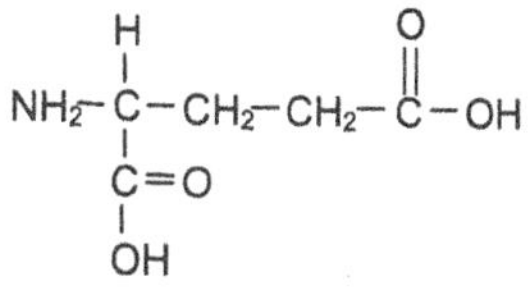

D
$NH_2-\overset{\overset{H}{|}}{\underset{\underset{OH}{|}}{\underset{C=O}{C}}}-CH_2-OH$

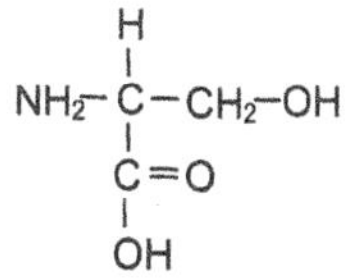

Q&A 2: Indicate which of the following are acids?

A
$\underset{OH}{CH_2}-\underset{OH}{CH}-C\overset{O}{\underset{H}{\lessgtr}}$

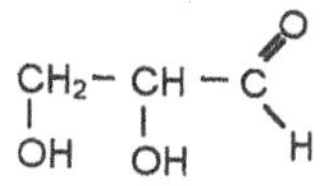

B
$CH_2-\underset{NH_2}{CH}-C\overset{O}{\underset{OH}{\lessgtr}}$

C
$CH_3-\underset{O}{C}-C\overset{O}{\underset{OH}{\lessgtr}}$

D
$CH_3-C\overset{O}{\underset{OH}{\lessgtr}}$

E
$\overset{O}{\underset{HO}{C}}-CH_2-\underset{O}{\overset{||}{C}}-C\overset{O}{\underset{OH}{\lessgtr}}$

F
$\overset{O}{\underset{HO}{C}}-CH_2-\underset{\underset{OH}{|}}{C}-C\overset{O}{\underset{OH}{\lessgtr}}$ with $\overset{O}{\underset{CH_2}{C}}\nearrow OH$

G
$\underset{HO}{\overset{H}{C}H}-\underset{O}{C}-\underset{OH}{\overset{H}{C}H}$

H
$\underset{HO}{\overset{H}{C}H}-\underset{O}{C}-\underset{O}{\overset{H}{C}H}-\underset{OH}{\overset{O}{P}}-OH$

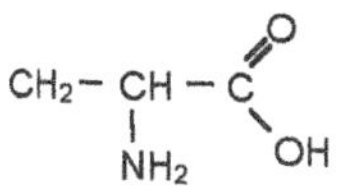

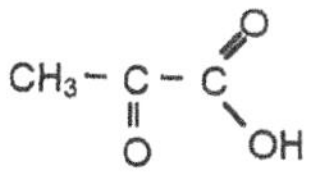

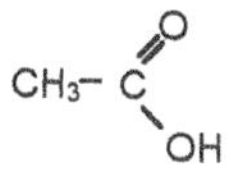

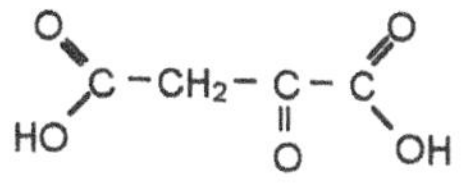

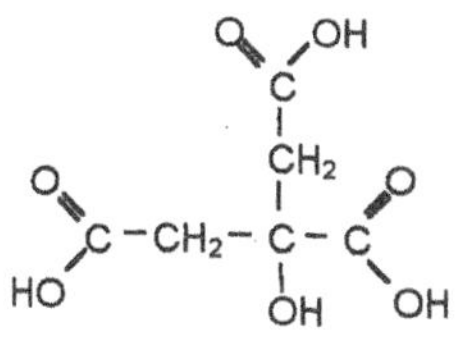

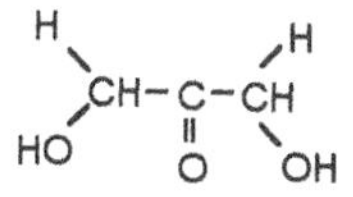

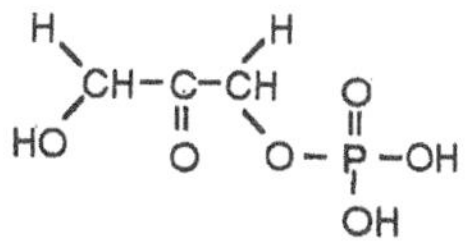

To form an ester bond, we simply condense (form water) the acid with the alcohol. In the following diagram, carboxylic acid is condensed with alcohol to form an ester bond and water:

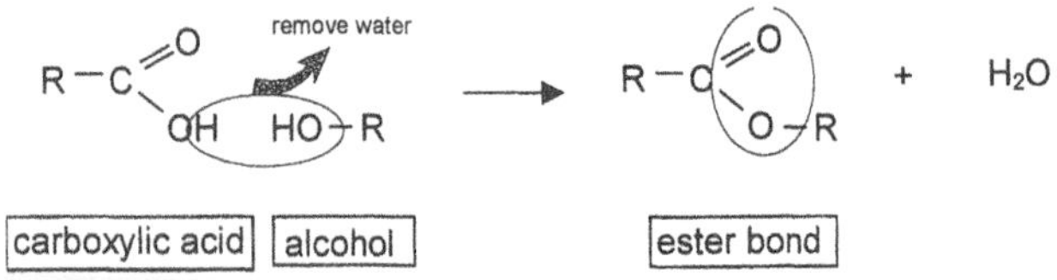

carboxylic acid   alcohol   ester bond

> **MAJOR POINT 6:** An ester is formed by condensing an acid with an alcohol.

Biochemical examples of ester bonds

Clearly from above, biochemicals that contain ester bonds must have basic structural elements of alcohols and acids. Examine the following molecule, glycerol. It comprises 3 carbons with an hydroxyl group at each carbon i.e. a poly-alcohol. Although this molecule rarely occurs biochemically, it is one of the most important structures to learn for biochemistry because it is the structural basis for many other biochemicals.

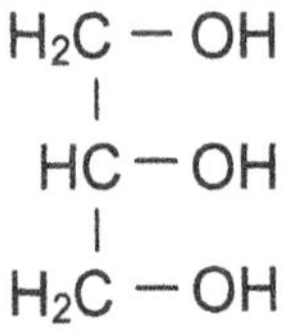

$$H_2C - OH$$
$$HC - OH$$
$$H_2C - OH$$

It is exciting because it has the potential to form 3 ester bonds, not just one. To form an ester bond, it needs an acid (or 2 or 3 acids).

> **Q&A 3:** Draw a carboxylic acid with 14 carbons (including the carbon of the carboxyl group) and a single bond between each carbon so that it is saturated with hydrogens. Remember that each carbon can only form 4 bonds.

When such a long uncharged chain forms would you expect it to be more water or more lipid soluble?

Answer: *Lipid soluble, because there is very little charge on such a molecule. Therefore a good name for such an acid would be a* **fatty acid.**

> **Q&A 4:** Draw a fatty acid with 20 carbons with double bonds between C5 and C6, C8 and C9, C11 and C12 and between C14 and C15 with carbon 1 being the carbon of the carboxylic acid.

This fatty acid is not saturated. It has double bonds so could theoretically take more H atoms to saturate the molecule fully. Since it is not saturated at more than 1 place (4 double bonds) it is called a polyunsaturated fatty acid.

> Nomenclature: A double bond between carbon atoms in a molecule is represented in its chemical name by 'ene'.

Therefore, something which has 4 double bonds is a tetraene (tetraenoic acid).

> **Q&A 5:** a What would a fatty acid with 3 double bonds be called? b. What would a fatty acid with 6 double bonds be called?

Esterify (i.e. form an ester bond between) the hydroxyl (OH) group on the first carbon of the glycerol molecule with the 14 carbon carboxylic acid you drew above. Do the same to the hydroxyl group of the second carbon of glycerol but this time esterify the 20 carbon polyunsaturated fatty acid to it.

Answer:

Given that the backbone for this molecule is glycerol and we have added 2 long carbon chains or acyl groups (an acyl group is just a chain of carbon atoms joined together and spare binding sites taken by hydrogen atoms) to it, what would be a suitable name for it?

Answer: *Diacylglycerol. This molecule is an important intracellular messenger.*

> **Q&A 6:** Note that the diacylglycerol still has an hydroxyl group free, hence another acid can be esterified to it. Esterify a carboxylic acid (16C atoms long) to the remaining hydroxyl group and name the molecule.

The phosphyryl group

Consider the following molecule, phosphoric acid.

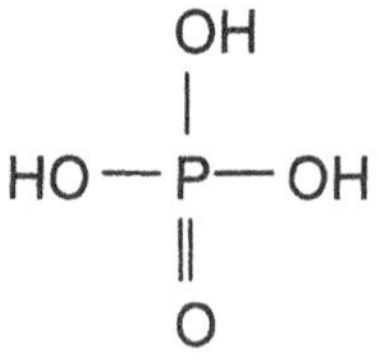

It is an acid because it can easily lose H^+ ions as illustrated below.

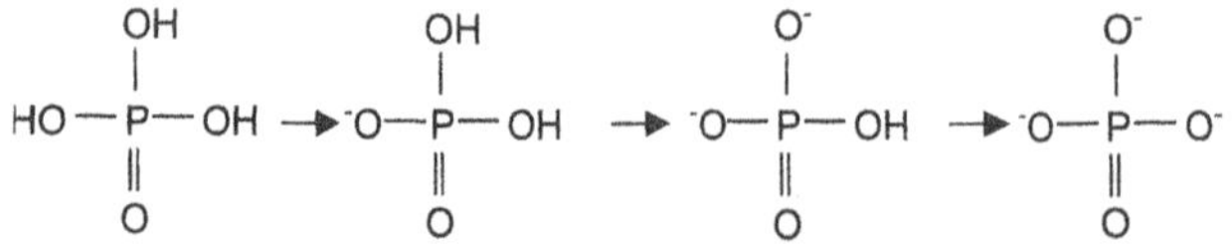

The last structure is the form this molecule normally takes at pH 7.4 (physiological pH).

> Nomenclature: When an acid loses its H^+ ions its name takes on an 'ate' form e.g. acetic acid becomes acetate, glutamic acid becomes glutamate, a carboxyl group becomes a carboxylate group.

Q&A 7: What would phosphoric acid and aspartic acid be called when they lose H^+ ions?

Nomenclature: Note that an hydrogen atom is made of a proton and an electron. An hydrogen ion (H^+) has lost an electron. Therefore an H^+ is really a proton. Often you will read about protonation and deprotonation. These refer to gaining or losing an H^+.

Two aspects of the phosphate molecule make it remarkable.

- It is relatively easy for the molecule to form 2 ester bonds (not just one, like carboxylic acids).
- It has a strong negative charge.

The phosphoester bond

Examine the following molecules: phosphoric acid (phosphate), on the left, and the sugar ribose on the right:

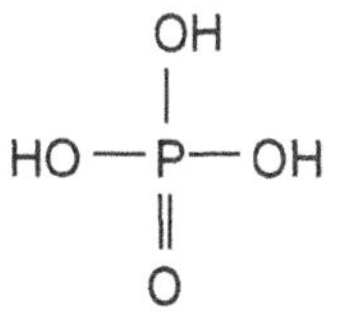

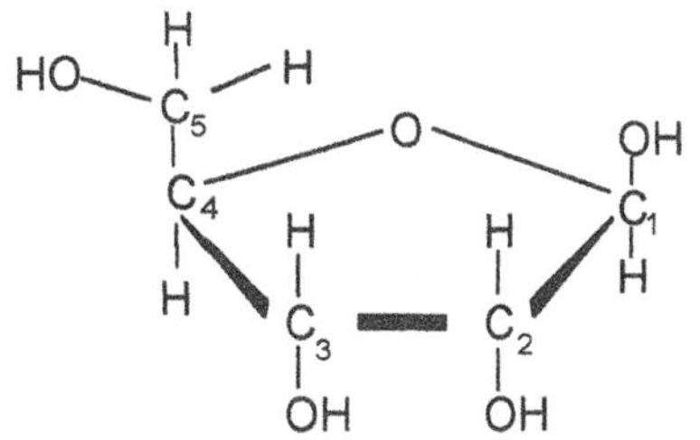

Ribose has a large number of hydroxyl groups, 1 at each of the carbon atoms except for C4, and therefore phosphoric acid can form an ester bond with ribose at any of these hydroxyl groups. Exciting, isn't it?

Esterify phosphoric acid to the hydroxyl group on carbon number 5 of ribose (ribose).

Answer:

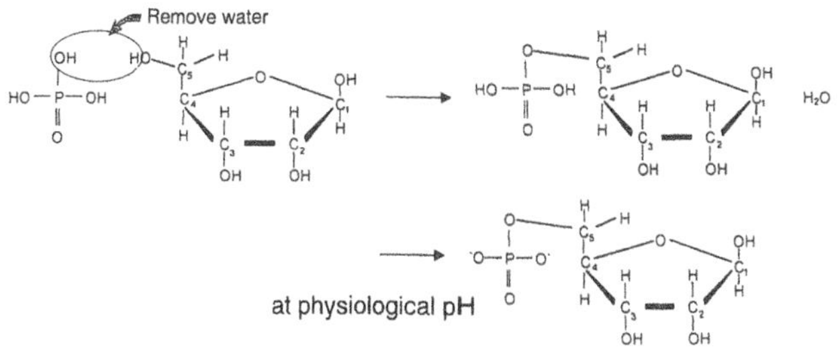

A phosphoester is formed by condensation (removing water), and at physiological pH, the phosphate group will have a negative charge (last diagram). The effect of this phosphorylation of ribose is to put a massive negative charge on it, making it highly water soluble and virtually impossible for it to pass across a lipid membrane. The new molecule is ribose monophosphate.

MAJOR POINT 7: A phosphoester bond is formed by condensation of phosphoric acid with an alcohol.

MAJOR POINT 8: Condensing a phosphate to a molecule (phosphoester bond) gives it a massive negative charge and hence makes it hydrophylic.

The phosphodiester bond

We have not finished with the above reaction. The phosphate can form another ester bond with the same ribose group. The closest hydroxyl group in this case being at C3:

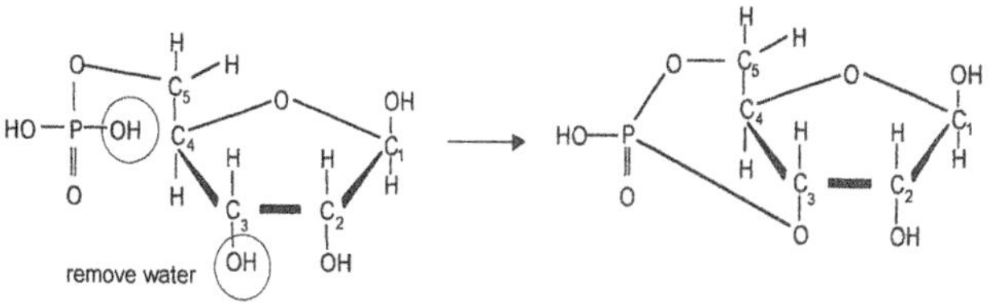

This produces a cyclic ribose monophosphate. The type of bond is now a phosphodiester bond because 2 ester bonds have been formed. The new phosphate ring, which has formed, makes this a very inflexible structure, and it still caries a negative charge.

Nomenclature: There is 1 phosphate (monophosphate) attached to the ribose and so it is a ribose monophosphate. However, the phosphate group forms a ring by attaching at 2 sites and so it becomes a cyclic monophosphate. Since the phosphate could have attached at other sites such as C_1 or C_2, we can give a fuller description of the molecule and so its full name is 3,5-cyclic ribose monophosphate.

Let us try some other examples. Could 2 ribose molecules be linked together by a phosphate bridge?. Of course they could by simply forming a phosphoester bond with an hydroxyl group of one ribose and another ester bond with an hydroxyl group of the second ribose, as illustrated:

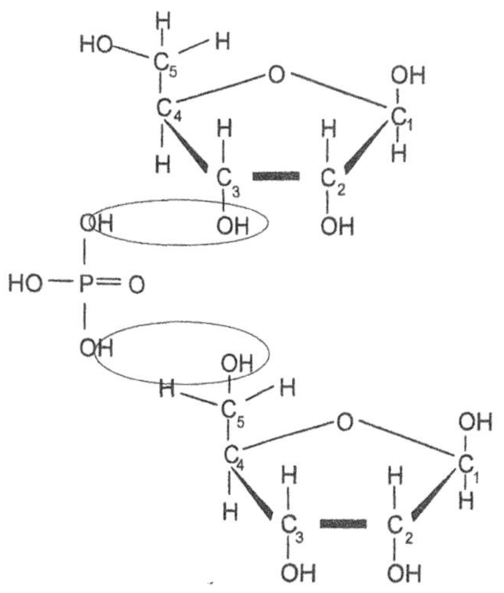

Q&A 8: Draw the ester linkages by condensing the phosphate with the 2 ribose molecules.

Although other bridges could have been formed (e.g. from C5 to C2, C1 or C5) this is a 5,3 bridge because it links C5 of one molecule with C3 of the other and is hence a 5,3 bridge. This particular formation is very important. **It is the backbone structure of RNA** and also DNA (well almost). For these molecules, there is a series of riboses, not just 2, and each is linked to the next via a phosphodiester bond between the C5 of one ribose to the C3 of the next ribose. It also means that DNA and RNA have a negative charge running down the outside of them due to the phosphate groups.

Reconsider a diacylglycerol such as was drawn earlier.

In this form, it has no charge and hence is fat soluble. Could it be made water soluble? Remember to do this a charge must be added to it. To add a charge to it we could esterify a phosphate to the free hydroxyl group, as illustrated:

Notice that the part of the molecule which is circled is charged and the rest is uncharged. This means that because it is a big molecule, one part is water soluble and the other part is lipid soluble. This is a phosphoglyceride and is a major component of cell membranes. (See Chapter 11)

Nomenclature: An enzyme that catalyses the breaking of an ester bond is an esterase. A well known example is acetylcholinesterase which breaks acetylcholine (a neurotransmitter) into acetic acid and choline.

Let us examine how acetylcholinesterase catalyses the breakdown acetylcholine. Water is added to the ester bond in acetylcholine to break the bond. Because the bond is broken by water we say that it is hydrolysed.

Consider the molecule we drew earlier, cyclic ribose monophosphate. There is an enzyme that catalyses the hydrolysis of 1 of the ester bonds of the <u>phosphodiester</u> (a diester because it forms 2 ester bonds) as illustrated.

Q&A 9: What would be a suitable name for an enzyme that catalyses the breakdown of a phosphodiester bond?

Nomenclature: An enzyme that catalyses the breaking of a phosphodiester bond is a phosphodiesterase. Such enzymes are very important in breaking cyclic adenosine monophosphate (cAMP) into adenosine monophosphate (AMP).

The amide bond – a cousin of the ester bond

We can extend the concept of an ester bond further. An amide bond is almost identical to an ester bond. In this case an acid is condensed to an amine (instead of an alcohol). To understand this we must know the structure of the functional group, an amine.

Draw an amine.

Answer:

As with an ester bond, the amide bond is formed by condensation; in this case the amine is condensed with an acid, as illustrated:

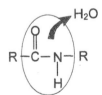

MAJOR POINT 9: An amide bond is formed by attaching an amine to an acid by removing water.

Nomenclature: A nitrogen attached to 2 H groups and 1 R group is a primary amine; 1 H group and 2 R groups a secondary amine; and to 3 R groups is a tertiary amine.

✱ Logic process: If you treat NH as O in chemical reactions then you find that they behave similarly, e.g. OH is equivalent to NH_2.

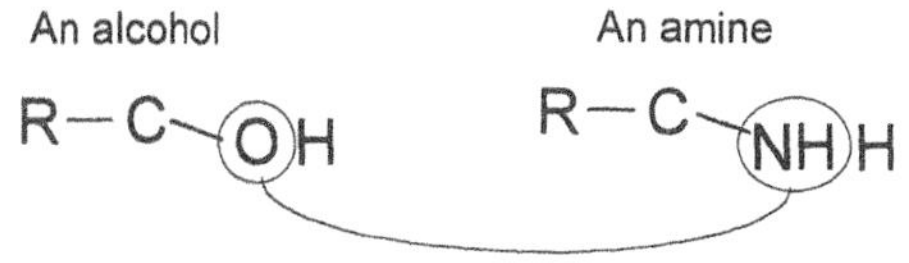

An alcohol   An amine

R—C—OH   R—C—NH—H

● an O is the same as an NH

Q&A 10: Indicate which of the following molecules have amine groups and circle the amine group.

A

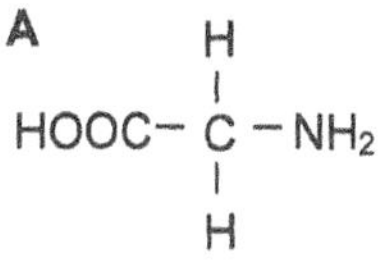

HOOC—C—NH₂ (with H above and H below)

B

O=C(OH)—CH₂—CH₂—CH(—COOH)—NH₂

C

O=C(NH₂)—CH₂—CH(—NH₂)—COOH

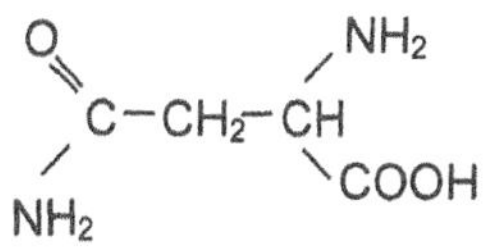

D

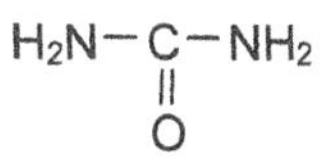

H₂N—C(=O)—NH₂

E

(purine ring structure)

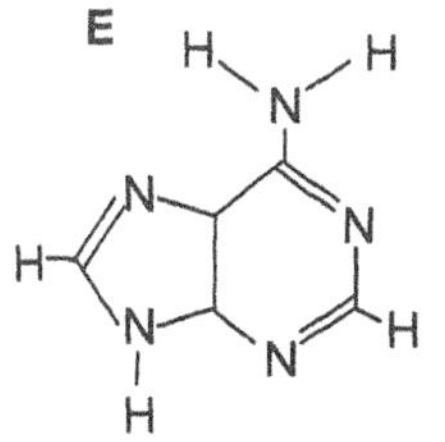

Q&A 11: Form an amide bond between the following 2 molecules. Remember remove water (an OH from the COOH group and an H from the NH_2 group). They are both amino acids, the first is glycine and the second is aspartic acid.

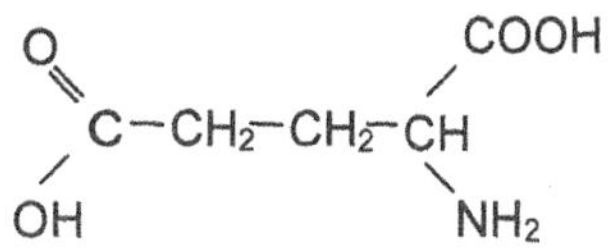

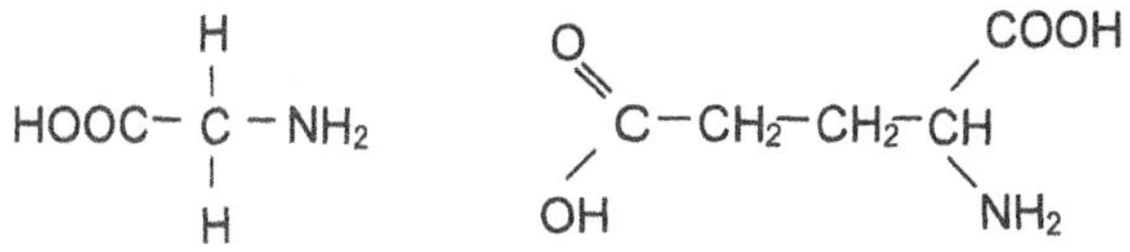

HOOC—C(H)(H)—NH₂   O=C(OH)—CH₂—CH₂—CH(—COOH)—NH₂

Nomenclature: When an amide bond occurs between 2 amino acids it is called an peptide bond. This bond is the basis of proteins.

The ether bond – cousin of the ester bond

How is an ether bond formed?

Answer: *An ether bond forms between 2 alcohols. As with ester and amide bonds, it involves a condensation reaction. The hydroxyl groups of the 2 alcohol molecules link to form an ether bond, and water:*

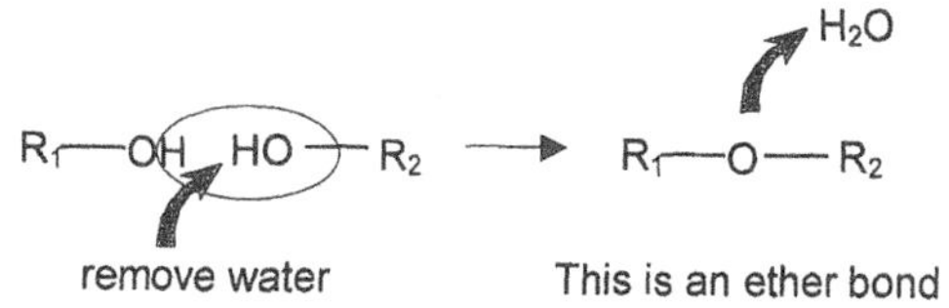

H_2O

R₁—OH   HO—R₂ → R₁—O—R₂

remove water   This is an ether bond

Major point 10: An ether bond is formed by attaching an alcohol to an alcohol and removing water.

Q&A 12: Indicate which of the following are alcohols and circle the 'alcohol' groups.

A

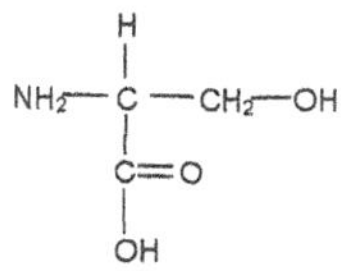

NH₂—C(H)—CH₂—OH with C=O and OH below

B

(ring with OH) CH₂—C(NH₂)(H)—C=O—OH

C

(glucose ring structure)

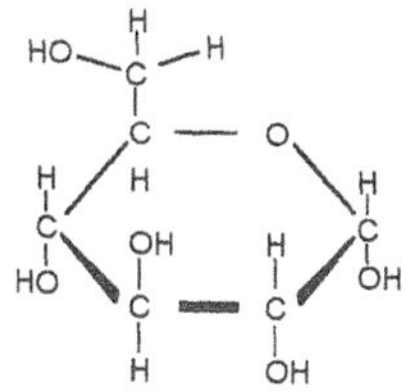

D

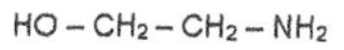

HO–CH₂–CH₂–NH₂

E

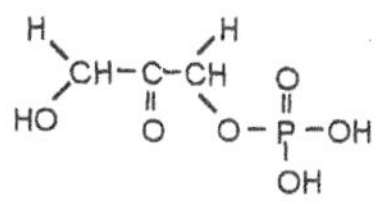

HO—CH—CH₂—O—P(=O)(OH)—OH

F

O=C(OH)—CH₂—CH₂—CH(NH₂)—C(=O)—N(H)—C(H)—COOH

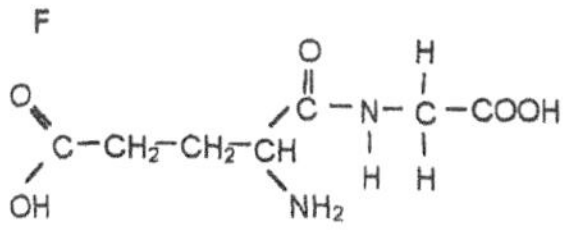

Q&A 13: Form an ether bond between the following 2 molecules.

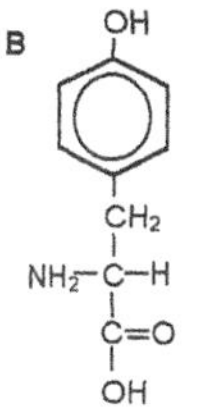

(Glucose ring structure) HO—CH₂—C(NH₂)(H)—C(=O)—OH

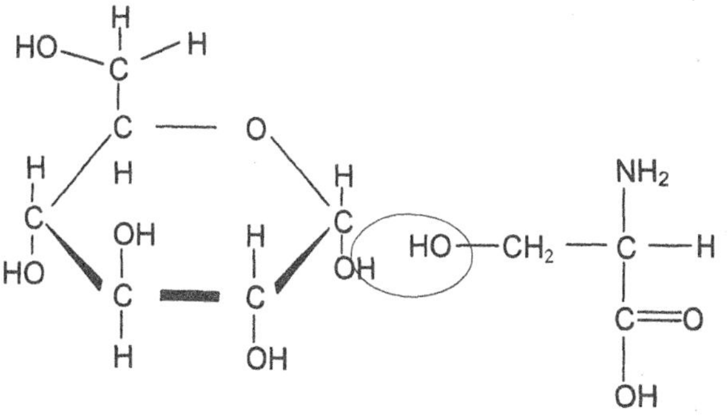

Glucose   Serine

Serine is an amino acid that occurs in proteins and is an important site of attachment for sugars to proteins.

Glycosidic bonds

You will probably have heard of the molecule glucose previously (see above). It is one of the most important molecules in biochemistry and is dealt with in detail in Chapter 3. Structurally, it is an exciting molecule because it has a large number of hydroxyl groups. This means that 2 molecules of glucose may be joined together via an ether bond. Even more importantly, a large number of different ether bonds can be formed. Some possibilities are C1→C4, C1→C1, C1→C6. Each of these is a completely different compound with different physical and chemical characteristics.

Q&A 14: List 3 other possible ether bonds that could be formed between 2 molecules of glucose.

Q&A 15: Draw an ether bond between C1 of 1 glucose molecule and C4 of the other.

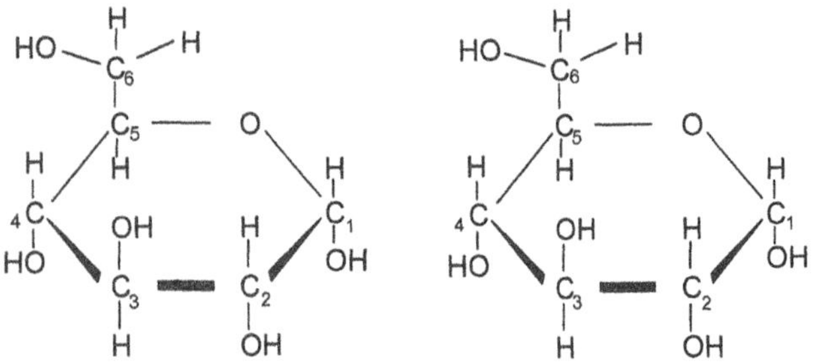

Nomenclature: An ether bond between 2 sugar units or a sugar and another alcohol is called a glycosidic bond. Because it involves an O in the final bonding, the very precise name is an O-glycosidic bond.

A similar bond between an amine and an hydroxyl group of a sugar is called an N-glycosidic bond. The following diagram illustrates how an N-glycosidic bond may be formed between an hydroxyl group of a glucose molecule and an amine group of an asparagine molecule. Asparagine is an amino acid that occurs in proteins and hence is another site at which sugars can attach to proteins.

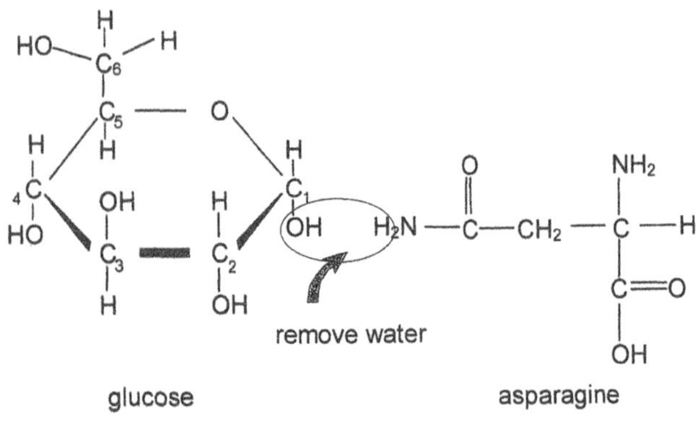

Note that the amine group here is really part of an amide bond.

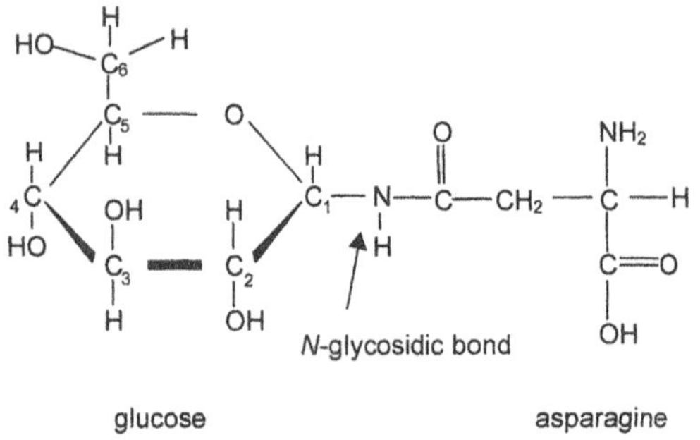

The linkage formed is an N-glycosidic bond. This is another way that sugars can attach to proteins.

Nomenclature: An N-glycosidic bond forms by the condensation of an amine and the hydroxyl group of a sugar.

Q&A 16: To test your skills, consider the following molecules: ribose, phosphoric acid and adenine Make an ester bond between C5 of ribose and phosphate, and then an N-glycosidic bond between C1 of ribose and the amine group of adenine (circled).

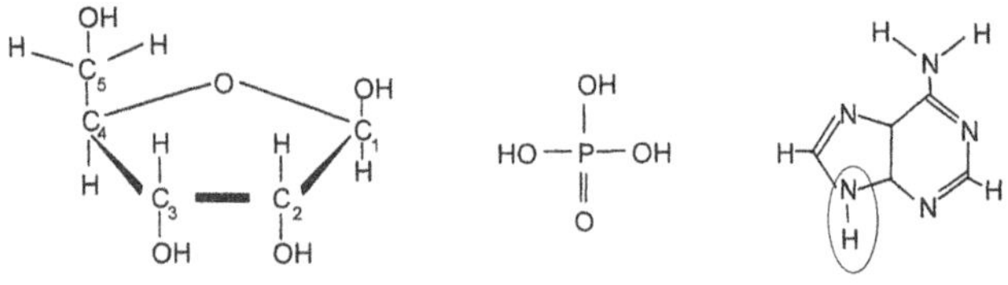

References and further reading

Cooke, R. and Kuntz, I.D. (1974) The properties of water in biological systems. *Annu. Rev. Biophys. Bioeng.* **3**:95-126.

Devlin, T.M. (1992) *Textbook of Biochemistry with Clinical Correlations,* 3rd edn. New York: Wiley-Liss.

Fersht, A.R. (1987) The hydrogen bond in molecular recognition. *Trends Biochem. Sci.* **12**:301-304.

Frieden, E. (1972) The chemical elements of life. *Sci. Amer.,* 227(1): 52-60.

Good, N.E., Winget, G.D., Winter, W., Conolly, T.N., Izawa, S. and Singh, R.M.M. (1966) Hydrogen ion buffers for biological research. *Biochemistry,* **5**:467-477.

Stillinger, F.H. (1980) Water revisited. *Science,* **209**:451-457.

Q&A Answers

1 The hydroxyl groups are circled.
- A (glycerol), B (choline), and D (serine) are all alcohols because they have an hydroxyl (OH) group.
- Note that D is also an acid as it has a carboxyl group (COOH)(boxed).
- Similarly, C (glutamic acid) is an acid, but in this case it has 2 acid groups boxed).

A

$$H_2C - \boxed{OH}$$
$$HC - \boxed{OH}$$
$$H_2C - \boxed{OH}$$

B

$$\boxed{HO} - CH_2 - CH_2 - \overset{+}{N} \overset{CH_3}{\underset{CH_3}{-}} CH_3$$

C

$$NH_2 - \overset{H}{\underset{\underset{OH}{|}}{\underset{C=O}{\overset{|}{C}}}} - CH_2 - CH_2 - \boxed{\overset{O}{\overset{\|}{C}} - OH}$$

D

$$NH_2 - \overset{H}{\underset{\underset{OH}{|}}{\underset{C=O}{\overset{|}{C}}}} - CH_2 \boxed{OH}$$

2   The acid groups are boxed. B–F have carboxyl acid groups whereas H has a phosphoric acid group.

A

$$CH_2 - CH - \boxed{C\overset{O}{\underset{H}{}}}$$
$$\underset{OH}{|} \quad \underset{OH}{|}$$

glyceraldehyde

B

$$CH_3 - CH - \boxed{C\overset{O}{\underset{OH}{}}}$$
$$\underset{NH_2}{|}$$

alanine

C

$$CH_3 - \overset{}{\underset{\|}{C}} - \boxed{C\overset{O}{\underset{OH}{}}}$$
$$\underset{O}{}$$

pyruvic acid

D

$$CH_3 - \overset{}{C} - \boxed{C\overset{O}{\underset{OH}{}}}$$

acetic acid

E

$$\boxed{\overset{O}{\underset{HO}{C}}} - CH_2 - \overset{}{\underset{O}{C}} - \boxed{C\overset{O}{\underset{OH}{}}}$$

oxaloacetic acid

F

$$\boxed{\overset{O}{\underset{HO}{C}}} - CH_2 - \overset{}{C} - \boxed{C\overset{O}{\underset{OH}{}}}$$
$$\boxed{\overset{O}{\underset{OH}{C}}} \; CH_2$$

citric acid

G

$$\overset{H}{\underset{HO}{CH}} - \overset{}{\underset{O}{C}} - \overset{H}{\underset{OH}{CH}}$$

dihydroxyacetone

H

$$\overset{H}{\underset{HO}{CH}} - \overset{}{\underset{O}{C}} - \overset{H}{\underset{O}{CH}} - \boxed{\overset{}{\underset{OH}{P}} - OH}$$

dihydroxyacetone phosphate

Important notes on these molecules

- A is glyceraldehyde. The group circled is an aldehyde not an acid group. It has an H group only not an OH. Can you see the similarity between this molecule and glycerol? Glycerol has an alcohol instead of an aldehyde group (circled).
- B is alanine. It is an amino acid because, besides the acid group, it has an amine group (circled).

- C is pyruvic acid. Many molecules are converted to pyruvic acid and then pyruvic acid is broken down (oxidised) further to CO_2 and H_2O to provide energy.
  See the relationship between alanine and pyruvate. Removal of ammonia (NH_3) and replacement with a ketone = O (circled) gives pyruvic acid.
- D is acetic acid, which is not so different from pyruvic acid. The C = O group is taken out of the middle. There is a chemical reaction in the body where this actually happens and pyruvate is converted to acetate.
- E is oxaloacetate. This is similar in structure to pyruvate except that it has 2 carboxyl groups (boxed).
  What would happen if we put an amine in place of the ketone group (circled)? We would form an amino acid. This is similar to alanine above. The amino acid so formed from oxaloacetate is called aspartic acid.
- F is citric acid that has 3 carboxyl groups. It can be made by adding acetic acid (D) (circled) to oxaloacetic acid (E) at the carbon atom (arrow).
- G is dihydroxy acetone. It is a ketone and the keto group is circled. Note that it is similar to glycerol except that glycerol has an hydroxyl (-OH) group to replace the keto (= O) group. Is dihydroxyacetone similar in structure to one of the earlier molecules shown in this group?
- H This was a trick question. Here the acid is phosphoric acid instead of carboxylic acid. Nevertheless, it is an acid. Its structure is dihydroxyacetone with a phosphate attached. Therefore it is called dihydroxyacetone phosphate – What a good name!
- The molecule dihydroxyacetone is related to is glyceraldehyde (A).

3

$$HO - \overset{O}{\overset{\|}{C}} \diagup\diagdown\diagup\diagdown\diagup\diagdown\diagup\diagdown CH_3$$

(Note: the zig-zag line represents a chain of CH_2 units)

4

$$HO - \overset{O}{\overset{\|}{C}} \diagup\diagdown\diagup\diagdown\diagup\diagdown\diagup\diagdown \overset{20}{CH_3}$$
$$\underset{1}{}$$

(Note: the carbons are numbered with carbon number 1 being the carbon of the carboxylic acid (-COOH).

5a triene (trienoic acid) is a fatty acid with 3 double bonds.
 b hexene (hexenoic acid) is a fatty acid with 6 double bonds.

6

$$H_2C - O - \overset{O}{\overset{\|}{C}} \diagup\diagdown\diagup\diagdown\diagup\diagdown CH_3$$
$$HC - O - \overset{O}{\overset{\|}{C}} \diagup\diagdown\diagup\diagdown\diagup\diagdown\diagup\diagdown CH_3$$
$$H_2C - O - \overset{O}{\overset{\|}{C}} \diagup\diagdown\diagup\diagdown CH_3$$

This is a triacylglycerol. This molecule is important because it is actually a fat and it is in this form that we store fat (it is uncharged) to be used later as an energy source.

7   Phosphate is phosphoric acid that has lost its hydrogens (the base form).

Aspartate is the base form of aspartic acid.

8

Note that the phosphate group has been drawn as it would appear at physiological pH (without an H$^+$).

9   A phosphodiesterase breaks one of the bonds in a phosphodiester by adding water which is also called hydrolysis.

10   The amine groups are boxed.

A

HOOC— C —NH₂

B

C-CH₂-CH₂-CH COOH NH₂

C

C-CH₂-CH NH₂ COOH

D

H₂N—C—NH₂

E

- A has an acid group (COOH). Therefore it is an amino acid. Its name is glycine.

- Similarly, B is an amino acid but has 2 acid groups. It is glutamic acid.

- C indeed has an amine group but the other amine is attached to a C=O grouping which in fact makes it an amide bond. This molecule is also an amino acid, asparagine.

- D is similar in that the amine groupings are attached to C=O which are amide bonds not amine. It is urea.

- E clearly has an amine group as illustrated (box). The NH group circled is also an amine group, but it is not obvious that it is so. In this case the nitrogen is attached to 2 carbons (or R groups) and 1H. This is called a secondary amine.

11 There are 3 possible combinations here. Did you get all 3?

12 The alcohol or hydroxyl groups are boxed.

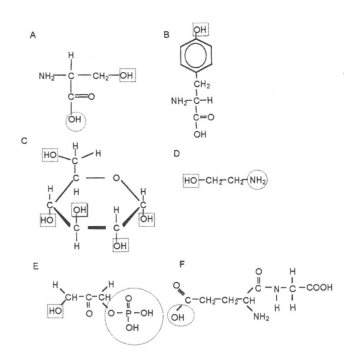

- A is an amino acid, serine. Note that the OH group circled is part of the carboxyl (COOH) group and is not an alcohol.

- B is also an amino acid (tyrosine). Here the alcohol group is attached to a phenyl group.

- C is really interesting. This is a sugar (glucose) and has <u>5</u> OH groups.
- D is an amine as well as an alcohol. If it did not have the amine group it would be ethanol. With the amine group (circled) it is ethanolamine.
- E is also an alcohol. We had this molecule earlier. It is dihydroxyacetone phosphate. The phosphate group (circled) has been esterified to dihydroxyacetone and hence there is now only 1 hydroxyl group.
- F This is a dipeptide with an amide bond. The OH group circled is part of the carboxyl group. For fun, indicate the amide bond.

13

- Note: This is one way sugar groups are attached to proteins.

14   There are several other possibilities e.g. C3→C3, C3→C4, C3→C6 etc. It is not possible to make an ether bond to C5 because it does not have an -OH group.

15   This ether bond between 2 sugars is an *O*-glycosidic bond.

16   Wow!!! This is fantastic!!! You have just created adenosine monophosphate (AMP).

Remember that the key to the structure of AMP it is the ribose group. It has a phosphoester bond to phosphate and an *N*-glycosidic bond to adenine.

3 Sugars

In this chapter you will learn:

- the structure of monosaccharides, disaccharides, polysaccharides and oligosaccharides
- the structure of glycerol, glyceraldehyde, glucose, mannose and galactose, glucosamine and galactosamine, *N*-acetylgalactosamine and *N*-acetylglucosamine, fructose, ribose and deoxyribose, AMP
- to recognise the importance of polyhydroxyl groups in sugars
- to understand the meaning of D-isomers and L-isomers
- the functional groups aldehyde and ketone
- about a chiral carbon
- how mutarotation forms an anomeric carbon and the à and á designation for sugars
- the basic structures of DNA and RNA
- the importance of the formation of an ester bond with phosphate and with sulphate
- the function of kinases
- *O*-glycosidic bonds and *N*-glycosidic bonds
- the structure and nomenclature of sugar acids.

General structure of sugars

Sugars, or saccharides, come under the more general name of carbohydrate. As the word carbohydrate suggests, they are hydrated carbon, or carbon with water added to it. As may be expected, carbohydrates have the general formulae $(CH_2O)_n$, where n is an integer of 3 or greater. The simplest carbohydrates are the sugars (monosaccharides) which have 3 to 9 carbon (C) elements.

Structurally, sugars have two important features:

1. more than one hydroxyl (OH) group (polyhydroxyl); and
2. a carbonyl group (C = O) as either an aldehyde group (-CHO) or a ketone group (-CO-).

The first of these points is important because sugars can be phosphorylated via an ester bond between one or more of the hydroxyl groups and phosphoric acid, or two sugars can be linked together by an ether bond.

Aldoses

The most commonly known sugar is the monosaccharide glucose. We understand its structure by perceiving it as a variation of glycerol. The aldehyde of glycerol is glyceraldehyde.

Glucose is glyceraldehyde with an extra 3 carbons and an extra 3 hydroxyl groups (or another glycerol group). The aldehyde group is indicated in the following figure.

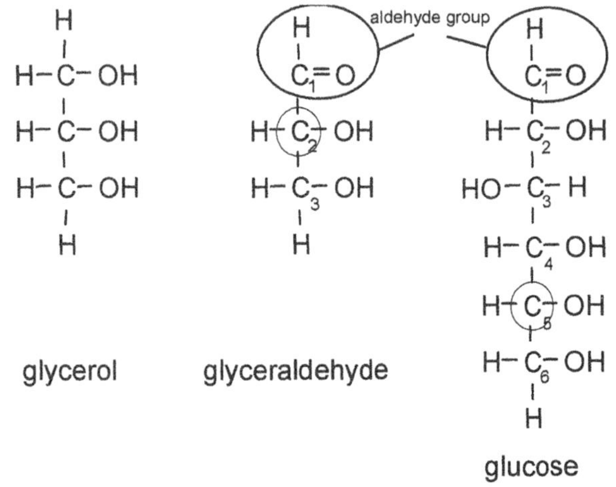

glycerol   glyceraldehyde   glucose

The difference between glycerol and glyceraldehyde is that molecular hydrogen (H_2) has been removed from the terminal C of glycerol to form the aldehyde group, as illustrated above. Glycerol is therefore dehydrogenated to give glyceraldehyde. C2 of glyceraldahyde and C5 of glucose (circled) determine if the molecule is a D or L isomer (see below).

Recall that the removal of hydrogen (H_2) is one form of oxidation. Therefore, oxidation of an alcohol, such as glycerol, gives an aldehyde. The general formula for an aldehyde is R-CHO where R is the symbol for any side chain. By convention, the numbering of the carbon atoms starts with the aldehyde group.

Most importantly, C2 of glyceraldehyde has four different groups attached to it. This means that there are two possible structures for this molecule, an L and a D form, as indicated below:

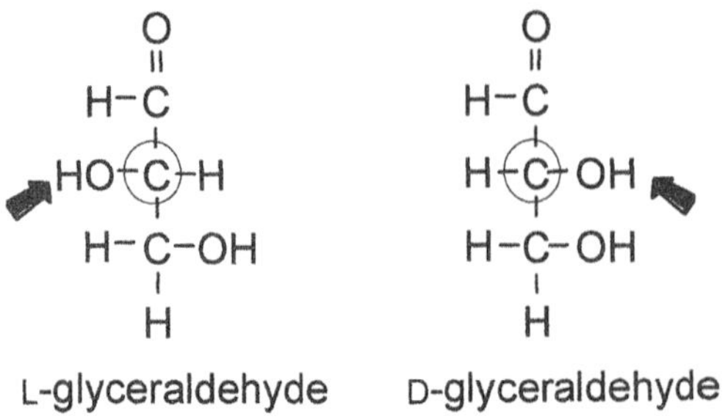

L-glyceraldehyde   D-glyceraldehyde

I hear you ask , 'How can I remember which is the L form and which is the D form?' If we draw glyceraldehyde in the standard notation, as shown above, the hydroxyl (-OH) group attached to the chiral carbon (see the arrows in the diagram above) can be drawn to the right or to the left. Which way it is drawn determines if it is the L or D form of glyceraldehyde. If the hydroxyl group is on the left-hand side then it is the L form of the molecule; if it is on the right-hand side, it is the D form of the molecule.(A good way to remember this is that when the hydroxyl group is on the left, it is the L form; L equaling left.)

Nomenclature: Any carbon atom with four different groups attached to it is called a chiral carbon atom. Chiral comes from the Greek 'kheir' which means hand. Look at your own hands and you see that they are the same but opposite in form. In other words, the two possible forms offers a 'handedness' to the molecule.

Recall that molecules are three dimensional and that four molecules bound to a carbon atom form a tetrahedron. The convention for representing molecules in flat projection is that the vertical bonds are going away from you, into the page and the horizontal bonds are coming out towards you. This can be remembered if you liken it to a stick figure with the arms coming out to hug you. Another way of representing the three-dimensional structure is to use dark bonds coming towards you and light bonds going into the page.

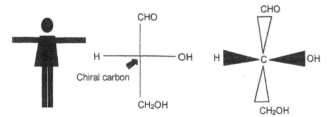

You might be wondering where the terms D and L come from. Glyceraldehyde has a polarising effect on light. If a solution of glyceraldehyde is placed in a beam of polarised light, the light is rotated. If it is a pure solution of L-glyceraldehyde then it rotates the light to the left (anti-clockwise) and if it is a pure solution of D-glyceraldehyde then it rotates the light to the right (clockwise). The L stands for levo meaning left and the D stands for dextro meaning right. However, this is the only molecule where D and L have this meaning. For all other molecules, including sugars and amino acids, the D and L terms apply to structure and not to the direction they rotate polarised light. Therefore, the direction that the D and L forms of these molecules rotate polarised light can only be determined experimentally. If it rotates polarised light to the right it is given the designation (+) and if it rotates light to the left then it is designated (-). There are some molecules for which the D form rotate light to the left (anti-clockwise), and for amino acids, the rotation of light can change direction depending upon the pH. Since a molecule may have more than one chiral

carbon, the D form of a molecule is the mirror image of its L form.

> **Major point 11:** For all molecules (other than glyceraldehyde), the D and L terminology refers to their structure based on the D and L forms of glyceraldehyde, not on how they rotate polarised light.

For sugars, the D and L forms are determined by the chiral carbon furthest away from the aldehyde group at C1 (or the ketone group at C2 for ketoses).

Examine the following glucose molecule more closely. Is it D-glucose or L-glucose?

$$
\begin{array}{c}
\text{O} \\
\| \\
\text{H--C}_1 \\
| \\
\text{H--C}_2\text{--OH} \\
| \\
\text{HO--C}_3\text{--H} \\
| \\
\text{H--C}_4\text{--OH} \\
| \\
\text{H--}\!\!\boxed{\text{C}_5}\!\!\text{--OH} \\
| \\
\text{H--C}_6\text{--OH} \\
| \\
\text{H}
\end{array}
$$

Answer: *Given that C5 is the chiral carbon furthest from the aldehyde group C1 and that the hydroxyl group on C5 (circled) is on the right, it is D-glucose.*

> **Q&A 1: Draw L-glucose.**

Nomenclature: An isomer of a molecule that is its mirror image is called an enantiomer. Enantio means opposite.

> **Major point 12:** L and D isomers are mirror images or enantiomers

> **Q&A 2: Can you predict which way D-glucose will rotate polarised light?**

Consider what would happen if one of the other hydroxyl groups of D-glucose reversed positions about its chiral carbon as illustrated:

$$
\begin{array}{ccc}
\begin{array}{c}
\text{O} \\
\| \\
\text{H--C}_1 \\
| \\
\text{H--C}_2\text{--OH} \\
| \\
\text{HO--C}_3\text{--H} \\
| \\
\boxed{\text{H--C}_4\text{--OH}} \\
| \\
\text{H--C}_5\text{--OH} \\
| \\
\text{H--C}_6\text{--OH} \\
| \\
\text{H}
\end{array}
&
\longrightarrow
&
\begin{array}{c}
\text{O} \\
\| \\
\text{H--C}_1 \\
| \\
\text{H--C}_2\text{--OH} \\
| \\
\text{HO--C}_3\text{--H} \\
| \\
\boxed{\text{HO--C}_4\text{--H}} \\
| \\
\text{H--C}_5\text{--OH} \\
| \\
\text{H--C}_6\text{--OH} \\
| \\
\text{H}
\end{array} \\
\text{D-glucose} & & \text{D-?}
\end{array}
$$

All that has changed is the position of the hydroxyl group on C4. This changes the whole nature of the molecule. It is now a completely different aldose: D-galactose.

Ring forms of sugar molecules

Sugar molecules of 5 or 6 carbons are quite flexible, and this flexibility brings the aldehyde group in close proximity to other hydroxyl groups on the same molecule. When this happens, the hydrogen of the hydroxyl group is transferred to the oxygen of the aldehyde group, and the oxygen from the hydroxyl group forms a bond with the carbon of the aldehyde group, as illustrated.

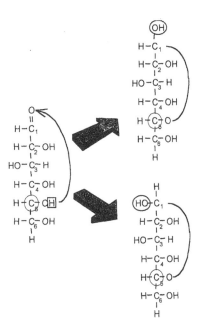

This reaction forms a stable ring structure. The hydroxyl group formed at C1 can be in one of two positions: either above or below the plane of the ring. This forms two possible ring structures: beta (β), where the hydroxyl group lies above the plane of the ring (illustrated on the left below) and alpha (α), where the hydroxyl group lies below the plane of the ring (illustrated on the right below). Both structures are D-glucose.

For glucose and other 5 and 6 carbon sugars in solution, there is a constant opening and closing of the ring, and as such there is constant interconversion between the two forms, α and β. This interconversion is called mutarotation.

The basis of this form of D-glucose is the 6-sided ring that is made from 5 carbons and 1 oxygen (C6 of the hexose lies outside the ring). This type of ring is called a pyran.

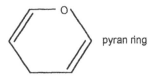

Since sugars containing 6 carbons more normally appear in a ring structure it is important to learn the structure of sugars in this form rather than the linear form. The following is a guide to drawing sugars in their ring form.

1. Draw a pyran ring, as above, but without the double bonds.

2. Number the carbon atoms clockwise starting from the oxygen (O), which you can imagine is zero.

3. The position of the hydroxyl groups at C2, C3 and C4 determine the name of the sugar. You will need to learn the positions of the hydroxyl groups at these carbons for specific sugars. (Note: C5 does not have an hydroxyl group.)

4. Starting at C2 determine whether the hydroxyl group is above or below the ring,. Do the same for C3 and C4. For example, in glucose the hydroxyl group is below the ring at C2 (down), above the ring at C3 (up), and below the ring at C4 (down); in galactose the hydroxyl groups are down-up-up, and in mannose up-up-down.

5. C6 sits above the ring if the sugar is a D enantiomer. (Ninety-nine per cent of sugars in biochemistry are in the D form so concentrate on this form. If you need to draw the L form, draw the D form and then draw its mirror image.)

6. The position of the hydroxyl group at C1 varies depending on where it was when the ring closed. If it lies below the ring it is called α and if it lies above the ring it is called β (remember a is less than b in the alphabet and therefore α lies below β).

> **Q&A 3:** The form of glucose on the left above is β-glucopyranose. What is the name of the form shown on the right?

> **Q&A 4:** Draw the structures of α-D-galactopyranose and β-D-mannopyranose, remembering that the location of the hydroxyl groups in galactose are down-up-up and in mannose they are up-up-down.

The sugars we have been discussing are aldoses, that is, sugars which contain an aldehyde group. If they have 6 carbons they are called aldohexoses, if they have 3 carbons they are called aldotrioses. The aldoses described so far are the 3 carbon sugar glyceraldehyde, and the 6 carbon sugars glucose, galactose and mannose. These are the essential ones to learn. You need to be able to recognise them when they are drawn and be able to draw them yourself, particularly in the pyranose form.

> **Nomenclature:** tri = 3; pent = 5; hexa = 6. Thus, sugars containing an aldehyde group and having 3 carbons are aldo<u>tri</u>oses, those with 5 are aldo<u>pent</u>oses, and those with 6 carbons are aldo<u>hexa</u>oses.

Ribose is also an aldose, but it has 5 carbons and is therefore an aldopentose. Having only 5 carbons, it forms a 5-sided ring, rather than the 6-sided ring seen with the aldohexoses. Due to opening and closing of the ring, mutarotation occurs giving both the β and α forms, as illustrated below. This is one of the most important molecules in biochemistry and therefore it is essential that you learn the structure of this molecule in its normal ring form.

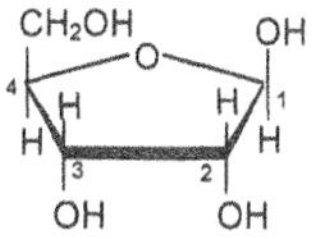

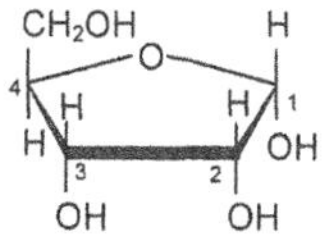

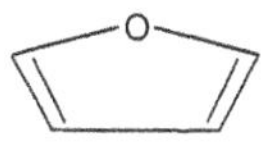

In learning its structure, note firstly that that the ring is now a 5-sided ring made of 4 carbons and an oxygen. This is called a furan ring. C5 of ribose sits above the ring, and therefore the sugar is in the D form. C2 and C3 have hydroxyl groups below the ring (down, down) which designate that it is ribose rather than another aldopentose. At C1, the hydroxyl group above the ring is the β form, and below the α form. The full name is of the molecule is β-D-ribofuranose on the left and α-D-ribofuranose on the right.

Examine the following forms of ribose, β-D-ribofuranose and α-D-ribofuranose. Both lack an oxygen atom at C2. What would be a good name for such forms of ribose?

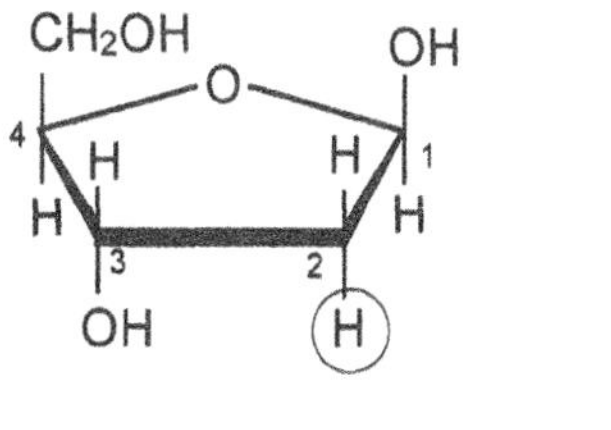

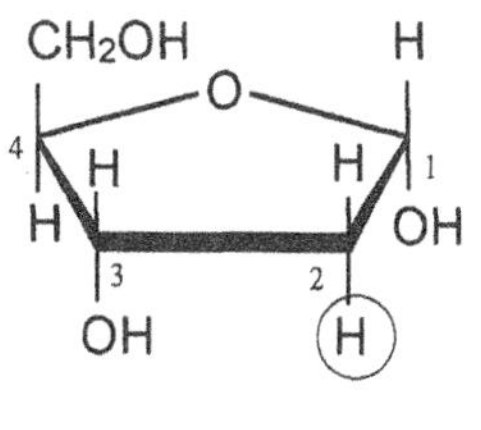

Answer: *These deoxygenated forms of ribose are called deoxyribose (technically β-D-deoxyribofuranose and α-D-deoxyribofuranose), which is the basis of deoxyribonucleic acid (DNA).*

In DNA, C5 of deoxyribose is connected to C3 of the deoxyribose of the next via a diphosphoester. This in turn has its C5 connected to C3 of the ribose of the next unit and so on, as illustrated. The bases are connected to C1.

(Note: the base is either adenine, guanine, cytosine or thymidine — A G C or T: see Chapter 13)

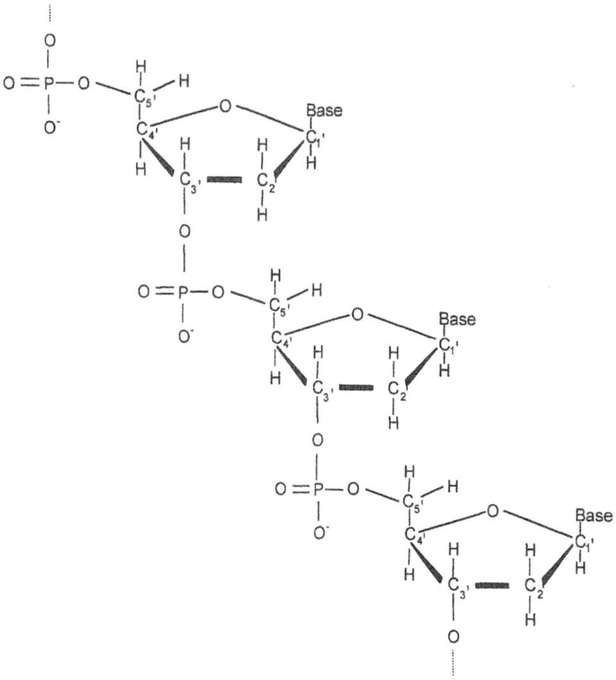

> **Nomenclature:** The carbons of the sugar ring are now given a prime designation C1'. This is because the atoms making up the ring of the base are numbered firstly, without the prime designation. The carbons of the ribose are then numbered and, to distinguish them from the atoms of the base ring (A, G, C, or T), they are given a prime (C') designation.
> The 5' to 3' designation associated with DNA or RNA refers to the direction of the molecule in terms of the phosphate linkage of one ribose to the next. The genetic code is read off the bases in this direction (5'→3').

Ketoses

There is another group of sugars called ketoses because they have a ketone in their structure, rather than an aldehyde. A ketone is formed by having a <u>carbonyl</u> group (C = O) on one of the non-terminal carbons. Examine the following molecules:

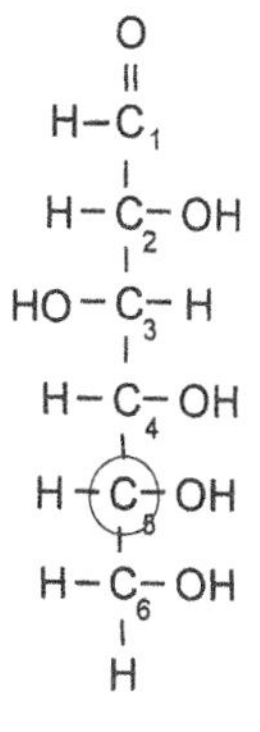

D-glucose 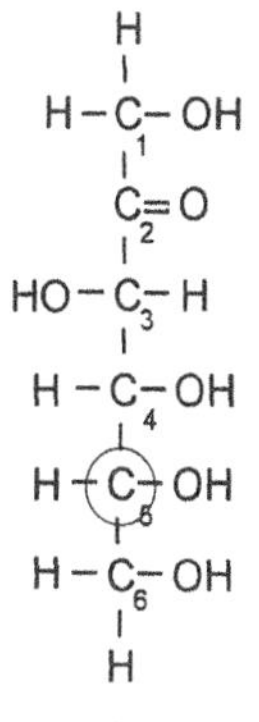   D-fructose

A ketone has the general structure R_1COR_2. D-fructose is a ketose, and is similar to D-glucose except that the carbonyl group (C=O) is on C2. (It is still the D form because the hydroxyl group on C5 is to the right.)

Like glucose, fructose normally exists as a ring form and the ring can open and close (mutarotation) in the same way as for the aldoses. Think about the way the ring was formed with glucose and how this might form a different type of ring with fructose. The ring now forms between C2 and C5, rather than C1 and C5 as it did for glucose. Consequently, a furan ring is formed because it contains 4 carbons and 1 oxygen.

Examine the two forms of fructose illustrated below. Both forms of fructose are in the D form because C6 sits above the plane of the ring (circled). Remember that there are two forms, α and β.

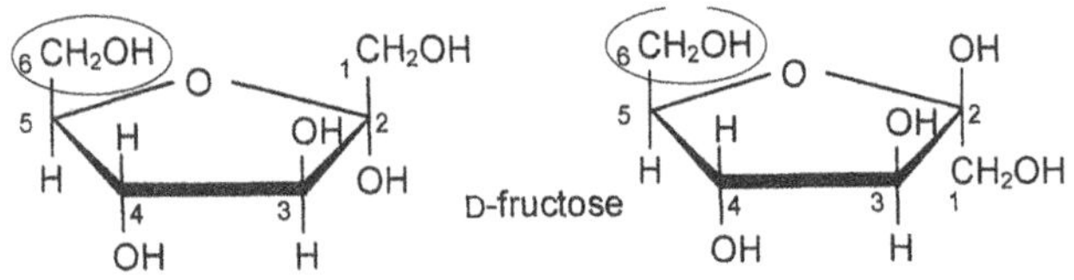

D-fructose

A   B

> **Q&A 5:** Which of the above is α-D-fructofuranose and what is the name of the other one?

Fructose is the only ketohexose (6 carbon ketosugar) which you must learn. It is an essential component in the metabolism of sugars. In particular, glucose is converted to fructose before it is oxidised (burnt) to produce energy.

To summarise, it is essential to learn: the pyran forms of the aldohexoses – glucose, galactose and mannose; the furan ring form of the aldopentose ribose; and the furan ring form of the ketohexose fructose. Learn these in the D form because for all intents and purposes, the L form does not occur in biochemistry.

Modified sugars

Amine substitutions

Some of the important sugars are modified by the substitution of other groups in their structure. One such group of sugars has an amine (NH_2) substituting for an hydroxyl (OH) group. The two most common amine substituted sugars of this type are galactose and glucose. This substitution occurs almost exclusively at C2. Therefore, you only have to worry about these two amine substituted sugars.

What would be suitable names for galactose with an amine substituted for the hydroxyl group at C2 and for glucose with an amine substituted at C2?

Answer: *Galactosamine and glucosamine. Although the correct terms are D-galactose-2-amine and D-glucose-2-amine, these molecules are so common that they are given these more trivial names galactosamine and glucosamine.*

Draw the structure of galactosamine and glucosamine in their a-pyranose forms.

Answer:

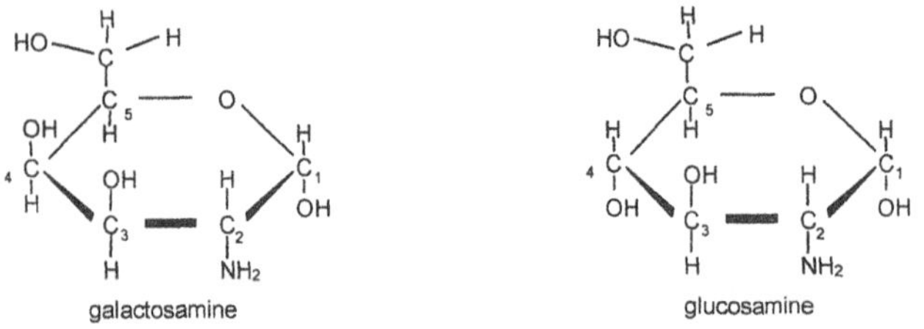

galactosamine   glucosamine

The amine in these sugars can be acetylated by forming an amide bond with acetic acid. Remember that an amide bond is formed between an amine and an acid by removal of water. (Note that to acetylate means to add acetic acid or acetate.) This process is illustrated in the following diagram where glucosamine is acetylated.

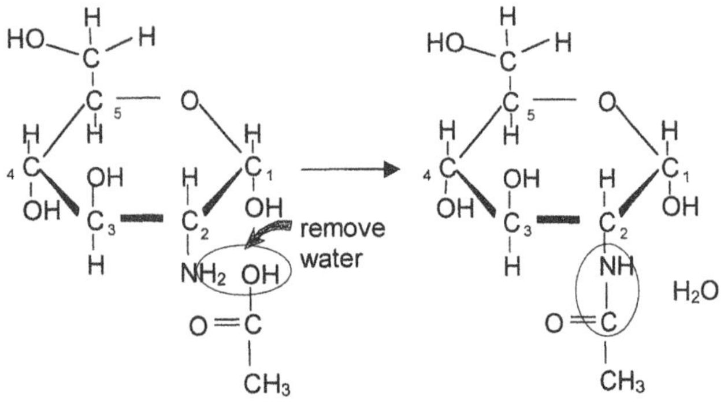

> **Nomenclature:** In the above diagram, glucosamine has been acetylated at the nitrogen group, and is therefore called *N*-acetylglucosamine (abbreviated to GlcNAc). The *N* designates where the acetylation occurred.

> **Q&A 6:** What is the structure of *N*-acetylgalactosamine, and what is its abbreviation?

Carboxyl substitutions

Other forms of substitution in sugars occur where an alcohol group (COH) or the aldehyde group at C1 is replaced by a carboxyl group (COOH). This can occur at either carbon, but most commonly occurs as a substitution of the alcohol group at C6.

> **Nomenclature:** When a carboxyl substitution in a sugar occurs at C6, the sugar has the 'ose' dropped from its name and uronic acid (-COOH) or uronate (-COO⁻) added to it. For example, glucose becomes glucuronic acid or glucuronate.

Two very important uronates occurring in biochemistry are D-glucuronate and L-iduronate (from the hexose idose) illustrated below.

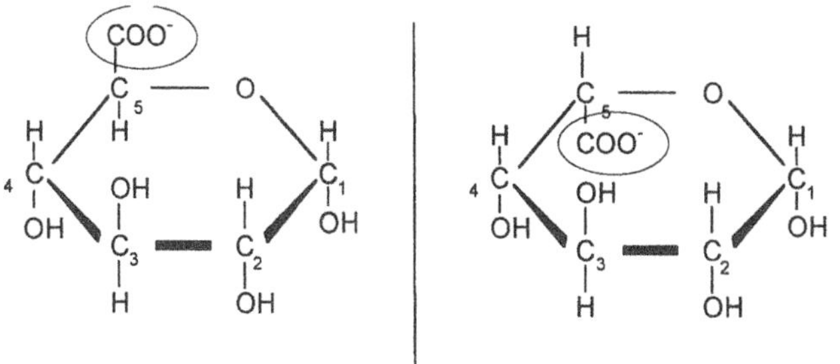

D-glucuronate L-iduronate

Do not panic about learning their structures. If you know the structure of D-glucose it easy to learn the structure of L-idose. Note the only difference between these two molecules is that the carboxyl group is above the ring for D-glucuronate and below the ring for L-iduronate. Notice that although C6 is below the ring, it is not the L-form of glucose because it is not the mirror image (imagine the vertical line is a mirror and you see immediately that these are not mirror images).

The significance or these uronates is that they carry a massive negative charge on the -COO⁻ that is not present in the aldoses. This is extremely important in polysaccharides made from these sugar groups (see Chapter 6).

Sugar esters

Consider how else sugars can be modified. Remember that we can add acids to alcohol groups (OH groups).

What type of bond is formed between an acid and alcohol?

Answer: *An ester bond. Two important acids that can be added to sugars are phosphoric acid and sulphuric acid.*

Sugar Phosphates

Let us examine the phosphorylation of sugars. Ribose monophosphate is a very important phosphosaccharide. In the following diagram, phosphoric acid is esterified to the sugar β-D-ribose at C5 to form ribose monophosphate:

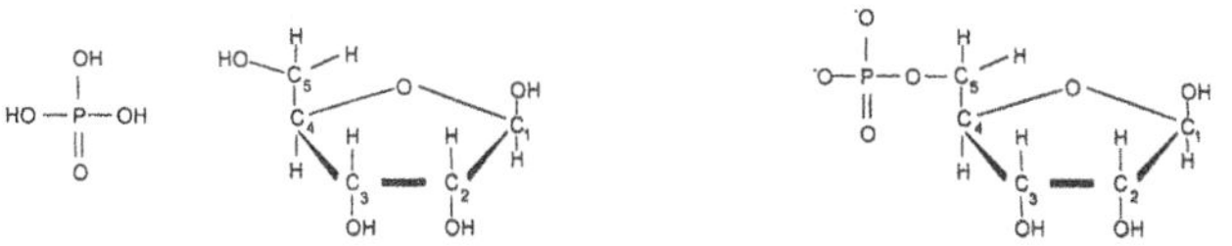

This reaction does two major things to the molecule: (1) it places a large negative charge on the molecule; and (2) it places it in a higher energy state (so it can do work when the reaction is reversed, releasing the energy stored in the phosphoester bond).

An enzyme (proteins which act as biological catalysts) which catalyses the attachment of a phosphate to a molecule is called a kinase. An enzyme which catalyses the attachment of a phosphate to a hexose such as galactose or glucose is called a hex-

ose kinase. This enzyme is found in muscle and fat cells. In the liver, the enzyme is more specific for glucose and is called glucose kinase. Both hexose kinase and glucose kinase phosphorylate at C6. In these reactions, the phosphate comes from adenosine triphosphate (ATP) which is converted to adenosine diphosphate (ADP).

> **Major point 13:** A kinase is an enzyme that catalyses the attachment of a phosphate (obtained from ATP) to a molecule

In a very important metabolic reaction, fructose 6-phosphate is converted to fructose 1,6-bisphosphate by esterifying a phosphate to C1 of fructose 6-phosphate. The name of the enzyme which catalyses the conversion of fructose 6-phosphate to fructose 1,6-bisphosphate is fructose 6-phosphokinase-1, which is simplified to phosphofructokinase-1. The '1' indicates that it is phosphorylated at C1.

Draw fructose (i.e. D-fructofuranose; the α or β designation does not matter).

Answer:

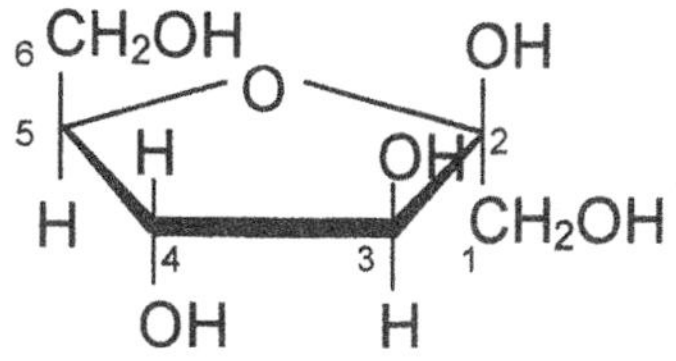

fructose

Next, draw fructose-6-phosphate by esterifying (removing water from) a phosphate to C6.

Answer:

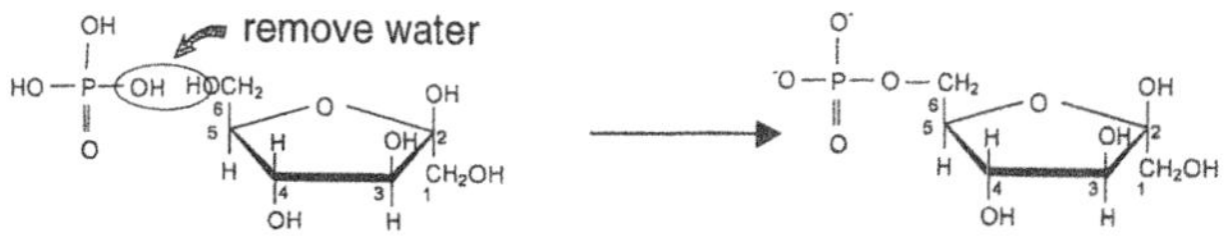

remove water

fructose → fructose-6-phosphate

> **Q&A 7:** Esterify a phosphate to C1 of fructose-6-phosphate to form fructose-1,6-bisphosphate.

> **Nomenclature:** Why is this structure called a <u>bis</u>phosphate and not a <u>bi</u>phosphate? <u>Bis</u> designates that the phosphates are attached to different carbons whereas <u>bi</u> (or di) indicates that 2 phosphates joined together (a phosphoanhydride) are linked to 1 carbon, as in ADP.

Sugar phosphates are important because: (1) they are metabolically active. If sugars act as a substrate in metabolic reactions, they do so as a phosphate ester; and (2) they do not cross the cell membrane very well (negatively charged) and are therefore trapped inside the cell. The most important reaction of sugar-1-phosphates is a reaction with a nucleoside triphosphate to yield a nucleoside diphosphate derivative. The general reaction is catalysed by NuDP-sugar pyrophosphorylase. UDP-sugars appear to be the most abundant nucleoside diphospho sugars in nature. (see Chapter 13)

Sugar sulphates

Sulphuric acid (H_2SO_4) can also be esterified to sugars. Once again the esterification of an acid to an alcohol involves dehydration, as illustrated.

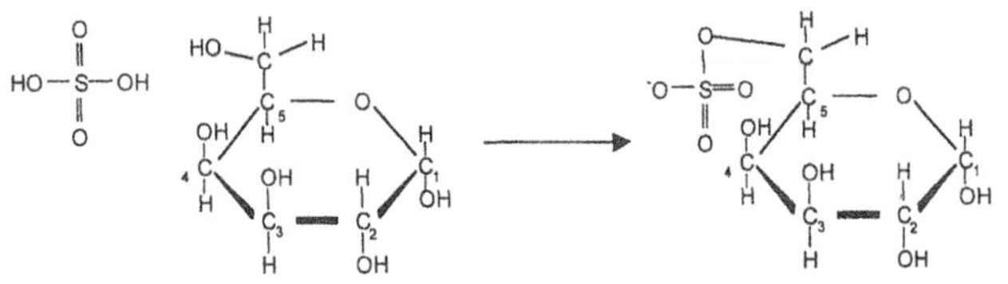

Here α-D-galactose becomes α-D-O-galactosulphate. It now carries a negative charge similar to the carboxylated sugars and as such they play an important role in polysaccharides that are major components of connective tissue and the extracellular matrix (see Chapter 6)

Nomenclature: The prefix '*O*' means that the glycosidic linkage to the acid is via an oxygen group.

Attaching sugars to other molecules

Oligosaccharides

One of the marvellous things about sugars is the multiple OH groups in their structures. This not only means that they can undergo condensation reactions with acids such as phosphate to form esters as above, but they can also be linked together by a condensation reaction between two hydroxyl groups.

Draw β-D-*glucopyranose and* α-D-*glucopyranose. Condense (form water) these at C1 of* β-D-*glucopyranose and C4 of* α-D-*glucopyranose.*

Answer:

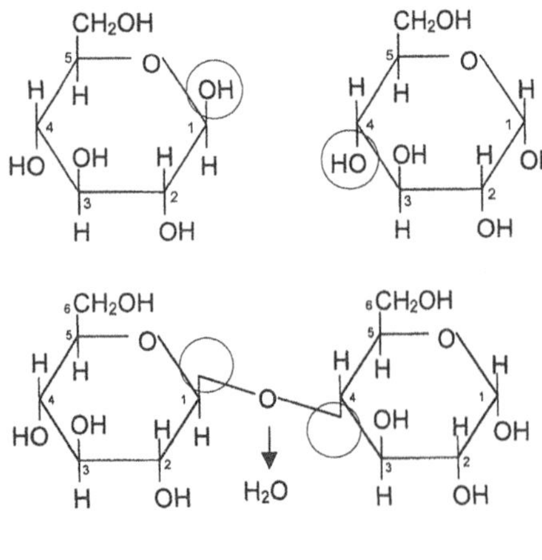

Two sugars have been linked together with an ether bond. The atom forming the bond is an oxygen group and hence the bond formed is called an *O*-glycosidic bond. The final structure has two sugars and hence is called a disaccharide.

Nomenclature: The structure above is called β-D-glucopyranosyl-(1→4)-α-D-glucopyranoside. Its trivial name is cellobiose, a component of cellulose.

Consider the endless possibilities by linking together sugars. Instead of connecting C1 to C4 as done above, you could connect C1 to either C1, C2, C3 or C6. Moreover, this connection could have been in the α or β configuration. Several other possibilities exist, e.g. we could connect C2 to C2, C3, C4 or C6. The most common linkages, however, are C1 to C4 and C1 to C1. Each of these linkages forms different molecules (disaccharides) with distinct properties. Comparatively, the number of possible molecules formed by joining two identical amino acids such as alanine together is only one! (See Chapter 4). This property of sugars is extremely useful. This variation in sugar linkages is the basis for the mechanism of cell–cell recognition. The number of variations of bonds formed between sugars compared with amino acids means that it is much easier to create the enormous variability required for identifying different cells by using sugars rather than amino acids.

Q&A 8: Draw α-D-glucopyranosyl-(1→4)-β-D-glucopyranoside. (Remember this is the a form (OH at C1) of glucose with a connection (glycoside bond) from carbon 1 to carbon 4 of the β form of glucose.)

The structure you have formed by answering this question is maltose and only differs from cellobiose (β-D-glucopyranosyl-(1→4)-D-glucopyranoside) in that it has an α linkage rather than a β linkage. However, its physical characteristics are enormously different. Note that it does no matter if the second sugar is in the α or β form, it is still maltose.

Nomenclature: We abbreviate the terminology such as α-D-glucopyranosyl-(1→4)-α-D-glucopyranoside to α-Glc-(1→4)-Glc. Remember in most biochemical situations we are dealing with the D forms of sugars.

Maltose (α-D-glucopyranosyl-(1→4)-D-glucopyranoside), which is a breakdown product of starch and glycogen, is a common disaccharide. The starch from barley is broken down to maltose and used to make a very popular beverage: whisky. Other common disaccharides are lactose (β-D-galactopyranosyl-(1→4)-D-glucopyranoside) found in milk, sucrose (α-D-glucopyranosyl-(1→2)-β-D-fructofuranoside) found in fruit and cellobiose (β-D-glucopyranosyl-(1→4)-D-glucopyranoside) which is the basis of cellulose.

Q&A 9: Draw the structures of lactose and sucrose.

Several medications have glycosidic bonds as main features of their structures. For example, various antibiotics, such as erythromycin, streptomycin and puromycin, have *O*-glycosidic bonds. Another interesting example is cardiac glycosides, such as digoxin, which is used to treat cardiac failure. The term glycoside means that there is a bond to a sugar and in this case the *O*-glycosidic bond is between a sugar and a steroid that is similar in structure to the oestrogens. Men who take them to improve the force of contraction of their hearts can develop breasts as a side-effect. Enzymes in the liver cleave the glycosidic bond releasing the steroid, which in turn acts like oestrogens and causes breast development.

Polysaccharides

Long chains of monosaccharides are called polysaccharides. Polysaccharides can either be homopolysaccharides, consisting of only one type of monosaccharide, or heteropolysaccharides, containing two or more different types. They may have molecular weights of up to several million and are often highly branched.

> **Major point 14:** A polysaccharide containing only one type of sugar it is called a homopolysaccharide. When they contain two or more different types they are called a heteropolysaccharide.

Illustrated below is a homopolysaccharide consisting of a series of glucose units that are joined together by α-1→4 linkages, except for one position where there is an α-1→6 linkage. The α-1→6 linkage causes a branch in the molecule, making the molecule globular rather than linear. Examples of such molecules are glycogen and starch, which are both storage forms of glucose; glycogen is used by animals to store glucose and starch is used by plants.

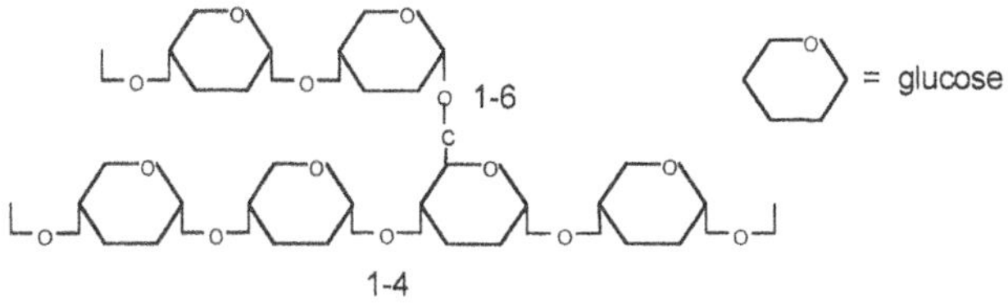

Long linear (unbranched) and branched arrays of sugars come under 2 general headings of structural (e.g. cellulose) and storage (e.g. starch) polysaccharides. As a general rule the structural polysaccharides have a β-linkage and the storage polysaccharides have an α-linkage. The β-linkage keeps the molecule linear whereas the α-linkage tends to fold the molecule, forming a globular rather than a linear structure.

N-glycosidic bonds

We have seen *O*-glycosidic bonds between sugars, which raises the obvious question: Can we have a glycosidic linkage to another atom? Remember the relationship between an amide bond and an ester bond. In this case we treated NH as being equivalent to an O. This makes an amide group (NH_2) equivalent to an hydroxyl group (OH). Therefore, when we condense an amine to an acid we have a structure that parallels the formation of an ester.

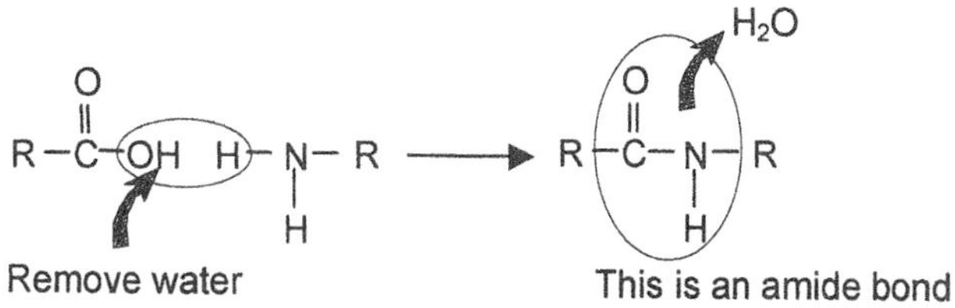

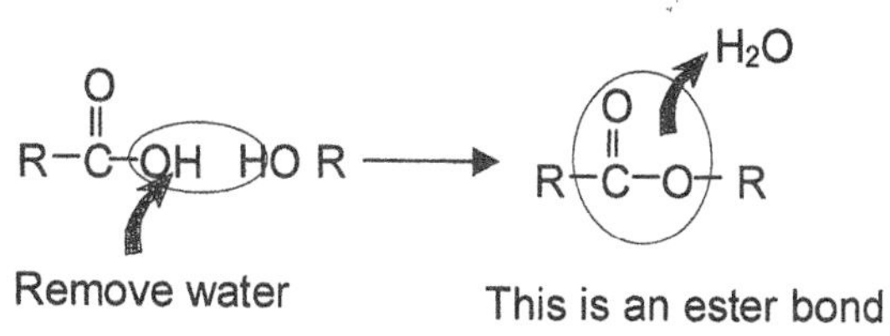

Treating NH as O again (NH_2 is equivalent to OH), we can form a bond between the hydroxyl (-OH) of a sugar and an amine (-NH_2) in the same manner that we formed a glycosidic bond between an OH of one sugar and that of another sugar (remove water).

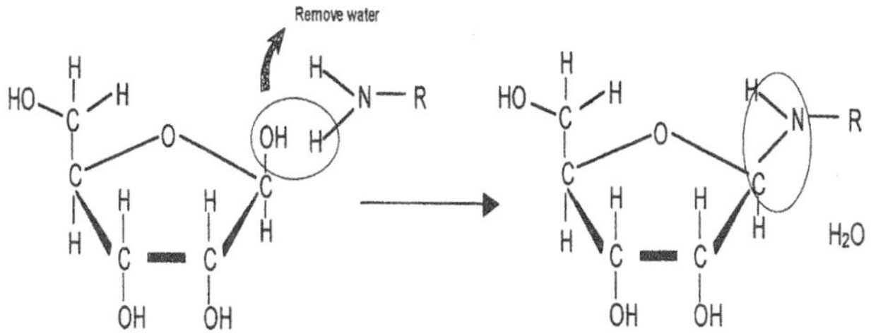

This is an *N*-glycosidic bond rather than an *O*-glycosidic bond because the linkage is through the nitrogen atom rather than an oxygen atom.

N-glycosidic bonds are very important in the linkage of ribose to purines or pyrimidines (the nucleosidic bases of DNA and RNA), and in the energy transfer nucleotides, ATP, ADP and adenosine monophosphate (AMP).

Examine the structure of AMP, focusing on the N-glycosidic bond by drawing the molecule.

Firstly draw ribose and then phosphorylate it at C5 to form ribose monophosphate. Then condense (remove water) adenine onto C1. Do not worry about the structure of adenine at this stage. We will learn how to remember its structure later. Concentrate on the N-glycosidic bond.

Answer: *See top of next page.*

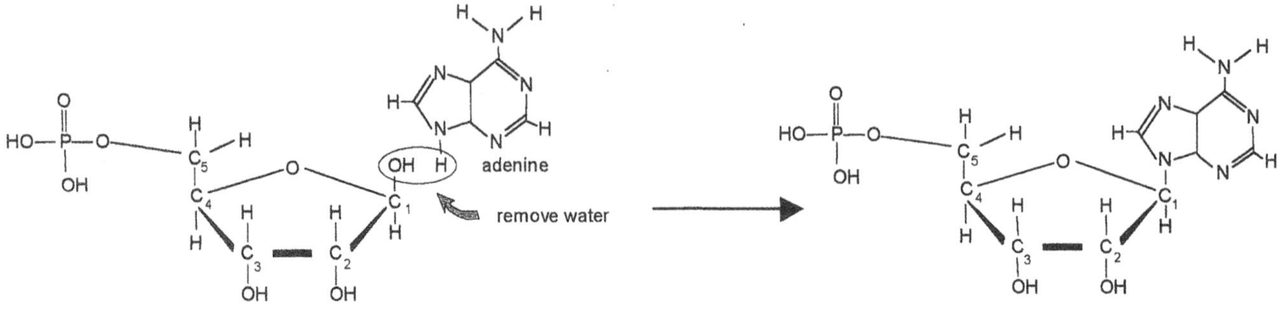

ribose monophosphate

adenosine monophosphate (AMP)

Q&A 10: Draw the variation: cAMP (cyclic 3',5' adenosine monophosphate). Hint – it is formed by connecting the phosphate attached to C1 of ribose to C3 of ribose by an ester bond.

Therefore, sugars can be attached to other sugars by a variation of the ester bond, the *O*-glycosidic bond, and to nucleoside bases by an *N*-glycosidic bond. These two strategies are used for linking sugars to proteins (see Chapter 6).

Attaching sugars to proteins

One of the very important modifications of proteins is in the attachment of sugar groups. As biochemists become more familiar with this, we are realising that it is an extremely significant process and hence understanding this process is essential. Glycosylation (sugar attachment) of proteins occurs via *O*-glycosidic bonds or *N*-glycosidic bonds. Proteins are made from amino acids (see Chapter 4), and therefore we are looking for amino acids that have an hydroxyl group in their R group or an amine group in their R group. There are several amino acids that fit this criterion, but only three are possible sites for sugar attachment.

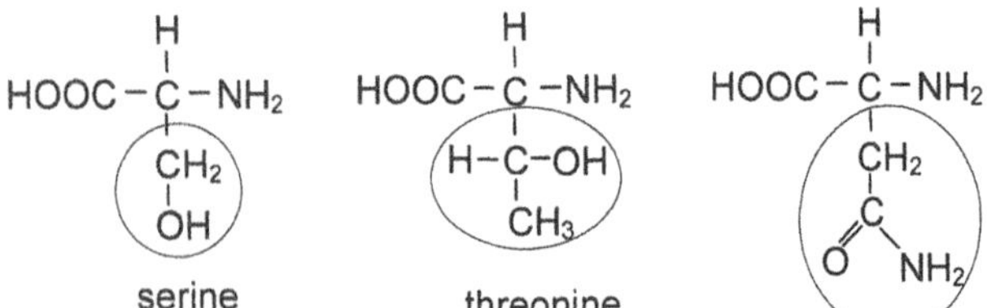

serine   threonine   asparagine

The R groups of the above amino acids are circled. Serine (S) and threonine (T) are possible *O*-glycosylation sites and asparagine is a possible *N*-glycosylation site.

Q&A 11: Glycosylate serine with *N*-acetylgalactosamine at C1.

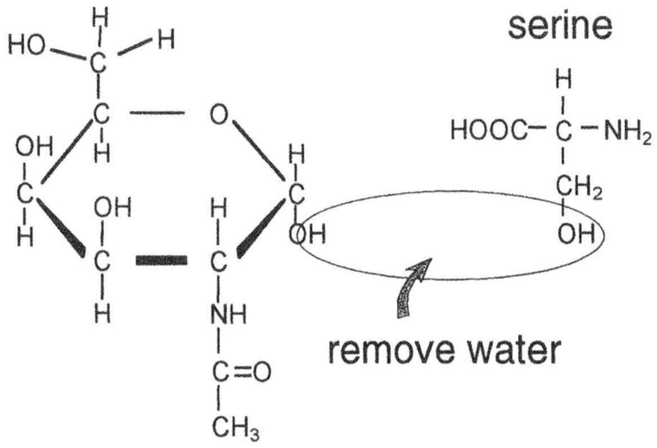

serine

N-acetylgalactosamine

Q&A 12: Glycosylate asparagine with *N*-acetylglucosamine at C1.

References and further reading

Aspinall, G.O. (1982) *The Polysaccharides, Vols. 1 and 2*. New York: Academic Press.

Cohen, J. (1995) Getting all turned around over the origins of life on earth. *Science*, **267**, 1265-1266.

Devlin, T.M. (1992) *Textbook of Biochemistry with Clinical Correlations*. 3rd edn. New York: Wiley-Liss.

Frieden, E. (1972) The chemical elements of life. *Sci. Amer.*, **227(1)**, 52-60.

Greenstein, B. and Greenstein, A. (1996) *Medical Biochemistry at a Glance*. Oxford: Blackwell Science.

Pigman, W. and Horton, D. (1972) *The Carbohydrates*. New York: Academic Press.

Roehrig, K.L. (1984) *Carbohydrate Biochemistry and Metabolism*. Westport, CT: AVI Publishing Company.

Q&A Answers

1 L-glucose is the mirror image of D-glucose. Note that C1 and C6 have not been rotated because these are not chiral carbons. If these (C1 and C6) had been drawn in their rotated forms, it is exactly the same structure.

The temptation is to simply switch the hydroxyl group on C5 from the right to the left. If you did this, **this is not the answer.** Remember, L-glucose is the mirror image of D-glucose.

D-glucose structure (C1 to C6) and L-glucose structure (C1 to C6)

D-glucose   L-glucose

2. No, you cannot predict which way D-glucose will rotate polarised light. The designation of D or L is structural for all molecules except for glyceraldehyde. The D designation for glucose is based on the D structural form of glyceraldehyde. In fact, it does rotate light to the right and to signify this it is given a + designation and hence is written as (+)-D-glucose.

3. The structure on the right is α-D-glucopyranose. It is glucose in the pyran form and the hydroxyl (-OH) group on C1 is below the plain of the ring, hence it is the α form.

4.

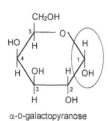

α-D-galactopyranose   β-D-mannopyranose

5. A is α-D-fructofuranose and the name of B is β-D-fructofuranose.

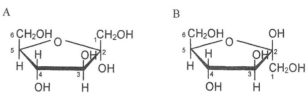

6. The structure of α-D-N-acetylgalactosamine is illustrated below. Its abbreviation is GalNAc.

structure of α-D-N-acetylgalactosamine

7. The phosphorylation of fructose-6-phosphate by phosphofructokinase gives fructose-1,6-bisphosphate.

fructose-1,6- bisphosphate

8.

disaccharide structure

9. lactose (β-D-galactopyranosyl-(1→4) -D-glucopyranoside) sucrose (α-D-glucopyranosyl-(1→2)-β-D-fructofuranoside).

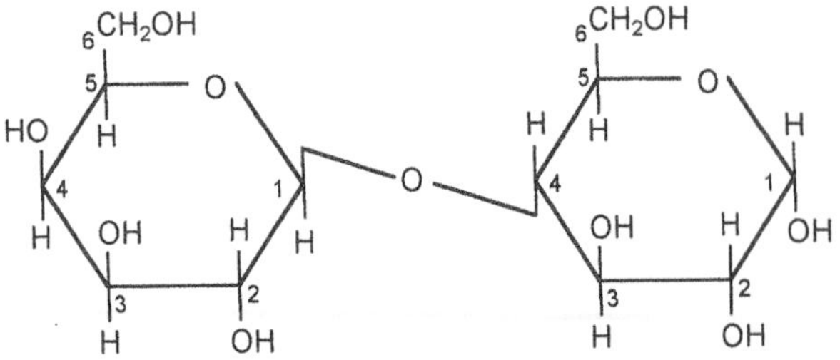

lactose

β-D-galactopyranosyl-(1→4)-D-glucopyranoside

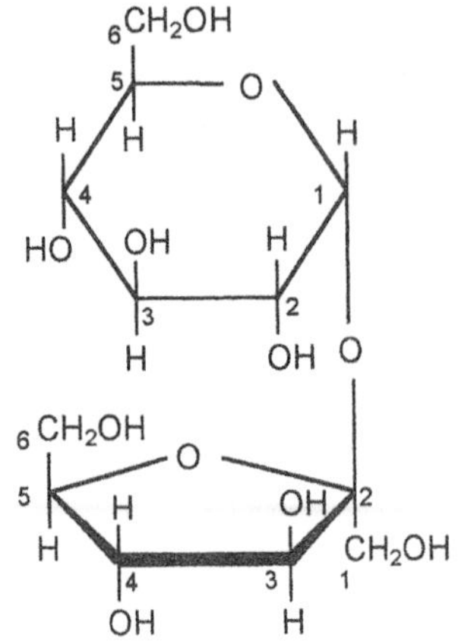

sucrose

α-D-glucopyranosyl-(1→2)-β-D-fructofuranoside

10.

adenosine monophosphate

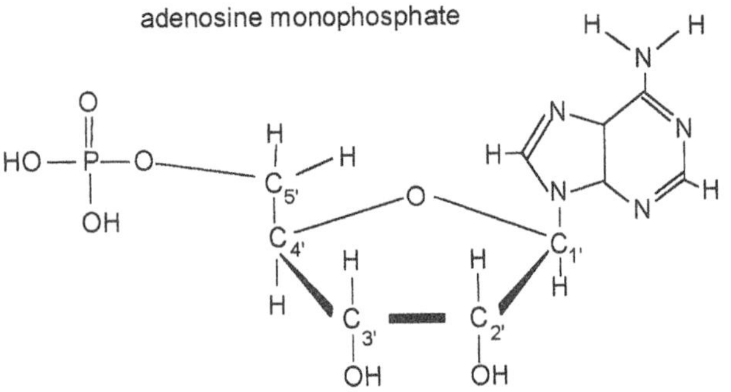

cyclic 3',5' adenosine monophosphate

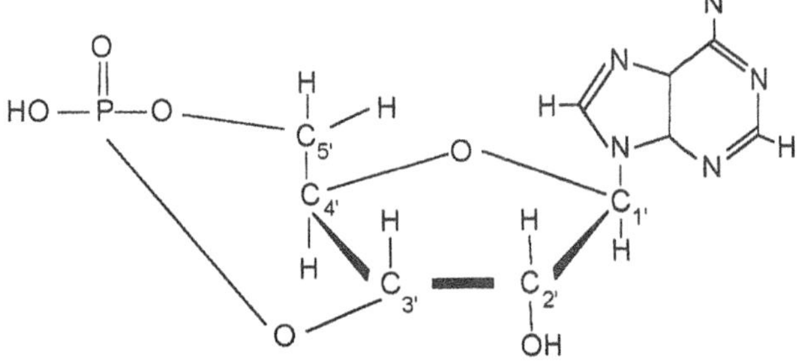

11. *N*-acetylgalactosamine is attached to serine by an *O*-glycosidic bond.

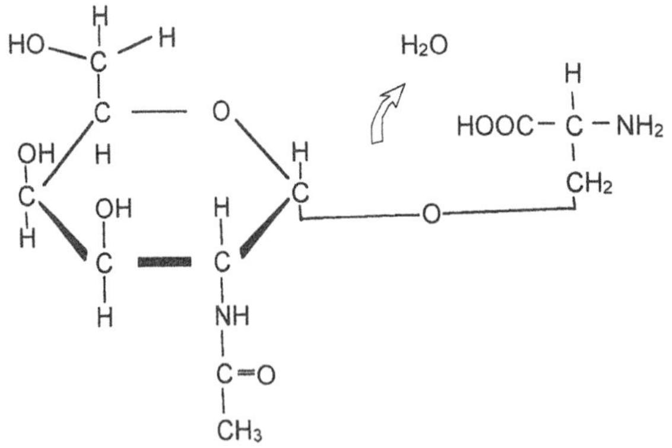

12.

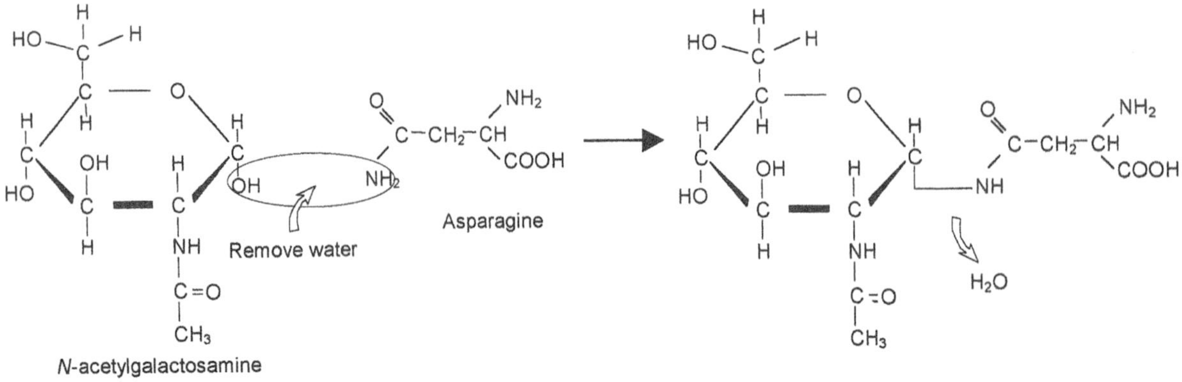

N-acetylgalactosamine

Asparagine

Remove water

4 Amino Acids and Their Functions

In this chapter you will learn:

- the functional groups: an amine and carboxyl groups
- to understand the general structure of amino acid
- the structures, names and single letter symbols for the 20 amino acids found in proteins
- how 2 cysteines may be oxidised to form the bridging amino acid cystine
- how the carbons of amino acids are named or numbered
- the terms ampholyte and zwitterion and how these relate to amino acids
- the structure of an amide bond and the special case a peptide bond
- special functions of amino acids (e.g. neurotransmitters) and structural relationships between amino acids
- how ketones are formed from amino acids by removing ammonia from the αC
- the synthesis of the bioactive amines dopamine, noradrenaline, adrenaline and serotonin.
- to understand the basis for Parkinson's disease and phenylketonuria
- that tyrosine, serine and threonine are phosphorylation sites in proteins
- the importance of decarboxylation in the formation of some active amines such as histamine
- how sugars may attach to the amino acids serine, threonine and asparagine

Basic structure and nomenclature of amino acids

The name amino acid suggests that these structures have an amine and an acid group. Indeed this is true; amino acids have an amino group and a carboxylic acid.

The structure of a typical L-amino acid is illustrated below. This type of amino acid is the basis of proteins.

$$HOOC - \overset{\overset{\displaystyle H}{|}}{\underset{\underset{\displaystyle R}{|}}{C}} - NH_2$$

Q&A 1: Draw the chemical structures of a carboxylic acid, and an amine group.

There is a central carbon that has bonds to an amine group, a carboxylic acid, an hydrogen and a variable R group. Since this central carbon has 4 different groups attached to it, it is a chiral carbon and hence there are 2 possible isomers, L and D. Nearly all amino acids in biochemistry are of the L-form (L for life). Note that this is the opposite of sugars, which nearly always occur as the D isomer. You need to learn their structure in this orientation. Remember that the L designation <u>has nothing to do</u> with the way they rotate polarised light, but is purely structural. This is based on the structure of L-glyceraldehyde.

$$OHC - \overset{\overset{\displaystyle H}{|}}{\underset{\underset{\displaystyle CH_2OH}{|}}{C}} - OH \qquad HOOC - \overset{\overset{\displaystyle H}{|}}{\underset{\underset{\displaystyle R}{|}}{C}} - NH_2$$

L-glyceraldehyde   L-amino acid

Although it is not important to remember this transposition, it is important to remember the amino acids in the L form. To do this, remember the notation that in the flat projection the horizontal arms are coming out to hug you – out of the page. Drawn the way illustrated, the mnemonic **co-r-n** forms. This is an L amino acid

$$HOOC - \overset{\overset{\displaystyle H}{|}}{\underset{\underset{\displaystyle R}{|}}{C}} - NH_2$$

Co   r   n

The next part is confusing so pay attention. To number the carbon atoms, C1 is the carbon of the carboxyl group, C2 is the chiral carbon and C3 (or more) is on the R group. However, the chiral carbon is also called the α-carbon. Hence the amino acids we commonly deal with are α-amino acids (they have a carboxyl group and an amine group attached to the α-carbon).

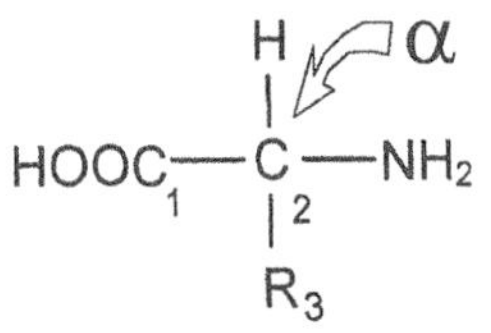

The ß carbon is C3 (the first carbon in the R group). With this system of naming, the carbon of the carboxyl group (C1) is referred to as the carbon of the α carboxyl group (the carboxyl group attached to the α C). At pH 7.4, which is the normal pH of the body, this has lost its proton (hydrogen ion) and has a negative charge with the electrons of the double bond shared across the two O atoms.

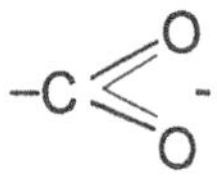

Also attached to the αC is an amine group that at pH 7.4 has an extra proton and so carries a positive charge.

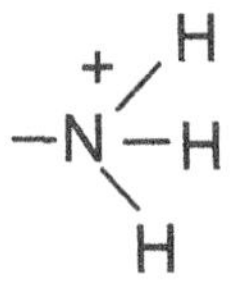

Therefore at normal pH, amino acids carry a positive charge on their amine group and a negative charge on their carboxyl group. These 2 charges cancel each other out and hence the net charge is dependent upon the charge carried on their R group.

| **Nomenclature:** Molecules which have both a negative and a positive charge are called <u>zwitterions.</u>

Note that at pH 7.4 it is possible for amino acids to lose an hydrogen ion (from $-NH_3^+$) or gain an hydrogen ion (onto $-COO^-$). A substance that easily loses an hydrogen ion is an acid and a substance that can easily take up an hydrogen ion is a base. Therefore, amino acids act as both an acid and a base.

| **Nomenclature:** Molecules that act as both an acid and a base are called <u>ampholytes.</u>

Note that in this book, most often the amino acids will be drawn with the carboxyl group as COOH and the amine group as NH_2. This is more convenient when dealing with the structure and bonding. However, from time to time it is important to emphasise the appearance of these molecules at pH 7.4 (physiological pH). Hence, be aware that the amino acids can be drawn in either form, but the more correct form is as the zwitterion.

The peptide bond

The most important reaction amino acids can undertake is the formation of an amide bond. <u>Learn this bond.</u> This forms between an amine group and an acid group and water is removed. An amide bond formed between 2 amino acids is given a special name – a peptide bond. Proteins and peptides (which are really small proteins) are formed from a string of amino acids linked together by amide bonds.

Major point 15: An amide bond between 2 amino acids is called a peptide bond.

Major point 16: Proteins and peptides are formed from a string of amino acids linked together by peptide (amide) bonds.

Both of the groups necessary for forming an amide bond occur in an amino acid (an acid -COOH and an amine – NH_2) and so 2 amino acids can be joined by an amide bond by removing water.

Form an amide bond between a carboxylic acid and an amine. Recall the similarity between this and an ester bond.

Answer:

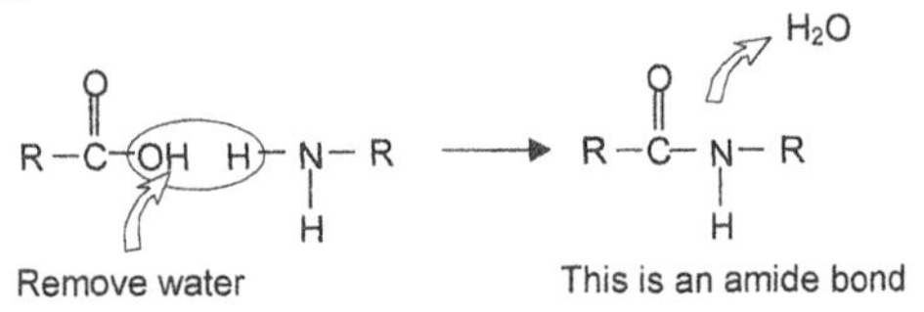

The structures and characteristics of the 20 amino acids found in proteins

There are 20 amino acids commonly found in proteins. You <u>must</u> learn these and practise them so that you know them. It is a daunting task, but persist and it will happen. Most texts group amino acids in a way that does not necessarily help to learn their structures. The best and most useful way of learning them is in terms of what they can really do. By this, I mean what types of reactions they take part in, and what role they play in proteins. You will also need to know their abbreviated single letter symbols. In this era of genetic characterisation and manipulation of proteins, it is essential that you learn these symbols.

Practising amino acids in context is not only important, it is fun. Any time you are reading about an amino acid and have forgotten its structure, look it up. This will be slow to start with, but very quickly it will be unnecessary to look it up, you will know it! The following is to help you learn about amino acids and place them in the broader context of biochemistry.

Glycine (G)

*Draw the structure of glycine (**G**)*

Answer:

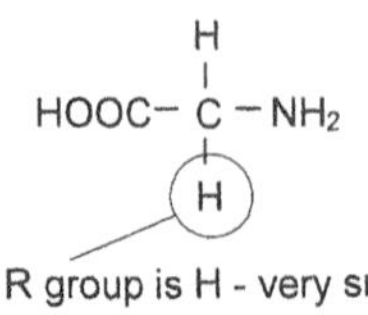

The R group is H - very small isn't it?

There are 2 main things to remember about glycine (**G**):

- the α carbon is not a chiral centre because it has two H groups attached to it. Therefore, there is no such thing as a D or L form of glycine (**G**).
- When it appears in proteins, the R group (H) provides little steric hindrance because of its size (very small). Proteins can bend or rotate easily where glycine forms part of their structure.

Glycine (**G**) is also a major inhibitory neurotransmitter in the brain. When released onto a neuron it hyperpolarises the neuron and hence decreases its activity.

Aspartic acid (D) and glutamic acid (E)

Note: glutamic acid (**E**) is the most important amino acid structure to learn.
Draw the structures of glutamic acid and aspartic acid.
Answer:

glutamic acid (E)   aspartic acid (D)

Note that their R groups (circled) contain a carboxylic acid. At normal pH (pH7.4), the carboxyl group carries a negative charge (COO⁻). Therefore, in proteins the R groups of aspartic acid (**D**) and glutamic acid (**E**) have a negative charge due to the carboxyl group, and are very water-soluble. This means that they are commonly found on the surface of proteins.

Glutamic acid (**E**) is a major excitatory neurotransmitter of the brain. It is released by a large number of neurons and is very important in a region of the brain which influences memory – the hippocampus. I wonder if you will remember this?

Nomenclature: When they have lost their proton from their R group, as found at physiological pH, they are given the names glutamate and aspartate. This signifies the base form. However, you should be aware that in common usage glutamic acid and glutamate are often used interchangeably, as are aspartic acid and aspartate.

An interesting feature of these molecules is that an amide bond can be formed with the carboxylic acid of the R group. We can do this with free ammonia – NH_3.

Redraw aspartic acid (D) and ammonia next to it. Form an amide bond between the carboxylic acid of the R group and ammonia by removing water.

Answer:

remove water

This gives us another amino acid — **Asparagine (N)**.

Do the same with glutamic acid (E). Form an amide bond between the carboxylic acid of the R group and ammonia by removing water.

Answer:

remove water

The amino acid **Glutamine (Q)** has been formed.
Not only can glutamic acid (**E**) be converted into glutamine (**Q**), but also the reverse can occur: glutamine (**Q**) can be converted into glutamic acid (**E**). Similarly, asparagine (**N**) can be converted into aspartate (**D**), by hydrolysing the amide bond.

If you remember the relationship between these molecules and the sequence: glutamine (**Q**) is converted to glutamic acid (**E**) and asparagine (**N**) is converted to aspartic acid (**D**), then you come up with a mnemonic **Q END**. This is one for billiard players.

★ **Logic process: Note that the R groups for both asparagine (N) and glutamine (Q) are quite polar, having a slightly positive charge. This is because the double bond to the O is really shared across the C to the N**

The small polar positive charge can be used to advantage when studying the function of proteins using a genetic technique called single point substitution. With this technique, the activity of the native protein is compared with one which has had one amino acid changed. If the change is subtle, and it has a large effect on the activity of the protein, then it becomes clear that the amino acid of interest is critical to the function of the protein. A subtle point mutation substitution is glutamine (**Q**) for glutamic acid (**E**), asparagine (**N**) for aspartic acid (**D**) (or vice versa). This alters the charge at this point in the protein but otherwise the R groups are essentially the same. If the protein retains its function then these amino acids are not critical for protein's function, but if the function is vastly altered then these amino acids are critical for the protein's function. Other subtle point mutations are glutamic acid (**E**) for aspartic acid (**D**) or asparagine (**N**) for glutamine (**Q**) (or vise versa). These retain the charge at that point in the protein, but alter the size of the R group affecting steric hindrance.

Ketone formation from amino acids by deamination

A very important reaction that some amino acids can undergo in the body is removal of ammonia (NH_3 = $-NH_2$ and $-H$) from the α-C. These 2 groupings are replaced by a ketone (= O).

Remove ammonia and replace it with a ketone group

This reaction is a feature of aspartic acid (**D**) and glutamic acid (**E**) and is really significant:

- it is how the body can isolate the carbon chain of amino acids so that it can be used as an energy source
- it is how ammonia, which is toxic, is removed.

> **Q&A 2:** Draw glutamic acid (**E**), circle the R group and label the α–carbon.
> Remove ammonia (NH_3) from the α-C and replace it with a ketone group (= O).

The molecule formed in Q&A 2 (α-ketoglutarate) is very important because it is a molecule in the tricarboxylic acid cycle (energy generating and carbon shuffling cycle in mitochondria, see Chapter 9). Therefore, by removing ammonia from glutamic acid (**E**) and replacing it with a keto group, the carbon skeleton of glutamic acid (**E**) can be metabolised for energy. Note that the reverse can occur as well, glutamic acid (**E**) can be formed from α-ketoglutarate by adding ammonia.

> **Q&A 3:** Draw aspartic acid (**D**), circle the R group and label the α–carbon.
> Remove ammonia (NH_3) from the α-C and replace it with a ketone group (= O).

The molecule formed in Q&A 3 (oxaloacetate) is also part of the citric acid cycle and is where aspartic acid (**D**) may enter the citric acid cycle; its carbon skeleton used for generating energy.

Alanine (A)

*Draw the structure of alanine (**A**)*

Answer:

You do not get much of a charge from this R group

This molecule is important in proteins because it has no charge on its R group.

- Its R group is an acyl group (just carbons and hydrogens not in a ring formation) and all amino acids with acyl groups for their R groups are uncharged.
- Regions of proteins with lots of these types of amino acids are lipid soluble regions of the protein.

It is also important as the basis for learning several of the other amino acids.

Similar to glutamate (**E**) and aspartate (**D**), ammonia can be removed from the αC.

*Remove ammonia from alanine (**A**) and replace it with a ketone (= O)*

Answer:

Fantastic!!! You have just made pyruvic acid.

Pyruvic acid is very important in metabolism. In the liver, pyruvic acid can be made into glucose for transport to the muscles, or it can be shuffled into the citric acid cycle so that its carbon skeleton can be used for making new molecules or for energy production. In muscles, the reverse process can take place, glucose is broken down to pyruvic acid to provide energy without using oxygen. The pyruvic acid in high concentration is toxic to the muscle and therefore an amine group from glutamic acid (**E**) can be transferred to pyruvate to form alanine (**A**). The alanine (**A**) is transported in the blood back to the liver where it is converted back to pyruvate by removing ammonia. The pyruvate is then made into glucose (gluconeogenesis) and transported back to the muscle. In this way the muscle not only removes the acid but also ammonia, both of which are toxic.

> **Major point 17:** Glutamic acid (E), aspartic acid (D) and alanine (A) can be deaminated and the carbon skeleton of the ketone formed can be used as an energy source.

*Redraw alanine (**A**) and then add a phenyl group (a benzene ring) to the R group*

Answer:

What name would be appropriate for this molecule?

Answer: Phenylalanine (**F**)

Phenylalanine (F) is an aromatic amino acid.

Its symbol (F) is the phonetic sound of the 'ph'
* it is very hydrophobic in proteins
* it is an essential amino acid in humans. It is very difficult to make the ring structure and hence we obtain this amino acid from our diet

> **Nomenclature:** Amino acids that we cannot synthesise ourselves are called essential amino acids. These must be obtained from our diet.

Examine the structure of a closely related amino acid,
Tyrosine (Y)

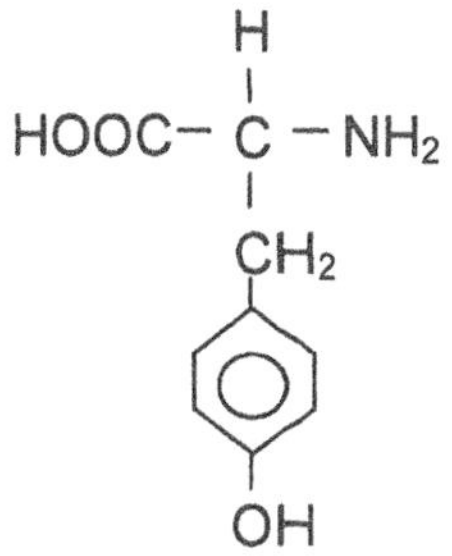

*What is the difference between this and phenylalanine (**F**)?*

Answer: *The difference is that it has an extra hydroxyl group (circled) attached to the phenyl.*

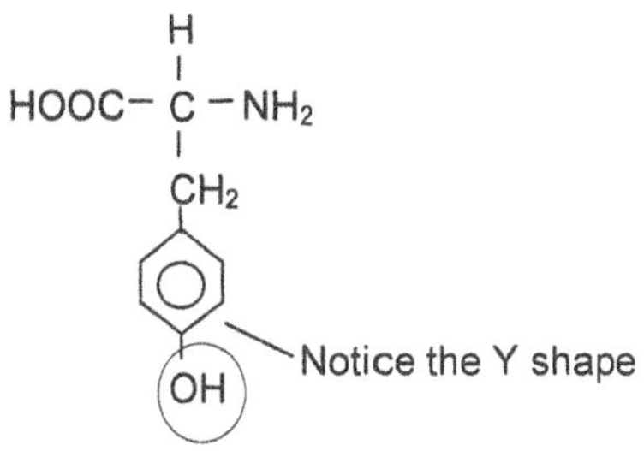

Notice the Y shape

To aid remembering its symbol (**Y**), notice that by adding the OH to the phenyl group of phenylalanine (**F**) a Y has been formed in its structure.

Because of the close structural relationship between phenylalanine and tyrosine (**Y**), you might expect tyrosine (**Y**) to be synthesised from phenylalanine (**F**) and this is indeed the case.

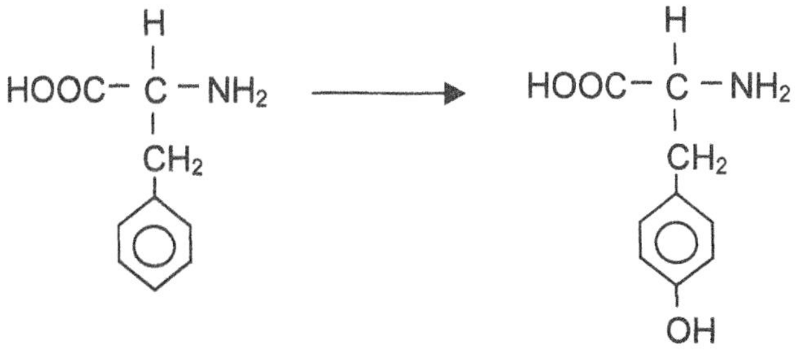

*An enzyme catalyses this hydroxylation of phenylalanine (**F**). Enzymes are often named after the reaction they catalyse and 'ase' is added to the end of their names. What would be a suitable name for the enzyme that catalyses this reaction?*

Answer: *Phenylalanine hydroxylase - an hydroxyl group has been added to phenylalanine (**F**).*

Clearly the hydroxyl group comes from somewhere. For this reaction to occur 2 additional substances are needed - oxygen and a co-enzyme named <u>tetra</u>hydrobiopterin. Tetrahydrobiopterin is converted to <u>di</u>hydrobiopterin with loss of H_2. One oxygen atom is added to the phenyl group of phenylalanine (**F**) to form tyrosine (**Y**) and one to the H_2 to form water. This is shown in the next diagram with the phenyl group drawn to show the H atoms.

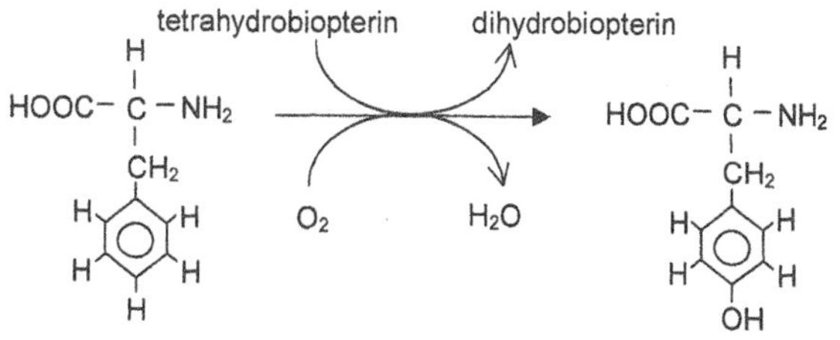

Phenylketonuria

Some people lack phenylalanine hydroxylase and cannot make tyrosine (**Y**). They therefore end up with an excess of phenylalanine (**F**) which is excreted into the urine after being modified into a ketone. This is called phenylketonuria. It is treated by dietary modification: not eating foods high in phenylalanine (**F**) e.g. cheese. Without treatment, mental retardation occurs. This is a genetic disease and a blood test, the Guthrie test, is used to screen newborn babies for this condition.

Tyrosine (**Y**) is very important
* in proteins, it is strongly hydrophobic.
* the aromatic ring allows tyrosine (**Y**) to absorb light at 280 nm. This is UV light. Since nearly all proteins contain tyrosine (**Y**), the amount of light absorbed at 280 nm by a protein is used as an indirect measure of protein concentration.
* it is the precursor for forming a group of neurotransmitters known as catecholamines. These include dopamine, noradrenaline and adrenaline.
* in proteins, its R group can be phosphorylated by forming an ester bond between the hydroxyl group and phosphoric acid (see below). This changes it from being hydrophobic (no charge) to hydrophilic (a massive negative charge). This causes the shape of the protein to change as the tyrosine (**Y**) tries to leave a hydrophobic region to move towards a more hydrophilic region of the protein. This changes the activity of the protein.

> **Nomenclature:** A catechol group is a benzene ring with one or more hydroxyl groups attached.

The synthesis of catecholamines

This is a multistep process governed by 3 enzymes. Firstly, an hydroxyl group is added in the meta position to the benzene ring of tyrosine (**Y**).

$$HOOC-\overset{\overset{H}{|}}{C}-NH_2 \longrightarrow HOOC-\overset{\overset{H}{|}}{C}-NH_2$$

Dihydroxyphenylalanine (phenylalanine with 2 hydroxyl groups) is formed. The name, dihydroxyphenylalanine is abbreviated to DOPA with the O standing for OH.

> **Q&A 4:** What would be an appropriate name for the enzyme involved in the conversion of tyrosine to DOPA? Remember that the enzyme that catalysed the conversion of phenylalanine (**F**) to tyrosine (**Y**) by adding an hydroxyl group was phenylalanine hydroxylase, and here we are adding an hydroxyl group to tyrosine (**Y**).

The next step in the reaction is the conversion of DOPA to dopamine by decarboxylating DOPA. Note that this reaction, a decarboxylation of the αC, occurs in some other amino acids to give active biological compounds (see below).

DOPA → **Dopamine**

Dopamine, the amine form of dihydroxyphenylalanine (DOPAmine), is formed.

> **Q&A 5:** What would be a suitable name for the enzyme that catalyses the decarboxylation of DOPA?

Dopamine is very important physiologically.

- It is a neurotransmitter in the brain where it has many functions, e.g. it is important in controlling blood pressure, increases in dopamine activity are thought cause Schizophrenia, Huntington's chorea, aggressiveness, and repetitive activity such as head swinging. Clinically, amphetamines are used to stimulate dopamine release to control attention deficit disorder. Used for social purposes they quickly create dependence and schizoid behaviour.
- It is also a hormone that controls the pressure of blood flow through the kidneys. If blood flow through the kidneys is decreased, dopamine is released and this causes a vasoconstriction of the kidney vessels which increases the pressure of blood flow in the kidneys.

> **Nomenclature:** A neurotransmitter is a chemical released by a neuron at a synapse to affect another post-synaptic neuron or organ (e.g. heart, lungs). Its effect is confined to the synapse and, unlike hormones, it is **not** released into the bloodstream. A hormone is a chemical released into the bloodstream by a gland in the body to affect organs or tissues elsewhere in the body.

Dopamine is then converted to noradrenaline by hydroxylating the β carbon

Dopamine → **Noradrenaline**

> **Nomenclature:** Another name for noradrenaline is norepinephrine and similarly, adrenaline is called epinephrine. Americans use epinephrine and norepinephrine.

> **Q&A 6:** What would be the name of the enzyme that converts dopamine to noradrenaline? Remember we have put an hydroxyl group onto the βC of dopamine.

Noradrenaline is an important hormone and neurotransmitter.

- It is released as a hormone by the adrenal medulla as part of the fight and flight response.
- It is the main neurotransmitter of the branch of the autonomic nervous system called the sympathetic nervous system.
- It plays several roles as a neurotransmitter in the brain.

Noradrenaline is then converted to adrenaline by adding a methyl group to the amine group of noradrenaline.

ethanolamine — phenyl — **Noradrenaline** → **Adrenaline**

In this case, you would never guess the name of the enzyme involved in this reaction unless you knew another name for noradrenaline. Note that noradrenaline can be broken down into the structures indicated above. Ethanol with an amine group = ethanolamine; and of course the phenyl group. Therefore noradrenaline is also a phenylethanolamine.

In forming adrenaline, a methyl (CH_3) group is transferred to the nitrogen. The name of the enzyme describes this reaction. It is phenylethanolamine-*N*-methyl transferase. Also known as PNMT. The *N* indicates that the transfer was to the nitrogen group.

Adrenaline is similar to noradrenaline in many of its functions.

- It is released as a hormone in the fight or flight response by the adrenal medulla.
- It is a major neurotransmitter in certain regions of the brain.

Major point 18: Tyrosine (Y), formed from the essential amino acid phenylalanine (F), is the precursor for the biologically active amines, dopamine, noradrenaline and adrenaline.

Tryptophan (W) and the formation of serotonin

In addition to phenylalanine (**F**) and tyrosine (**Y**), there is another aromatic amino acid, **tryptophan (W)**. Aromatic amino acids have a ring structure in their R groups. For tryptophan (**W**) the ring is an indole and so it is also known as an indolamine.

To help remember its symbol, you can see a W in its ring structure.

It is important in proteins because its R group (circled)

- is hydrophobic
- absorbs UV light at 280 nm.

Like tyrosine (**Y**), most proteins contain tryptophan (**W**) and its absorbance of light at 280 nm is used as an indication of protein concentration. It is another of the essential amino acids because it is very difficult for our bodies to make the complex ring structure.

Similar to the conversion of tyrosine (**Y**) to noradrenaline, it can be hydroxylated and decarboxylated to give a biologically active substance 5-hydroxytryptamine or serotonin.

Firstly, the R group is hydroxylated at carbon 5 of the ring structure. Numbering in the ring starts with the nitrogen group.

Q&A 7: What is the name of the enzyme that catalyses the hydroxylation of tryptophan (**W**)?

Next, 5-hydroxytryptophan is decarboxylated to form 5-hydroxytryptamine (serotonin).

Q&A 8: What is the name of the enzyme that catalyses the decarboxylation of 5-hydroxytryptophan?

5-Hydroxytryptamine (serotonin) is a very important neurotransmitter in the brain.

- It is involved with spinal reflexes, sleep-wake cycle, flow of sensory afferents and habituation.
- Decrease in serotonin has been implicated in the depressive phase of some manic depressive disorders. Low levels of noradrenaline are probably involved as well.
- Lysergic acid diethylamine (LSD) is an inhibitor of serotonin receptors. Use of this drug causes vivid hallucinations.
- Clinically, a drug called Prosac, which activates serotonin receptors in the brain, has been used to give people a sense of well-being.

Decarboxylation of amino acids and the formation of gamma amino butyric acid and histamine

The reactions described above in the formation of dopamine and serotonin were decarboxylations. Other amino acids are changed into biologically active components by decarboxylation: removing -COOH from the α-carbon. One such amino acid is glutamic acid (**E**)

Q&A 9: Draw the structure of glutamic acid (**E**) and circle the α-carboxyl group. There are 2 carboxyl groups to choose from. One belongs to the α-carbon and the other is attached to the γ-carbon (gamma-carbon).

Glutamic acid is decarboxylated at the αC.

Glutamic acid

Q&A 10: What is the name of the enzyme that catalyses the decarboxylation of glutamic acid (**E**)? Remember the enzyme is usually named after the reaction it catalyses and then 'ase' is added to its name.

To work out the name of the product we need some extra information:

- Butyric acid is $CH_3CH_2CH_2COOH$.
- The carbon with the carboxyl group (-COOH) has higher priority now than the carbon with the amine group (-NH_2). Therefore the carbon with the carboxyl group becomes the α-carbon (formerly the γ-carbon) and the carbon with the amine group becomes the γ-carbon (formerly the α-carbon).

Therefore we have butyric acid with an amine group on the γ-carbon. Its name describes this – γ-amino butyric acid (gamma-amino butyric acid: GABA).

In summary, decarboxylation of glutamic acid (**E**) by glutamic acid decarboxylase forms γ-amino butyric acid.

GABA is very important. It is a major inhibitory neurotransmitter in the brain. Most neurons in the brain are continually receiving GABA input that keeps them quiet. People who have anxiety are often prescribed benzodiazapines (e.g. Valium) which enhance the action of GABA on the brain.

It is ironic that a major inhibitory neurotransmitter in the brain is formed from a major excitatory neurotransmitter – glutamic acid (**E**).

Note that GABA has both an amine group (-NH_2) and an acid group (-COOH) and therefore is still an amino acid. However, it is not an α amino acid and therefore is not used for making proteins.

A second amino acid that undergoes a similar decarboxylation to form an important biologically active product is the amino acid **histidine (H)**.

The R group of histidine (**H**) is circled and the ring structure is known as an imidazole.

Histidine (**H**) is decarboxylated to give histamine.

| Histidine | Histamine |

Q&A 11: What would be a suitable name for an enzyme that catalyses the decarboxylation of histidine (**H**)?

Histamine is a very important molecule for our body's defence against infection.

- It leads to a strong inflammatory response. It causes dilatation of arterioles (blood vessels) which gives the typical redness associated with inflammation, and it makes capillaries leaky which leads to the swelling associated with inflammation.
- Histamine is released from: mast cells, a group of cells found in high numbers just under epithelial surfaces; basophils which are one of the types of white blood cells; and by platelets which are part of the clotting system in the blood.
- People who suffer from strong allergy responses release too much histamine.

Histidine (**H**) is very important in proteins – the R group can change its charge at around physiological pH (7.4).

If the pH is below this, then it gains an extra proton and carries a positive charge, and if it is above this, then the proton is lost and it has no charge. It is the only amino acid in proteins whose charge on the R group is changed around physiological pH. Hence small changes in pH around physiological pH can alter the charge (water solubility) of histidine (**H**) and this can of course alter the shape, and hence the function, of the protein.

add a proton (H^+)

Histidine

The charge is actually shared across these two amines

Serine (S) and threonine (T): hydroxyl-containing amino acids

Two amino acids besides tyrosine (**Y**) have hydroxyl groups: serine (**S**) and threonine (**T**)

The first of these, serine (**S**), is alanine (**A**) with an hydroxyl group added to it. Threonine (**T**) is alanine (**A**) with an hydroxyl and a methyl group added to it

| alanine | serine | threonine |

The R groups are circled. Their symbols just take on the first letter of their names.

Phosphorylation of serine, threonine and tyrosine

The hydroxyl groups make these 2 of the most important amino acids in proteins. Having an hydroxyl group means that they can form an ester bond. They form the ester bond with phosphoric acid.

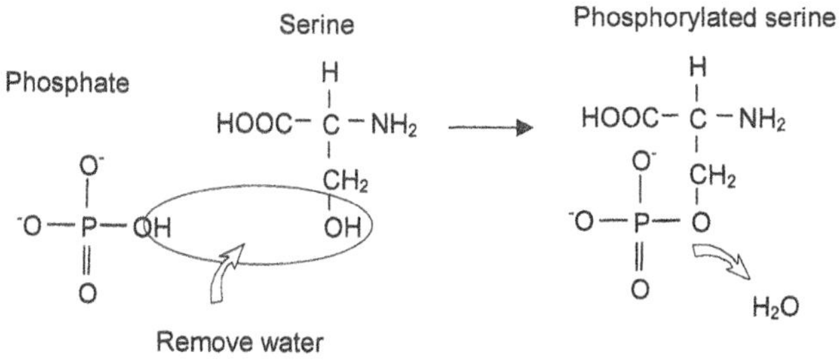

This is really important because it means that where the protein had no charge at a serine (**S**) or threonine (**T**), once phosphorylated it has a massive negative charge. This changes the shape of the protein and affects its function. In this way, they are similar to tyrosine (**Y**).

Major point 19: In proteins, serine(S), threonine(T), and tyrosine(Y) can be phosphorylated. Hence their charge changes from neutral (or slightly polar) to negative, which changes the shape and function of proteins.

Glycosylation of amino acids

Remember what else can be done with hydroxyl groups – ether bonds can be formed by condensing it with another hydroxyl (-OH) group. This is really exciting because it means that sugars, which have multiple hydroxyl groups (see Chapter 3), can be attached to proteins at serine (**S**) or threonine (**T**) by an ether

bond. Since it is a sugar being attached, the type of bond formed is an *O*-glycosidic bond.

Note that such sugar attachments do **not** occur with tyrosine (**Y**).

Another amino acid to which sugars can attach is asparagine (**N**) which we have covered already. In this case an N-glycosidic bond is formed (attaches to the nitrogen) as illustrated at the bottom of the previous page.

Major point 20: Serine (S), threonine(T) and asparagine (N) are possible attachment sites for sugars onto proteins.

Cysteine (C) and Methionine (M): Sulphur-containing amino acids

Two amino acids both contain sulphur. However, this is the only real similarity. Their functions are quite different.

Cysteine (C) (sis-tay-een)

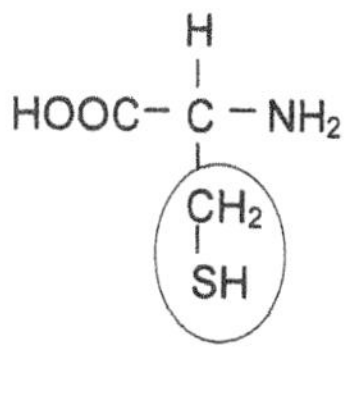

It is alanine (**A**) with a sulphydryl (-SH) group. This is really important in proteins because 2 of them can be oxidised to form a disulphide bridge. Such bridges are important for determining the shape of a protein.

Attach the sugar N-acetylgalactosamine to serine.

Answer:

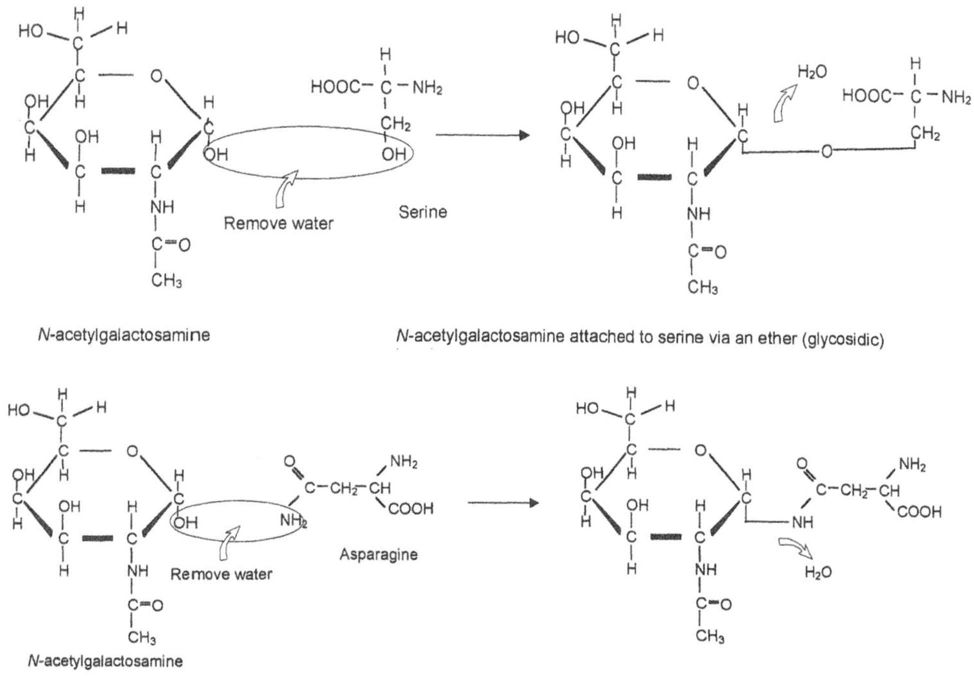

N-acetylgalactosamine

N-acetylgalactosamine attached to serine via an ether (glycosidic)

N-acetylgalactosamine

Oxidise 2 cysteine (C) groups together. Here oxidation occurs by removal of hydrogen.

Answer:

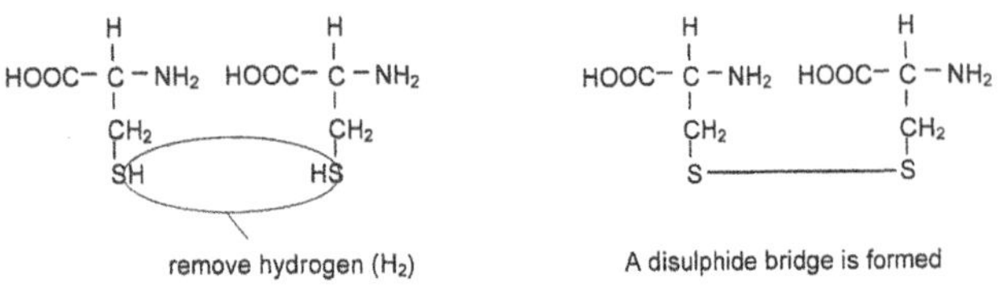

remove hydrogen (H₂)

A disulphide bridge is formed

The new molecule is **cystine** (sis-teen). In cartoons of proteins these are often represented in the structure as illustrated.

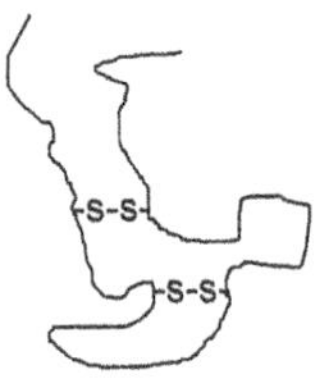

In asthma, the mucous formed is very viscous because the protein component of the mucous makes a large number of disulphide bonds. *N*-acetylcysteine (cysteine (**C**) with acetic acid condensed onto the amine group, an amide bond) is given as treatment.

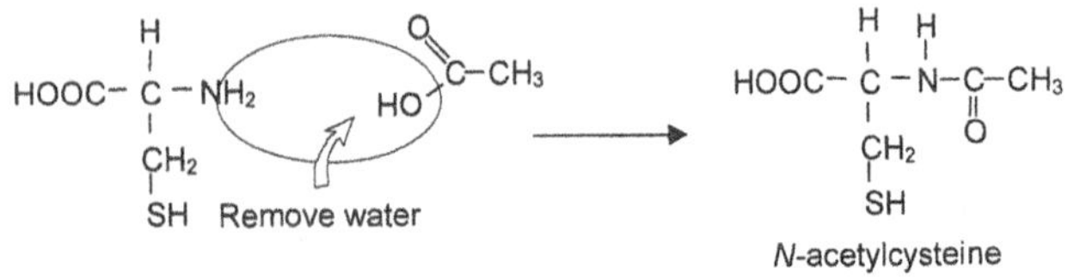

N-acetylcysteine

This binds to the cysteine (**C**) groups (oxidation) in proteins and hence preventing cross linking. The result is a less viscous mucous which enables the person to cough it up.

Methionine (M) is the other sulphur-containing amino acid.

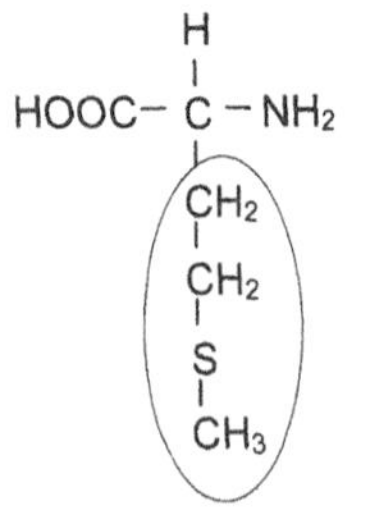

What is particularly noticeable about this group?

Answer: *The methyl group (-CH₃) sticks out like a sore thumb.*

This is very important because

- methionine (**M**) is used in proteins to transfer the methyl group. Therefore, it is inevitably found in enzymes called methyl transferases.
- it is very non-polar in proteins and hence is lipophilic.

Lysine (K) and Arginine (R): Amino acids carrying a positive charge in their R groups

The symbol for **lysine** is K. One would think it should be L, but this is a lie; it is K. The K looks a little bit like a 4 rather than a 2, 3, or 5 and hence it has 4 carbons in its R group. The charge is carried by an terminal amine.

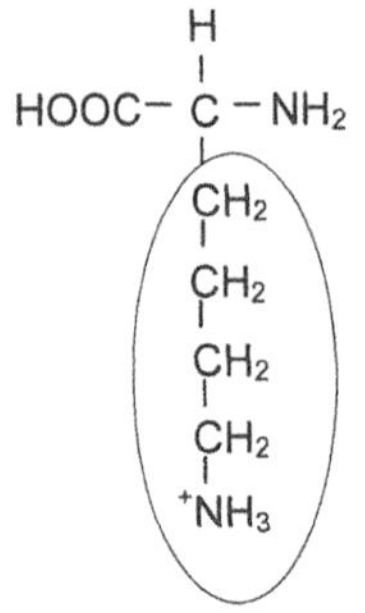

The symbol for **Arginine** (R) is a phonic for the beginning of its name. It contains a guanidinium group. This structure appears in the nucleoside base guanine, which is found in DNA. Like lysine (**K**), arginine (**R**) has 4 carbons in its R group, but remember one is part of the guanidinum group. Therefore you have C,C,C,N,C.

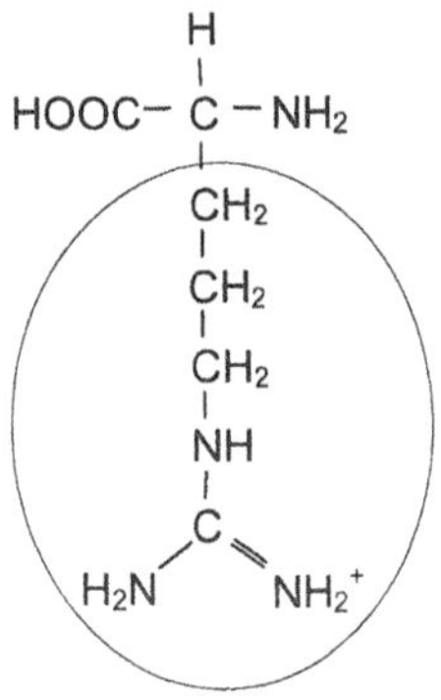

One of the important roles for arginine (**R**) is that it is used in making urea. This occurs in the liver and is how our bodies remove ammonia which in high levels is toxic to us. Cleaning agents which kill bacteria often have high levels of ammonia.

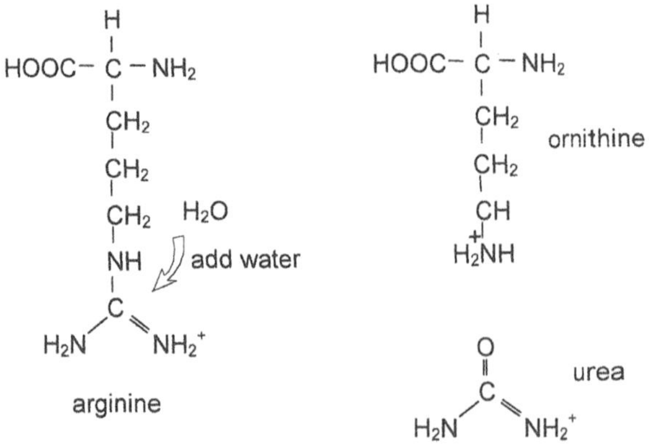

arginine

ornithine

urea

Amino acids that carry no charge

In addition to alanine (**A**) (see above), some other amino acids also carry no charge.

Valine (V), leucine (L) and isoleucine (I)

HOOC— C —NH₂ (valine)

HOOC— C —NH₂ (leucine)

HOOC— C —CH₃ (isoleucine)

valine

leucine

isoleucine

The R groups are circled. Notice the similarity. Valine (**V**) can be remembered because its R group is an inverted V in structure.

These and alanine (**A**) are important in proteins because they are uncharged (hydrophobic), hence often are in the internal regions of proteins away from the surface.

Proline (P)

Proline (**P**) is the last amino acid to learn.

HOOC— C —NH₂⁺

proline

This is a very interesting amino acid because the R group circles around forming a bond with the amine group. This forms a ring and so the structure looks like a P lying on its face for proline (**P**). The ring makes the amino acid very rigid and so when it appears in proteins, it puts a bend in their structure.

References and further reading

Almers, W. (1994) How fast can you get?. *Nature* **367**:682-683.

Baker, P.J., Britton, K.L., Engel, P.C., Farrants, G.W., Lilley, K.S., Rice, D.W. and Stillman, T.J. (1992) Subunit assembly and active site location in the structure of glutamate dehydrogenase. *Proteins* **12**:75-86.

Barrett, G.C>, ed. (1985) *Chemistry and Biochemistry of the Amino Acids*. New York: Chapman and Hall.

Cohen, J. (1995) Getting all turned around over the origins of life on earth. *Science* **267**:1265-1266

Devlin, T.M. (1992) *Textbook of Biochemistry with Clinical Correlations*, 3rd ed. New York: Wiley-Liss.

Frieden, E. (1972) The chemical elements of life. *Sci. Amer.* **227(1)**:52-60

Greenstein, B. and Greenstein, A. (1996) *Medical Biochemistry at a Glance*. Oxford: Blackwell Science.

Heiser, T. (1990) Amino acid Chromatography: the "best" technique for student labs. *J. Chem. Educat.* **67**:964-966.

Kamtekar, S, Schiffer, J.M., Xiong, H., Babik, J.M. and Hecht, M.H. (1993) Protein design by binary patterning of polar and nonpolar amino acids. *Science* **262**:1680-1685.

Meister, A and Anderson, M.E. (1983) Glutathione. *Annu. Rev. Biochem.* **52**:711-760.

Rizo, J. and Gierasch, L.M. (1992) Constrained peptides: Models of bioactive peptides and protein substructures. *Annu. Rev. Biochem.* **61**:387-418.

Snell, K. (1979) Alanine as a gluconeogenic carrier. *Trends Biochem. Sci.* **4**:124-128.

Q&A Answers

1

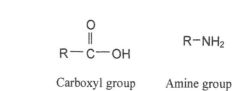

Carboxyl group   Amine group

2 Ammonia can be removed from the αC of glutamic acid (**E**) (the R group is circled) and replaced with a ketone group giving α-ketoglutarate (α-ketoglutaric acid). The name indicates that there is a ketone on the αC of glutaric acid.

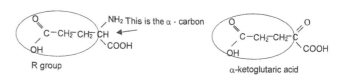

3 Removal of ammonia from aspartic acid (**D**) and formation of a ketone results in the formation of oxaloacetate (oxaloacetic acid).

R group

oxaloacetic acid

4 Tyrosine hydroxylase catalyses the formation of an hydroxyl group on tyrosine (**Y**) to form DOPA.

5 DOPA-decarboxylase is the enzyme that catalyses the removal of a carboxyl group from DOPA to form dopamine.

6 Dopamine-β-hydroxylase is the enzyme that catalyses the addition of an hydroxyl group to the βC of dopamine to give noradrenaline.

7 Tryptophan hydroxylase is the enzyme that catalyses the conversion of tryptophan (**W**) to 5-hydroxytryptophan.

8   5-hydroxy tryptophan decarboxylase is the name of the enzyme that catalyses the decarboxylation of 5-hydroxytryptophan to 5-hydroxytryptamine.

9

$$O = C(OH) - \underset{\gamma}{CH_2} - \underset{\beta}{CH_2} - \underset{\alpha}{CH} \underset{NH_2}{\overset{}{}} \boxed{COOH}$$

10   Glutamic acid decarboxylase removes the α-carboxyl group from glutamic acid (**E**) to form γ-amino butyric acid (GABA)

$$O = C(OH) - \underset{\alpha}{CH_2} - \underset{\beta}{CH_2} - \underset{\gamma}{CH_2} - NH_2$$

11   Histidine decarboxylase catalyses the decarboxylation of histidine to produce histamine.

5 Peptides, Proteins and Enzymes

In this chapter you will learn:
- the basic structure of peptides and proteins
- the structure and the 3-dimensional shape of a peptide bond
- to understand what is meant by the N and C terminus of a protein
- the difference between primary, secondary, tertiary, and quaternary structure of proteins
- to understand the importance of hydrogen bonding and its role in the formation of secondary structures in proteins
- to understand the difference between helical structure, random structure, parallel sheets, and antiparallel sheets
- to realise the importance of glycine, proline, and cysteine in the formation of tertiary structure
- to understand the difference between conjugated versus simple proteins, and the definitions of holoprotein, apoprotein, and prosthetic group
- to understand subunits of proteins and what dimers, trimers etc. are
- to understand the principles used for determining protein structure including sequencing
- about naturally occurring peptides, their synthesis and functions
- how glutathione is synthesised and know about its functions
- about the general properties of binding proteins as transporters and receptors
- about the properties of haemoglobin including its genetics and haemoglobinopathies
- to study enzyme function and classification
- to understand the difference between equilibrium and rate constants

- about the concept of transition state in enzyme catalysed reactions
- about the role of co-factors in enzymatic reactions
- the principles for studying enzyme activity, in particular the Michaelis Menten equation
- to understand the concept of steady state and how this differs from equilibrium
- to know the difference between competitive and non-competitive inhibitors of enzyme activity and how this affects the Lineweaver Burk plots of enzyme activity
- to know how the activity of enzymes can be controlled in the body

Overview of peptides and proteins

There are about 100 000 proteins in the human body. Their functions and locations are enormously varied.

> **Q&A 1: List 6 functions of proteins in the body.**

With just these 6 functions, it is obvious that they must differ in their structure and hence the obvious question arises, 'What are they made of, how do they differ, and how do they work?' Proteins are made from strings of amino acids linked together by amide bonds between the α carboxyl and the α amine groups. When amide bonds occur between amino acids they are called peptide bonds.

> **Q&A 2: Why were the α carboxyl and α amine groups specifically mentioned above rather than simply the carboxyl and amine groups?**

The size of these chains varies from about 3 to more than a thousand amino acids. Smaller chains are called peptides (up

Draw 3 general amino acids and link them with amide bonds. Remember that an amide bond is a variation of an ester bond. Just treat the NH of -NH$_2$ as the O of -OH.

Answer:

to about 40 amino acids), and slightly larger chains are called polypeptides (up to 100 amino acids) and larger chains are called proteins. Very small peptides are named after the number of amino acids e.g. a tripeptide has 3 amino acids in its structure. This division is arbitrary and there is a continuous spectrum from peptides to proteins. The amino acids used for forming these chains are the 20 α L-amino acids mentioned earlier and cystine. The sequence of amino acids determines the structure and the properties of the peptide or protein. Some general observations about proteins are given for interest (not for learning):

1. There is no sequence or partial sequence of amino acids common to all polypeptides.
2. Every possible combination of successive amino acids has been detected.
3. Proteins with different functions have very different sequences.
4. Proteins with similar functions have similar sequences.
5. The same proteins serving the same function in different species have extensive similarities.

> **Nomenclature:** Some proteins also contain other compounds within their structure. These are called conjugated proteins. The protein part is called an apoprotein and the non-protein part a prosthetic group. Combined, they are called a holoprotein. Proteins with no prosthetic group are called simple proteins.

Some simple proteins, which are found in the blood, are albumins and globulins. Examples of conjugated proteins found in the blood are haemoglobin (the prosthetic group is haeme), and lipoproteins (the prosthetic group is lipid).

Structure of proteins

Primary structure

The order of the amino acids in a protein is called the primary structure. It is nothing more than this, just the sequence of amino acids! Clearly, in a polypeptide there are two ends. At one end is the *N*-terminal amino acid, which has a free amine group, and at the other end is the *C*-terminal amino acid, which has a free carboxyl group.

> **Major point 21:** The amino acid sequence of a peptide or a protein (primary structure) is always listed from the amino acid with the free amine group to the amino acid with the free carboxyl group, i.e. from the *N*-terminus to the *C*-terminus.

Secondary structure

The secondary structure of a peptide or protein centres on the 3-dimensional shape of the peptide bond.

Draw a peptide bond between two amino acids, circle the amide bond, indicate the αCs and name the 4 atoms that form the peptide bond.

Answer:

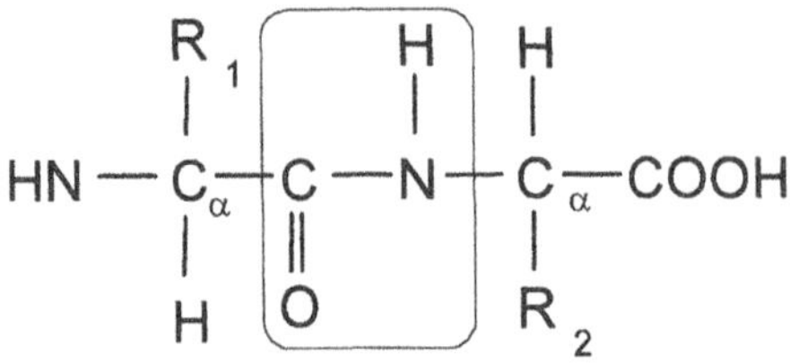

Focus on the 4 atoms of the peptide bond, the C, N, O and H. There are 4 main features to remember about the orientation of these molecules. The first, and most important, is that the double bond between the C and the O is actually shared across the bond between the C and the N. This is because there is a tendency at physiological pH (about 7.4) for the N group to form 4 bonds rather than 3 e.g. NH_4^+. Therefore, the bond between the C and the N has the characteristics of a double bond. This leads to the next 3 points. It is very difficult to rotate around this bond (C-N) because of its double bond nature, the O and the H are in the *trans* position, and all 4 molecules lie in the same plane.

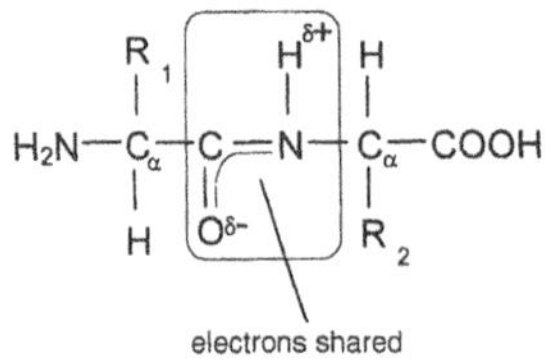

electrons shared

the peptide bond is planar with the O and H groups *trans*　*trans*

Another very important point to note is that the slight unpairing of electrons between the C and the O due to the sharing of the double bond with the N, means that the O has a slightly negative charge δ- and the electron of the H atom spends most of its time in the bond between the N and the H which gives the H atom a slightly positive charge δ+. This means that these atoms are able to take part in hydrogen bonding very readily.

If the peptide bond is fixed in a plane this means that the two αC atoms can be either *trans* or *cis* to each other.

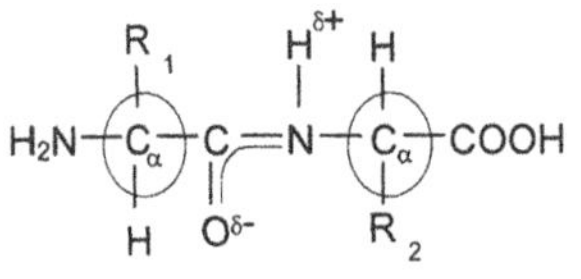

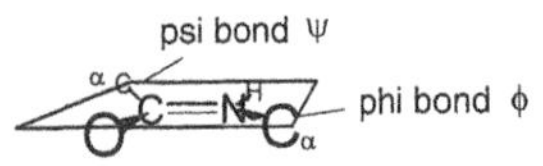

psi bond ψ

phi bond φ

The two αCs are *trans*

The αCs are *trans* and lie in the same plane as the 4 atoms of the peptide bond. These two bonds are single bonds and there is free rotation about them. Although this does not change the position of the αCs, it changes the position of the groups attached to them relative to the plane of the peptide bond. The angle of rotation about the bond between the N and αC is called the φ (phi: *fie*) angle and between the C and the αC the ψ (psi: *sigh*) angle. To remember CCψ (*see, see, sigh*) go together. For longer chains, if the φ and ψ angles are 180° then the molecule is as stretched out as possible and the distance between the two αC atoms is only 7.23Å.

The R groups project out of the plane and are *trans* to this plane.

H₂N—C$_\alpha$—C=N—C$_\alpha$—COOH

The two R groups are *trans*

The amino acids in a polypeptide or protein <u>could</u> stretch out to be as far away from each other as possible, but this is not the case because it does not provide the lowest free energy. The amino acids in the chain twist around the ϕ and ψ bonds until they find the lowest state of free energy. There are two things that influence the final shape of the chain: hydrogen bonding and the R groups of the amino acids. Hydrogen bonds occur because of the dipole effect at the C=O and N-H of the peptide bond. They also occur because some of the R groups have similar dipole groups, e.g. serine (**S**) with an -OH group. Clearly the dipoles within 1 peptide bond cannot interact with themselves and therefore the interaction must be with other amino acids elsewhere in the chain. This can occur by interactions with nearby amino acids above or below it in the chain or with distal amino acids. If there is a bend somewhere in the polypeptide chain, this enables amino acids a long way apart in the sequence to come near to each other. As a result of these interactions, 3 basic types of secondary structure are identified in polypeptide chains: helical structures; sheet structures; and random structures.

Helical structures

These occur when there is strong hydrogen bonding between the carbonyl (C=O) group and the amide (NH) group of the nearest amino acid with which it can interact. This is not the next amino acid in the chain because the two R groups are trying to position themselves as far away from each other as possible (*trans* to the peptide bond plane) and hence the carbonyl and amide groups are too far away to interact. In fact the next amino acid with which it can form stable hydrogen bonds is 3 amino acids away in the sequence. This occurs because the amino acid chain twists in a spiral to bring the carbonyl group of 1 amino acid adjacent to the amine group of the amino acid 3 amino acids away form it. Therefore each amino acid will form hydrogen bonds with 2 other amino acids: one with an amino acid 3 positions above it in the chain and the other with an amino acid 3 positions below it in the chain.

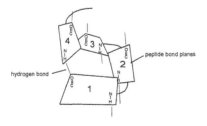

This forms a helical structure and the most common conformation is a right-handed helix where $\phi = -48°$ and $\psi = -57°$. This results in a very stable arrangement because there is little or no crowding of atoms and the C=O dipole and the N-H dipole are optimally orientated for intrachain hydrogen bonding which makes the molecule stable. This is called the α structure for packing amino acids in a protein. Since it is a helical structure, it is also called an α helix. <u>A β helix does not exist.</u> An α helix can occur as a left- or right-handed helix and this gives the protein its optical activity (whether it rotates polarised light to the right or the left) in solution. A left-hand α helix is rare.

> **Nomenclature:** A right-handed helix is based on your right hand. Make a semi-fist with your right hand such that the thumb points upwards, and then the right-handed helix spirals upwards twisting in the direction of your fingers.

If a protein is made up predominantly of α helices, then it is a fibrous protein. The α-keratins found in hair, scales, wool, hooves, and nails are good examples.

Sheet structures

When a bend occurs in a protein, very distant amino acid sequences in the protein can become aligned with each other. This enables the formation of hydrogen bonds between adjacent chains. If this occurs in an ordered array, a sheet is formed and $\phi = -139°$ and $\psi = +135°$. Large numbers of these chains causes a pleated sheet. Such sheets are called β structures or β sheets. The bend in the protein can occur so that the chains are in the same orientation (N terminus to C terminus aligned with another strand from N terminus to C terminus). This is called a parallel sheet. They can also align so that one chain is orientated N terminus to C terminus and the other oriented C terminus to N terminus. This is called an antiparallel sheet. This is more stable because the carbonyl bond (C=O) aligns more closely to the amide bond than is possible with the parallel sheets. Silk is a very good example of a β sheet and comprises many sheets laminated on top of each other.

Random structure

Random structure, as the name suggests, occurs as the amino acids have no ordered arrangements. These are usually in the regions of a protein between α helixes and β-sheets. The α helix, β sheet and random conformation are all secondary structures of proteins. A protein may contain a mixture of secondary structures or a predominance of one type. Globular proteins contain a mixture of secondary structures, and in fibrous proteins, one type of secondary structure predominates.

> **Major point 22:** α structures are helical and involve H bonding within the amino acid chain. β structures are sheet structures and involve H bonding between chains.

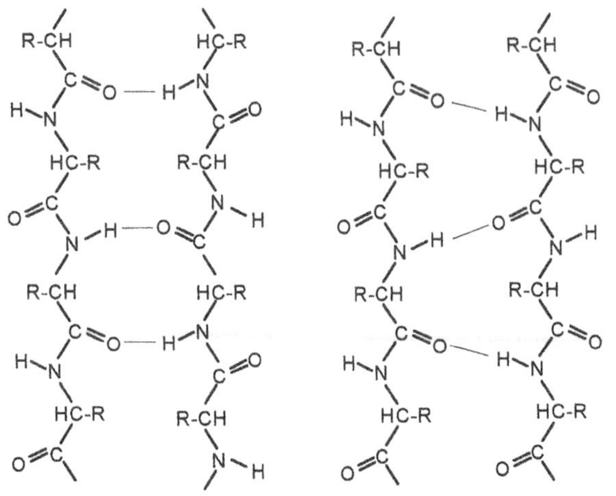

antiparallel   parallel

β structure

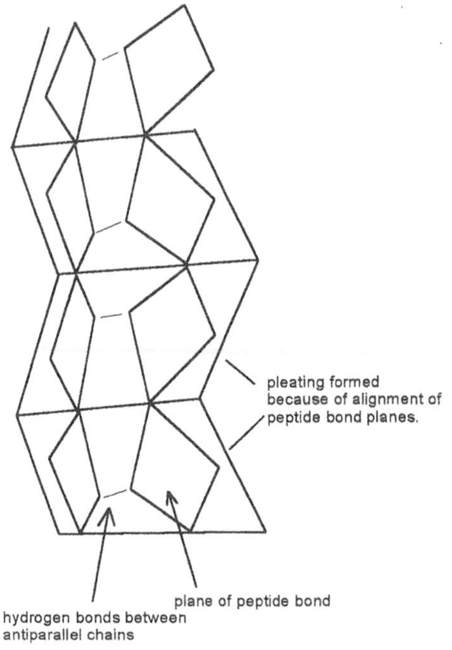

pleating formed
because of alignment of
peptide bond planes.

plane of peptide bond

hydrogen bonds between
antiparallel chains

Tertiary structure

The tertiary structure of a protein is the overall shape of the protein. The shape of a protein is critical to its function and hence there is much research being carried out on the shape of proteins. In terms of shape, one generally wants to know 2 things: the order that secondary structures occur along the length of the protein; and then the way these secondary structures are positioned relative to each other. For example, an imaginary protein might have the following structure - 6 helices (usually named alphabetically (A-F) separated by random chain regions with an antiparallel β-sheet between helix D and E. Helix C could lie very close to helix F and quite a long way from helix B or D. Of particular interest are hydrophobic regions (usually in the centre of the protein) and hydrophilic regions, usually to the outside.

There are 3 amino acids that have considerable influence over the tertiary structure of a protein, these are cysteine (**C**), proline (**P**) and glycine (**G**).

*Draw the structures of cysteine (**C**), proline (**P**) and glycine (**G**) and explain why they may influence the shape of proteins.*

Answer:

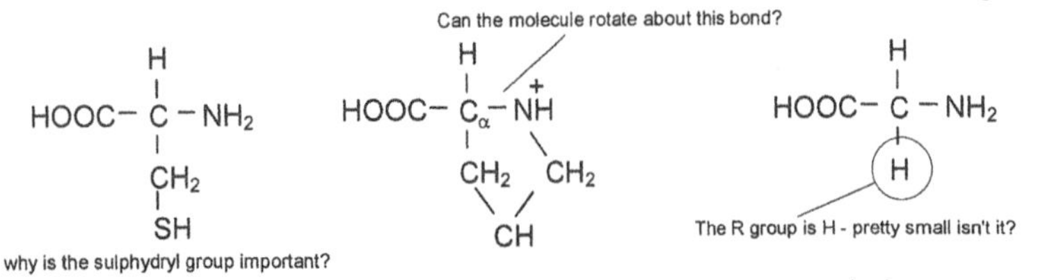

why is the sulphydryl group important?

cysteine   proline   glycine

Can the molecule rotate about this bond?

The R group is H - pretty small isn't it?

Cysteine (**C**) is extremely important because it can be oxidised (dehydrogenated) to form disulphide bridges (cystines). This causes 1 part of the protein to be covalently bonded to a distal part of the chain. A protein may have several disulphide bonds. I am sure that you have seen such disulphide bridges symbolised in cartoons of proteins.

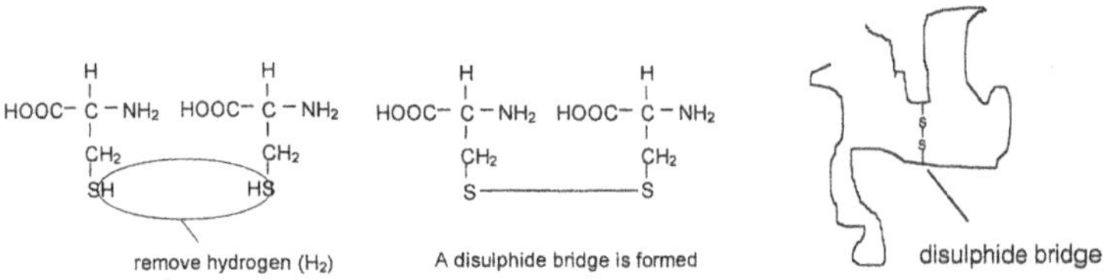

remove hydrogen (H_2)   A disulphide bridge is formed   disulphide bridge

The keratin in hair has many disulphide bridges. In permanent waving of hair, the disulphide bonds are broken with a perming solution, and the hair is curled with rollers. When the disulphide bonds reform with the hair in rollers, they reform in different positions from normal. As a result, the protein is now in a different shape from normal and the hair is permanently waved.

Proline (**P**) forces proteins or peptides to have a bend in their structure. The bond between the αC of proline (**P**) and the amine group cannot rotate freely around the φ bond because it is part of a ring. Almost the opposite is true for glycine (**G**). Here the R group is very small and gives no hindrance to rotation about the φ and ψ bonds. Larger R groups limit the rotation around these bonds. The consequence is that proteins and peptides are very flexible where glycine (**G**) occurs. Therefore proline (**P**) in a protein puts a fixed bend in it, and glycine (**G**) is a flexible site in a protein.

> **Major point 23:** Proteins and peptides may have internal disulphide cross-linking at cysteine (C) groups. They have great flexibility at glycines (G) and must bend where a proline (P) occurs.

Quaternary structure

The quaternary structure of proteins is the final assemblage of protein chains to form a physiologically active protein, e.g. haemoglobin is the assemblage of 4 protein units, and 2 of the units are coded for by one gene and the other 2 from a different gene. The number of subunits in a protein leads to the protein being described as a monomer, dimer, trimer, tetramer etc. (1,2,3,4,etc).

Q&A 3: Would a monomer have quaternary structure?

Summary

1. Primary structure – the sequence of amino acids from the N-terminal end.
2. Secondary structure – the geometrical orientation of sections of the polypeptide chain including the plane of the peptide bonds, the α helix, the β sheet and random structure
3. Tertiary structure – The order and relationship of secondary structures in a protein and the complete 3-dimensional arrangement of these.
4. Quaternary structure – The interactions between 2 or more polypeptide chains which may or may not be coded by different genes.

Determination the amino acid sequence of peptides and proteins

Note that it is not the purpose of this book to give detailed practical instructions about protein chemistry, but rather to discuss the principles of protein analysis based on what we know about protein structure. Please refer to supplementary reading list.

In general terms, the shape of a protein is determined by its primary sequence and therefore, many protein chemists and molecular biologists focus their attention on determining the amino acid sequences of proteins and from this information model the 3-dimensional structure. To determine the primary sequence, certain preliminary steps are carried out.

Considering the primary, secondary, tertiary and quaternary structure of proteins, what characteristics of a protein might help you determine its sequence/structure?

Answer

- **The size of the protein**

The mass of proteins is estimated usually by using SDS-PAGE (sodium dodecylsulphate-polyacrylamide gel electrophoresis) and comparing the position of the protein of interest with known molecular weight markers. For SDS-PAGE the protein is mixed with an SDS buffer that saturates the protein with sulphate groups (negative charges). Covered with negative charges, all proteins tend to be linear in shape and hence can be separated purely on the basis of size. If it is known that large molecular weight proteins are to be separated then low

density (4-6%) polyacrylamide gels are made, but if they are small, then a denser gel would be used (12%). Large molecular weight proteins cannot move through dense gels easily and hence they do not separate if a high-density gel is used. The reverse is true for low molecular weight proteins. If there is a mixture of proteins to be separated, then a gradient density gel is used to separate the proteins. For SDS-PAGE, the samples are placed in wells at the top of a gel and a current is run through the gel.

Q&A 4: How would the electrodes be connected positive (anode) to the top and negative (cathode) to the bottom or vise versa? Would the high molecular weight proteins travel further towards the bottom than the low molecular weight proteins? Explain your answers.

Nomenclature: Because proteins are big, the precise molecular weight is not a concern and only an approximate molecular weight is required. The size of proteins is measured in kilodaltons (kDa). One dalton is the mass of atomic hydrogen (mass of H = 1Da).

- **The ratios of the different amino acids**

The protein is degraded by acid hydrolysis in 6NHCl for 12–36 hours at 100–110°C (but this destroys tryptophan (**W**)) and the ratio of the amino acids that formed the protein is determined. This is useful because the ratio of the amino acids can be compared with that from known proteins and this may give a clue to its structure.

- **The N and C terminal amino acids. If it has more than 1 N or C terminal amino acid then it has quaternary structure: several chains joined together.**

The N-terminal amino acid is determined by binding it to dansyl chloride. Dansyl chloride forms a covalent bond with the NH_2 terminal. The tagged protein is then hydrolysed to its constituent amino acids (6N HCl, 110°C, 18–36h). Only the N-terminal amino acids will be dansylated and these can be identified by separating them using column chromatography and comparing them against dansylated standards.

The C terminal amino acid can be determined by treating intact polypeptide with hydrazine (NH_2NH_2). Hydrazine reacts hydrolytically at the carbonyl grouping (C=O) at each peptide bond resulting in acyl derivatives of each residue except the C-terminus which has an α-carboxyl (-COOH) group rather than a peptide linkage. The amino acids are then analysed by column chromatography and compared with standards. The ones without hydrazine are the C-terminal amino acids.

- **The type and amounts of prosthetic groups if it is a conjugated protein**

- **The presence of any interchain disulphide bonds**

One common method for determining interchain disulphide bonds is to use mercaptoethanol to reduce the disulphide bond. The sulphydryl groups formed can then be protected by reacting them with iodoacetate.

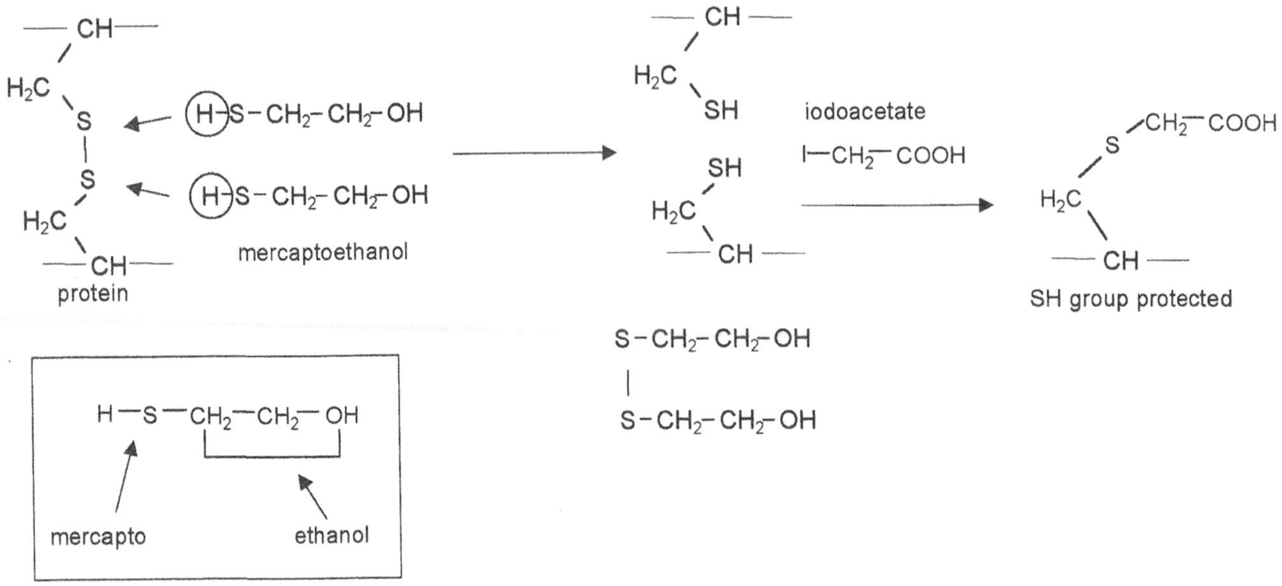

mercaptoethanol

protein

iodoacetate

$I-CH_2-COOH$

SH group protected

$S-CH_2-CH_2-OH$

$S-CH_2-CH_2-OH$

$H-S-CH_2-CH_2-OH$

mercapto   ethanol

• The amino acid sequence

This is achieved by Edman degradation on automatic machines called cyclic sequenators. In Edman degradation phenylisothiocyanate is bound to the N-terminal amino acid at pH 9 and then treated with trifluoroacetic acid (gentle acid hydrolysis). This causes a cyclisation and release of the N-terminal amino acid. The second amino acid then becomes the N-terminal amino acid and the cycle is repeated by returning to alkaline conditions. The phenylthiohydantoin separated is determined spectrophotometrically by comparing it with standards.

Only about 20 to 25 amino acids can be sequenced this way and therefore there is a problem in sequencing a whole protein which could be several hundred amino acids long. To overcome this problem, the protein sample is divided into 2 and each is treated with a different endopeptidase, which cuts the protein into smaller fragments (peptides). The peptides of each sample are separated and sequenced. This gives overlapping sequences and by knowing the N-terminal amino acid of the whole protein, the peptides can be regrouped to give the whole of the protein sequence.

To illustrate the principle, imagine the alphabet is an unknown sequence. We know that the N-terminal letter is A. The alphabet is then cleaved by 2 different peptidases.

ABCDE-FGHIJKL-MNOPQRSTU-VWXYZ

ABC-DEFGH-IJKLMN-OPQRS-TUVWXYZ

Each 'peptide' is separated and sequenced. This gives a pool of peptides

IJKLMN   ABCDE   DEFGH   MNOPQRSTU
TUVWXYZ   ABC   OPQRS   FGHIJKL   VWXYZ

and 2 of these peptides begin with A. These are lined up next to each other

ABCDE

ABC

The next 'peptide' will begin with DE (DEFGH)

ABCDE

ABCDEFGH

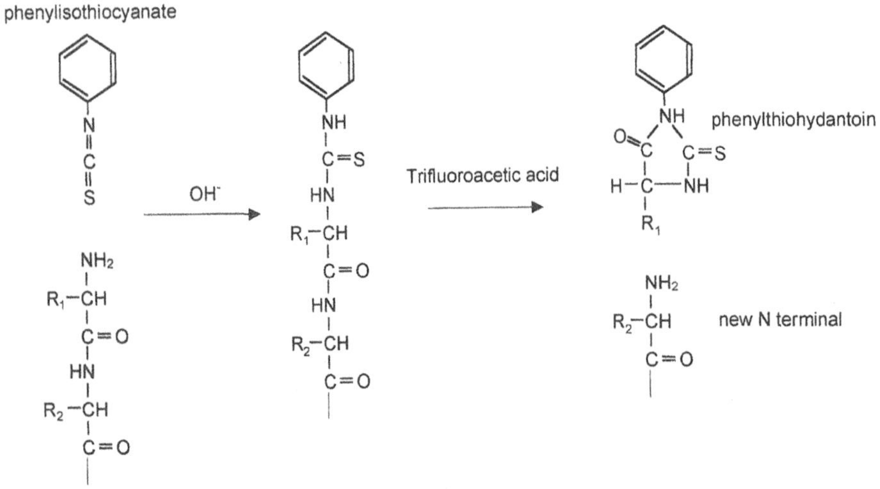

phenylisothiocyanate

Trifluoroacetic acid

phenylthiohydantoin

new N terminal

The next 'peptide' will begin with FGH (FGHIJKL) etc

Commonly used peptidases are chymotrypsin, which cleaves on the C-terminal side of aromatic amino acids (tyrosine (**Y**), phenylalanine (**F**), tryptophan (**W**)) and trypsin, which cleaves on the C-terminal side of basic amino acids (arginine (**R**), lysine (**K**)).

> **Q&A 5:** Assume the following peptides were cleaved from a trypsin and a chymotrypsin digestion respectively. Reconstruct the protein assuming that the N-terminal amino acid was G
> Trypsin cleavage - GPF AYK LHWIR GVEFMK DGSFTQR
> Chymotrypsin cleavage - IRAY GVEF KDGSF TQRGPF MKLHW

Molecular biology is also used for sequencing, and indeed is now the most common method used for full protein sequencing. This has the advantage that very little of the protein is needed and the protein does not have to be fully purified. The steps are that the protein is partially purified and the first 8 amino acids (approx.) at the N terminal end are sequenced using an automatic sequenator. These 8 amino acids are compared with an international protein/DNA sequence library available on the Internet, e.g. GENEBANK. If lucky, a match occurs in terms of molecular weight which means that someone has already isolated and sequenced the same protein or a similar protein. If this is the case, then further sequencing is probably unnecessary. If no match is found, then a cDNA molecule that codes for the N terminal amino acid sequence is made. This is called a cDNA primer. RNA is extracted from the cells that produce the protein and the RNA is probed with the cDNA primer. The cDNA primer attaches to the RNA that codes for the protein because it has a complementary sequence to the RNA sequence. DNA complementary to the rest of the RNA sequence is then synthesised using a polymerase chain reaction (PCR) method. This results in the synthesis of large amounts of DNA that codes for the protein. The DNA is then sequenced and the protein sequence derived from the DNA sequence. Although this sounds complicated, sequencing DNA is a relatively simple procedure.

- **The tertiary and quaternary structure of the protein**

This used to involve (and still does sometimes) purifying enough of the protein to form crystals of the protein. The crystal was then bombarded with X-rays (X-ray crystallography). The path of the X-rays is altered by high mass proton cores. The pattern of X-rays formed indicates the relative positions of atoms in the molecule and from this, the 3-dimensional structure can be modelled. Although this technique is still used, most modelling is now done by computer. Enough proteins have now been modelled that computer programs have been made to predict the most likely tertiary structure from the primary structure (amino acid sequence). The power of this technique is enhanced by the ability to use molecular biology to change 1 amino acid in the primary sequence. If the model of the tertiary structure is correct, then certain changes will not alter the shape/function where other changes will e.g. substituting proline (**P**) for glycine (**G**) will force a fixed bend in the molecule where it was previously very flexible. Identical techniques are used for determining the quaternary structure of a protein.

Naturally occurring peptides and their synthesis

Synthesis of peptides from preproteins

Many hormones or neurotransmitters are peptides. Hormones are chemicals produced by one cell type or organ in the body and released into the bloodstream where they carried to and affect a distant organ or tissue. A good example is insulin, which is a large peptide. Neurotransmitters are chemical substances released by neurons from their axon terminals to affect a neighbouring cell in very close proximity to the axon terminal. A good example is enkephalin, which is one of the body's own opiates or endorphins, used for dulling pain. The first neurons that were shown to produce and release peptides were in the part of the brain that integrates with the hormone system of the body: the supraoptic and paraventricular nuclei of the hypothalamus. These release the octapeptides (8 amino acids) oxytocin and vasopressin into the blood through their nerve endings located in the posterior pituitary. This was followed by the discovery of 3 hypothalamic hormones, thyrotropin releasing hormone (TRH) luteinizing hormone releasing hormone (LHRH) and growth hormone releasing inhibiting hormone (somatostatin). It is now known that these peptides are widely spread in neurons outside the hypothalamus.

In theory a cell could synthesise peptides by taking individual amino acids and sequentially joining them together until the peptide required is built, or by synthesising a large protein which contains the peptide sequence and then cleaving the peptide sequence from the protein. It is the latter that occurs. Peptides are encoded within larger protein sequences and sometimes more than one copy is encoded within the protein sequence. These large protein precursors are given names like preproenkephalin. Preproenkephalin was so named because enkephalins were discovered before the precursor molecules. The first precursor molecule to be discovered was called proenkephalin, but then a still larger precursor was discovered and hence the name preproenkephalin. It is the preproprotein which is hydrolysed (cleaved) by specific peptidases into the bioactive species.

Synthesis and function of glutathione

Remembering that the main function of glutathione is that it is easily oxidised (removal of hydrogen), how could this molecule be oxidised? Hint, 2 glutathiones are required for this to happen and recall how cysteine groups are oxidised.

Some peptides are not derived from proteins, but are synthesised by sequentially joining amino acids together, e.g. glutathione, carnosine, and tyrocidin A. Such non-protein derived peptides usually have abnormal structural properties. Glutathione (glutamylcysteinylglycine) has a peptide bond through the γ-carboxyl group. Carnosine (β-alanylhistidine) contains a β-amino acid and tyrocidin A contains α D-amino acid (D-Phe). Here, the synthesis and function of glutathione is used as an example.

Glutathione is ubiquitous in aerobic life forms and can be in very high (millimolar) concentrations with the liver having the highest concentrations. Its special feature is that it can be oxidised readily and as such, is an oxygen free-radical scavenger. It is a tripeptide comprising glutamate (**E**), cysteine (**C**), and glycine (**G**) with glutamate forming an amide bond with cysteine through its γ-carboxyl group rather than its α-carboxyl group. To synthesise glutathione, cysteine is firstly attached to glutamate and then glycine is attached to cysteine.

In each step, the synthesis of a peptide bond requires energy and this comes from the conversion of ATP to ADP.

Draw glutamate and cysteine. Name the carbons of glutamate α, β and γ. Form a peptide bond between cysteine and the γ-carboxyl group of glutamate. Name the enzyme that synthesises this new peptide, γ-glutamylcysteine.

Answer:

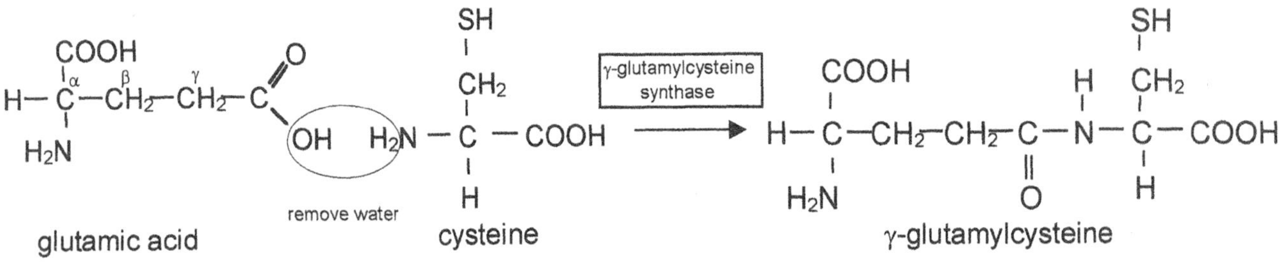

glutamic acid — remove water — cysteine → γ-glutamylcysteine (γ-glutamylcysteine synthase)

To complete glutathione, draw glycine and attach it to the cysteine of γ-glutamylcysteine via a peptide bond and name the enzyme that synthesises glutathione.

Answer:

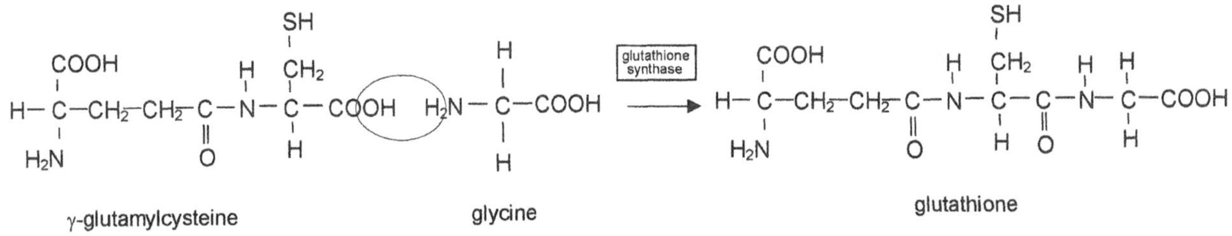

γ-glutamylcysteine + glycine → glutathione (glutathione synthase)

Answer:

Glutathione is oxidised by forming a disulphide bond between 2 glutathione molecules.

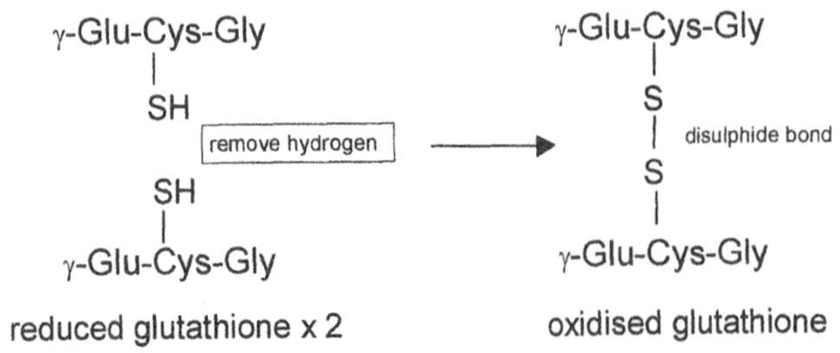

reduced glutathione x 2 — remove hydrogen → oxidised glutathione (disulphide bond)

Reduced glutathione is symbolised as GSH and oxidised GSSG. Therefore GSH + GSH → GSSG.

Glutathione acts as a co-enzyme in many reactions, e.g. for transhydrogenases and peroxidases. When it is oxidised, many highly reactive and toxic compounds are reduced. One of its most important functions is to reduce hydrogen peroxide (H_2O_2) to water. About 5% of mitochondrial O_2 consumption generates H_2O_2. If H_2O_2 were not reduced, it would form a superoxide anion $O_2^{\cdot-}$ and hydroxyl radical $OH^{\cdot}$. Both have unpaired electrons and therefore are highly reactive. Another function is to enable disulphide bonds in proteins to be broken. This is important in protein synthesis, degradation, activation and inactivation depending upon the circumstances. These reactions are catalysed by transhydrogenases. Reduced glutathione is regenerated by oxidising NADPH$^+$ to NADP.

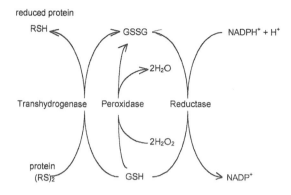

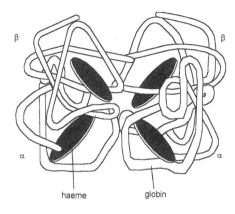

Binding proteins

There are a number of binding proteins in the body whose general role is to transport substances, or they act as receptors. One good example is albumin, which is found in high concentrations in the blood. It has non-specific binding properties for lipids and hence is one of the main transporters of lipids in the bloodstream. Lipoproteins are the other main transporters of lipid. A most fascinating transport molecule is haemoglobin and this will be dealt with in detail.

Haemoglobin

It seems straightforward that oxygen can be picked up in the lungs and transported to the tissues where it is unloaded, but of course this is not the case. The problem is the dual role: firstly to bind strongly to oxygen in the lungs and then to bind weakly to oxygen in the tissues so that it can unload it. Haemoglobin is the molecule that has the properties to make this possible. It has an additional advantage in that it transports some carbon dioxide back to the lungs for release. The mechanisms of O_2 and CO_2 transport are entirely different.

Haemoglobin is the major protein (70%) in erythrocytes (2.8×10^8 molecules per cell). It is a protein complex made from 4 subunits (a tetramer) held together by hydrogen bonding. Normal haemoglobin has 2 α subunits and 2 β subunits ($\alpha_2\beta_2$). Each subunit consists of a prosthetic group, a haeme group which actually binds oxygen, and a protein part called globin. Alpha chains are made of 141 amino acids and the β chains of 146. For naming purposes, each chain is divided into helical and non-helical areas. The helical areas are designated A-H. Amino acids can be designated by number or according to helical region, e.g. C5 is amino acid number 5 of helical region C. The haeme molecules are in clefts deep inside the molecule. These clefts are hydrophobic portions of the molecule.

> **Q&A 6: How many haeme groups would a molecule of haemoglobin have?**

Oxygen binding

Haemoglobin depends on the presence haeme groups to bind oxygen. Each haeme consists of a planer organic molecule (a porphyrin known as protoporphyrin IX) with 4 N atoms co-ordinated to a ferrous ion, Fe^{2+}. The Fe^{2+} is also co-ordinated to an N atom located above the plane of the porphyrin group. This N atom belongs to a histidine residue of the globin (protein) part of the molecule (histidine 93 or F8). When oxygen is not bound to haemoglobin, the Fe^{2+} sits 0.75Å above the plane. As O_2 enters the hydrophobic pocket in the globin part of the molecule on the opposite side of the plane, it causes the Fe^{2+} drop into the porphyrin plane.

This tiny movement has massive effects. As Fe^{2+} moves into the plane, it pulls histidine 93 of the protein backbone with it. This small movement triggers a sequence of intramolecular rearrangements that are transmitted to the other protein subunits. Changes occur to the contacts between the haemoglobin subunits, $\alpha1\beta2$ and $\alpha1\beta1$. The $\alpha1\beta1$ contact change is slight, rotating the β unit by about 4° and shifting it about 1Å. Movement of the $\alpha1\beta2$ contact is large: a rotation of about 13.9° with displacements of atoms up to 5.7Å. The consequence of these changes is that the binding of O_2 to the first haeme group makes it easier for O_2 to gain access to, and bind to the other haeme groups in an adjacent subunit. The ratio of binding constants for binding the 4 O_2 molecules to haemoglobin is 1:4:24:9. This fascinating effect is what enables haemoglobin to take up O_2 so readily in the lungs and reverse this process in the tissues. When a substance non-covalently binds to a protein and increases or decreases the activity of the protein, it is called an allosteric effect. Here, it is the non-covalent bonding of O_2.

This allosteric effect is seen in the binding curve of haemoglobin, which is sigmoidal rather than hyperbolic. A hyperbolic binding curve would occur if there were no allosteric effect. If the binding curve were shifted a small amount to the left, as shown below, then O_2 would bind better by about 2%, but the amount of oxygen released to the tissue would be much less (~40% of O_2 released to tissue instead of ~70%). Therefore, despite the fact that the haemoglobin can carry more oxygen, it has in fact been a disadvantage to the body because it cannot release it to the tissues.

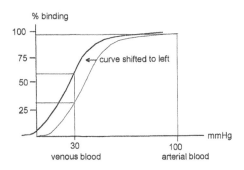

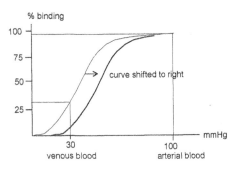

Fetal haemoglobin, which is made from slightly different protein subunits (see below), has a binding curve shifted to the left. This increased binding allows fetal haemoglobin to strip haemoglobin from the mother (adult haemoglobin). To compensate for the decreased ability to release O_2, the levels of Hb are almost twice as high in a fetus as in an adult.

> **Q&A 7:** In high altitude, where the partial pressure of oxygen is less, would it be an advantage to shift the oxygen-binding curve to the left?

> **Major point 24:** Interactions between spatially distinct sites of a molecule are termed allosteric interactions. In proteins they serve as a controlling factor for the activity of the protein.

Haemoglobin can also transport H^+ and CO_2 and these also have allosteric effects that alter the binding of O_2. In the tissues, where there are high CO_2 levels, CO_2 becomes bound

to Hb. It does not bind to the haeme group, but to the N-terminal end of the 4 globin groups to form carbamino compounds. This is possible because the N-terminal amino groups of haemoglobin have a pK between 7 and 8 and therefore, at physiologically pH, there is a relatively high proportion of un-ionised amine which is capable of forming a carbamino compound.

$$-NH_2^+ \rightleftharpoons -NH_2$$

$$-NH_2 + CO_2 \rightleftharpoons -HN-C{\overset{O}{\underset{O^-}{}}} \quad H^+$$

carbamino group

In oxyhaemoglobin, the 4 terminal carboxyl termini are free to move, but in deoxyhaemoglobin they are anchored. The carboxyl terminus of the α1 chain interacts with the terminal amino of the α2 chain and the arginine side chain which is the is carboxyl terminal residue is linked to an aspartate on the β chain. The carboxyl terminus of the β1 chain is linked to a lysine side chain on the α2 chain Finally the imidazole side chain of the carboxyl terminal residue (histidine 146) of the β1 chain interacts with and aspartate on the same chain. These pairs are held by salt links. This tends to cause the N termini to bind CO_2 as carbamino groups.

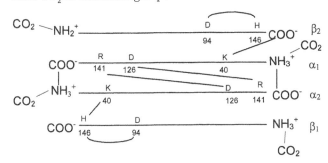

The changes in affinity of haemoglobin for CO_2 is called the Haldane effect.

The acidity $[H^+]$ also affects this interaction and the binding of O_2. It makes sense that certain R groups of the globin part of the molecule will tend to be protonated at lower pH (more H^+ present in the tissues) and deprotonated at higher pH (less H^+ present in the lungs). The most likely candidates are the R groups of amino acids that already have a pK around 7. These are histidine (**H**), cysteine (**C**) and the terminal NH_2 groups. Experimentally, it turns out to be histidine 146 on the β chain, which can be readily protonated in the deoxy form of haemoglobin. This is the terminal amino acid of the β chain. It occurs because at low pH, histidine 146 tends to be protonated and hence carry a positive charge. This attracts the negatively charged aspartate 94 on the same β chain with which it forms a salt bridge. Formation of the salt bridge promotes the release of O_2 from haemoglobin. In the lungs, $[H^+]$ decreases as CO_2 is eliminated and O_2 binding to haemoglobin increases. Therefore, changes in pH affect the oxygen binding curve (shifts it to the right in the tissues and to the left in the lungs). This is known as the Bohr effect. This was named after the degree of interest it promoted in students of biochemistry. The Haldane effect (shifts in CO_2 binding) is physiologically more important than the Bohr effect which relates to O_2 binding.

Higher level

Genetics of haemoglobin

What is extremely important to realise is that the ability for haeme to bind oxygen depends on its interaction with the flexible globin part of the molecule and it is this that allows oxygen to bind more or less readily. It is therefore quite conceivable that changes to the globin structure will alter its flexibility and hence oxygen binding ability. Structural changes to the globin chains are caused by genetic disorders and are grouped together as haemoglobinopathies.

Adult haemoglobin is made from combinations of 4 globin chains designated as α, β, γ, δ. It is mostly HbA (97.5%) which comprises 2 α-globins and 2 β-globins ($\alpha_2\beta_2$). The remaining haemoglobin is made up of 2% haemoglobin A_2 ($\alpha_2\delta_2$), and 0.5% fetal haemoglobin or haemoglobin F ($\alpha_2{}^G\gamma_2$, or $\alpha_2{}^A\gamma_2$). The genes for the α-globin are actually part of a gene cluster which is located on chromosome 16 and similarly the genes for β globin exist as a gene cluster on chromosome 11. We have 2 chromosome #16s and therefore we get proteins being produced as products from the genes of both chromosomes. The same is true for chromosome #11.

In a gene cluster, there is a sequence of expression of the genes during development. As one gene is switched on another gene is switched off. In the first few weeks of life, haemoglobin is produced in the yolk sac and is a tetramer made from 2 ζ chains which are encoded within the a cluster and 2 ϵ chains transcribed from the β cluster. This is rapidly changed to the formation of 2 α-globins from 2 genes in the a cluster, $\alpha1$ and $\alpha2$, which are identical in their coding regions and therefore produce a single protein. Expression of these genes continues for the rest of life. The fetal haemoglobin genes are expressed on the β cluster. These are a pair of genes, $^G\gamma$ and $^A\gamma$, which differ in their coding sequence by only 1 amino acid at position 136: glycine for $^G\gamma$ and alanine for $^A\gamma$. This means that the globin from the $^G\gamma$ gene is more flexible than that from the $^A\gamma$. The γ genes are turned off around the time of birth and the β-globin production takes over.

Haemoglobin F is absolutely necessary because the fetus needs a substance that binds oxygen more strongly than the mother's haemoglobin so that it can strip oxygen off the mother's haemoglobin. This has a disadvantage in that it is also more difficult to release the haemoglobin. This is overcome in a fetus by having more haemoglobin (about 210 g/L blood compared with about 120 g/L in adults).

This excess haemoglobin is broken down into bilirubin when the baby is born and hence the jaundice observed in newborn babies.

Many mutations of these genes have been discovered, but we will concentrate upon the mutations known as thalassaemia. Thalassaemia results from a frameshift mutation. This occurs when there is an insertion or deletion of a small number of nucleotides that are not a multiple of 3. When this occurs, the amino acid sequence beyond the point of frameshift is completely garbled. In thalassaemia, this frameshift occurs near the origin (5' end) and leads to a virtually unrecognisable protein product that bears no resemblance to globin. If this occurs on the β gene, the person produces little β globin and a relative excess of α globin. The result is an abundance of α globin tetramers instead of the normal $\alpha_2\beta_2$ tetramers. This is known as β thalassaemia. An equivalent condition, known as α-thalassaemia, occurs when α-globin is not produced. The severity of the condition varies considerably because the defective gene may be on one chromosome and not on the other and therefore some normal globin is still produced. Added to this, for α-thalassaemia there are 2 genes on each chromosome coding for α-globin and only 1 of these may be affected by the defect. If both β genes are defective there will be no haemoglobin A whatsoever produced (thalassaemia major). If only 1 of the genes is defective then some haemoglobin A will be produced and the disease is generally asymptomatic and is known as thalassaemia minor.

Ligand gated ion channel receptors

You will be aware that most chemical signals in the body are transported in the bloodstream (hormones) or pass across the extracellular fluid (neurotransmitters) to their target tissue. Therefore, most of these substances are water soluble. To affect the target cells, they must somehow change the inside of the cell. The problem is how to gain access to the inside of the cell when they are water soluble and the cell membrane is a lipid. The answer is that they cannot. Instead, a protein is incorporated within the cell membrane. These proteins have regions of water soluble (charged and polar) amino acids and regions of lipid soluble (uncharged) amino acids. The water soluble regions of the protein stick out from the membrane and the lipid soluble amino acids lie within the cell membrane and commonly are α-helices. Therefore, these proteins wend their way in and out of the membrane with some water-soluble parts lying on the outside of the cell and some on the inside of the cell. The different domains are named as follows. In the illustration below we have an N-terminal extracellular domain, a C terminal intracellular domain and 5 intramembranous domains called (M1 to M5). The regions between the intramembranous domains are called loops e.g. the M1-M2 loop.

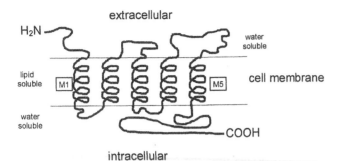

Q&A 8: Is the M2-M3 loop intracellular or extracellular?

Q&A 9: If the protein had 4 membrane spanning regions instead of 5 would the C terminus be intracellular or extracellular.

For one group of receptors, binding of a ligand (neurotransmitter or hormone) to the receptor protein on the outside of the cell changes its shape on the inside. This affects other proteins on the inside and the cell responds without the hormone or neurotransmitter ever reaching the inside of the cell.

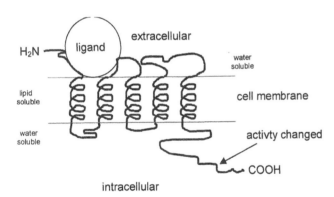

In some cases, a cluster of transmembrane proteins is grouped together. When the ligand binds to this cluster, the cluster changes shape or reorganises itself in such a way as to form a hole or channel in the membrane. These channels allow ions to pass across the membrane in accordance with electrochemical gradients and this changes the ionic environment inside the cell causing it to respond to the stimulus. Such clusters of proteins, which form a pore, are called ligand gated ion channel receptors.

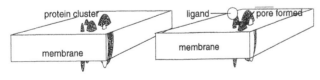

ligand gated ion channel

Receptors for acetylcholine, serotonin, gamma-amino butyric acid, and glycine are of this type. Each comprises a cluster of 5 membrane-spanning protein subunits. Each of the 5 subunits has 4 membrane spanning regions and hence the N

terminal and the C terminal ends of the protein are extracellular. Forming the channel opening when the ligand binds requires long-range allosteric interactions.

Enzymes

These are protein catalysts – increase the rate of a reaction without itself undergoing a change. Enzymes alter the rate and not the equilibrium constant of a reaction.

Consider the reaction $a+b \rightleftharpoons ab$ at equilibrium we might have $5a+7b \rightleftharpoons 20ab$. The equilibrium constant indicates the proportion of the substrates and products when equilibrium is reached. The equilibrium constant is the product (mathematical) of the concentration of the substrates divided by the product (mathematical) of the concentration of the products. So for $5a+7b \rightleftharpoons 20ab$ and $2a+2b \rightleftharpoons 6c +3d$ the equilibrium constants K would be

$$K = \frac{[5a][7b]}{[20ab]} \text{ and } K = \frac{[2a][2b]}{[6c][3d]} \text{ respectively.}$$

These can be measured empirically. However, the equilibrium constant gives no idea of the rate of the reaction. The rate is completely independent of the equilibrium. The reaction $a+b \rightleftharpoons ab$ might take 1000 years to reach its equilibrium $5a+7b \rightleftharpoons 20ab$ or a few seconds.

Major point 25: Enzymes increase the rate of a reaction, and do not alter the equilibrium.

Enzymes are distinguished by their specificity and many show absolute specificity for their substrates, including stereospecificity. A simplistic view of how they work is that they bring the atoms of the reactants close together and in the correct orientation for the reaction to take place, and hence this is not left to chance. The reaction is often enhanced by drawing electrons away from or adding electrons to the interacting atoms and this also increases the possibility that the reaction will take place. Therefore, there is a transition state when going from substrate to product and this is at a much higher energy level. Without an enzyme, the substrates would never reach this transition state at the higher energy level and hence never become products, even though the products are at a lower energy state than the substrates. It is like a big hill and the enzyme lowers the hill so that the substrates easily climb over to roll down the other side to become products. Proteins are ideal for being enzymes because they can make pockets for binding specific reactants and have enough flexibility to bring them into the right position for the reaction to take place.

Cofactor or coenzyme-dependent enzymes

Optimum activity of many enzymes depends on the co-operation of non-protein substances called cofactors. The molecular partnership of an enzyme and its cofactor is called a holoenzyme. The protein without the cofactor is called an apoenzyme and exhibits low activity (perhaps none at all). Cofactors may be inorganic ions Zn^{2+} Mg^{2+} Mn^{2+} Fe^{2+} Cu^{2+} K^{2+} and Na^{2+} or organic cofactors. Organic cofactors are called co-enzymes and are often derived from vitamins. An example of a common cofactor is flavin adenine dinucleotide FAD or flavin mononucleoside (FMD) which is derived from riboflavin (Vitamin B_2).

The cofactor either activates the enzyme by changing its 3-dimensional shape, thus providing a shape that maximises the binding and interaction of the enzyme with the substrate, or it actually participates in the overall reaction as another substrate (coenzymes mainly). In this case, it commonly acts as a donor or acceptor of a particular chemical group (group transfer agents).

Enzyme nomenclature

Some enzymes have non-descript names such as pepsin, renin, lysozyme but most end with the term -ase on the basis of the type of reaction it catalyses e.g. acetylcholine esterase, alcohol dehydrogenase.

> **Q&A 10:** What would the name be for the enzyme that catalyses the oxidation of alcohol by removing hydrogen?

> **Q&A 11:** What would be a suitable name for the enzyme that catalyses the breaking of the ester bond in acetylcholine?

This random name allocation led to some enzymes having more than one name and so a systematic approach was adopted in 1965 and revised in 1972. This divided the enzymes into 6 main groups, and then further into subclass, sub-subclass, and a serial listing of the specific enzyme in its sub-subclass. This leads to an enzyme being defined by 4 numbers, the first of which is one of the main classes.

For example, the enzyme 4.1.1.22 (histidine carboxylase) is a lyase (4). The 4.1 indicates it is a carbon-carbon

Main class	Type of reaction catalysed
1. Oxoreductases	Oxidation-reduction reactions of all types
2. Transferases	Transfers an intact group of atoms from a donor to an acceptor
3. Hydrolases	The hydrolytic (H_2O) cleavage of bonds
4. Lyases	The cleavage of bonds by means other than hydrolysis or oxidation
5. Isomerases	Interconversion of various isomers
6. Ligases	Bond formation due to the condensation of 2 different substances, with energy supplied by ATP

lyase; the 4.1.1 indicates it is a carbon lyase of C-COO-bond and the 4.1.1.22 indicates that it cleaves the C-COO-bond in histidine. 4.1.2.X indicates it cleaves a carbon-aldehyde bond; and 4.2.X.X indicates that it is a carbon-oxygen lyase.

Enzyme kinetics

Enzymes are the most efficient catalysts known. Biochemists are interested in the rates of enzymatic reactions at different pH levels, temperatures, concentrations of reactants, products and inhibitors because by studying these parameters, a great deal can be learnt about the nature of a particular enzyme and the biochemical pathway involved. This is called enzyme kinetics.

In studying enzyme kinetics, the focus is on how quickly the substrate is converted into product. You could imagine that at the beginning of the reaction, when there is substrate alone, the substrate is converted to product quickly. As the product builds up and the substrate decreases, the reaction slows and eventually stops at equilibrium.

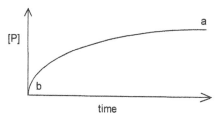

The rate of the reaction is the change in [P] (or [S]) with time. At point a on the graph there is no new product made because the reaction has reached equilibrium. At point b the reaction is very fast. Therefore there is an obvious problem: how to measure the reaction rate when the rate is changing. The answer is simple, you don't measure all points just the initial rate of enzyme activity (point b). A method for doing this was developed by Michaelis and Menten. This resulted in a formula for the initial reaction rate before the levels of product begin to affect the reaction rate.

The Michaelis Menten equation

$$v = \frac{V_{max}[S]}{K_m + [S]}$$

Where v is the initial velocity of the reaction. This can be measured experimentally by the rate of disappearance of the substrate or the rate of appearance of the product during the first few minutes of a reaction. Whether product or substrate is actually measured depends upon how easily either can be measured. V_{max} is the fastest the reaction could possibly occur. For any particular set of conditions it is a constant. K_m is a special rate constant called the Michaelis constant. [S] is the substrate concentration at the initial part of the reaction. Of course this changes as the reaction proceeds, but this is by only a small amount in the initial part of the reaction. Hence, the known substrate concentration at the beginning of the reaction approximates the substrate concentration when the experimental data are taken, i.e. in the first few minutes of the reaction.

Learning the derivation of the Michaelis Menten equation is higher level learning. However, it is very important to read through it to understand better the principles behind it.

Higher level

Derivation of the Michaelis Menten equation

Consider a substrate converting to a product under the control of an enzyme **S → P**. Distinct steps must occur in this process. To begin with the enzyme must bind to the substrate.

Do you think an enzyme and substrate is different from the enzyme and substrate joined together?

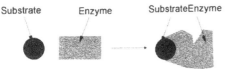

Substrate   Enzyme   SubstrateEnzyme

Obviously, once the enzyme has joined to the substrate it is different from the substrate and the enzyme alone. The product is then released from the EnzymeSubstrate complex to give product and regenerate what?

EnzymeSubstrate

Product

Enzyme

Product is released and the enzyme is regenerated

In terms of developing a model for studying enzyme kinetics, it is important to realise that each of these reactions takes time and hence each has a rate constant associated with it. In addition, sometimes when the substrate binds to the enzyme, it will not be converted to product, but simply separate from the enzyme to regenerate enzyme and substrate alone. This will also have a rate constant associated with it. Theoretically, at least, the product could combine with the enzyme and be converted back to an enzyme substrate complex. This is extremely unlikely because enzymes are designed to release product, not to bind it, and furthermore we are dealing with the initial part of the reaction when there is practically no product. This results in the following equation.

$$E + S \underset{k_3}{\overset{k_1}{\rightleftharpoons}} ES \underset{k_4}{\overset{k_2}{\rightleftharpoons}} E + P$$

In this case, the *k*s are rate constants (<u>not equilibrium constants</u>). The rate constant k_4 can be ignored because

this reaction practically never takes place in the initial part of the reaction.

The next step is based on the assumption that steady state is reached quickly. This is not to be confused with equilibrium, and indeed lots of product would be formed during this phase. Steady state means that *ES* is being formed as quickly as it is being broken down. The rate of formation of *ES* can be expressed as

$$v_{formation} = k_1[E][S]$$

Write an equation which represents the rate of breakdown of ES. Remember that ES is being broken down to E + P and E + S.

$$v_{breakdown} = k_2[ES] + k_3[ES]$$

Since steady state is assumed, the rate of breakdown of ES equals the rate of formation. Write an equation for this.

$$k_1[E][S] = k_2[ES] + k_3[ES]$$

Now the equation is rearranged to bring the rate constants together. Firstly, rearrange the right hand side of the equation

$$k_1[E][S] = [ES](k_2 + k_3)$$

Then divide both sides of the equation by k_1

$$[E][S] = [ES]\left[\frac{k_2 + k_3}{k_1}\right]$$

This is a great idea because the rate constant term is just a constant and therefore, it can be substituted by a simpler form, K_m, which is called the Michaelis constant.

substituting

$$\left[\frac{k_2 + k_3}{k_1}\right] = K_m \quad \text{where } K_m \text{ is the Michaelis constant}$$

then at steady state in the initial part of the reaction

$$[E][S] = [ES]K_m$$

The question now is can we determine K_m. To do this, we must be able to measure quantities within the reaction mixture. Once the reaction starts, some of the free enzyme is converted to enzyme bound to substrate so we cannot measure free enzyme or enzyme bound to substrate. However, we do know the amount of enzyme that was first used. The amount of free enzyme at any time [E] in the reaction is therefore the amount of enzyme first put into the mixture $[E_t]$ minus the amount converted to ES. Free enzyme can then be expressed as

$$[E] = [E_t] - [ES]$$

Substitute this into the equation $[E][S] = [ES] K_m$ to give

$$([E_t] - [ES])[S] = [ES] K_m$$

Mathematically, it is an advantage to have like terms grouped on one side of an equation and you will notice that [ES] appears on both sides. To do this both sides are divided by [S], which yields

$$[E_t] - [ES] = \frac{[ES]K_m}{[S]}$$

and then divide both sides by [ES] to give

$$\frac{[E_t]}{[ES]} - \frac{[ES]}{[ES]} = \frac{K_m}{[S]} \quad \text{or} \quad \frac{[E_t]}{[ES]} - 1 = \frac{K_m}{[S]}$$

This can be rearranged to give

$$\frac{[E_t]}{[ES]} = \frac{K_m}{[S]} + 1 \quad \text{or} \quad \frac{[E_t]}{[ES]} = \frac{K_m + [S]}{[S]}$$

We now need to obtain an alternative expression for $[E_t]/[ES]$ since ES cannot be measured easily. This required a particularly clever piece of logic by Michaelis and Menten. They considered an extreme hypothetical condition. Such a strategy is often used in maths to solve equations. If all the enzyme is saturated with substrate then none will be free and hence $[E_t] = [ES]$. The velocity of the reaction will be the maximum possible and therefore $V_{max} = k_2[E_t]$. In any other situation the velocity of the reaction $v = k_2[ES]$. (This is a mathematical way of saying how quickly the product is formed.

$$\text{Therefore } [E_t] = \frac{V_{max}}{k_2} \text{ and } [ES] = \frac{v}{k_2}$$

From these 2 expressions we can obtain the ratio $[E_t]/[ES]$

$$\frac{[E_t]}{[ES]} = \frac{V_{max}/k_2}{v/k_2} = \frac{V_{max}}{v}$$

This can be substituted into the earlier equation and solving for v. It is a good idea to solve the equation for v because the velocity of the reaction can be measured over a short period of time.

$$\frac{V_{max}}{v} = \frac{K_m + [S]}{[S]} \quad \text{or} \quad \frac{v}{V_{max}} = \frac{[S]}{K_m + [S]} \quad \text{or} \quad v = \frac{V_{max}[S]}{K_m + [S]}$$

This is the Michaelis Menten kinetic equation.

To derive it, there are 2 key points to remember:
- Steady state is assumed and therefore the rate of formation of ES = the rate of breakdown of ES
- The special situation where all of the enzyme is attached to substrate $E_t = ES$ and when this occurs the rate is maximum V_{max}

> **Major point 26:** The Michaelis Menten equation states
> $$v = \frac{V_{max}[S]}{K_m + [S]}$$

Measurement of K_m and V_{max}

The Michaelis Menten equation is useful for comparing different enzymes or the activity of the same enzymes when various inhibitors are included in the reaction. For such comparisons, it is important to establish values for K_m and V_{max}. This is done experimentally by measuring the initial reaction rate using a variety of different substrate concentrations.

Imagine that a huge amount of substrate was used in the reaction. Common sense would say that if this were the case, then the reaction would progress as fast as possible i.e. V_{max}. Using the Michaelis Menten equation this is also true. If [S] is very big then K_m is <u>relatively</u> small and hence $K_m + [S]$ is almost the same as [S]. Hence

$$v = \frac{V_{max}[S]}{K_m + [S]} \text{ when } [S] \text{ is very large } v = \frac{V_{max}[S]}{[S]} = V_{max}$$

Therefore, we can approximate V_{max} by using a high substrate concentration.

Consider the situation when [S] is the same as K_m.

$$v = \frac{V_{max}[S]}{K_m + [S]} \text{ when } K_m = [S] \text{ then } v = \frac{V_{max}[S]}{2[S]} = \frac{V_{max}}{2}$$

When the velocity of the reaction is half of the maximum reaction rate ($\frac{1}{2}V_{max}$) then $K_m = [S]$.

When the velocity is measured experimentally for different [S] the curve drawn using the Michaelis Menten equation results in a rectangular hyperbola.

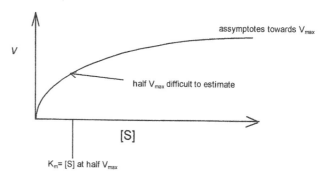

A better estimate of V_{max} can be achieved by a mathematical manipulation of the Michaelis Menten equation. Simply make the reciprocal of the equation.

$$v = \frac{V_{max}[S]}{K_m + [S]} \text{ goes to } \frac{1}{v} = \frac{K_m + [S]}{V_{max}[S]}$$

rearranging

$$\frac{1}{v} = \frac{K_m}{V_{max}[S]} + \frac{[S]}{V_{max}[S]} = \frac{K_m}{V_{max}}\left(\frac{1}{[S]}\right) + \frac{1}{V_{max}}$$

This is a straight line of general form y = mx + b, where y = $1/v$ and x = $1/[S]$

If $1/v$ is plotted against $1/[S]$, a straight line is formed and this is called a Lineweaver Burk plot. Where the graph cuts the y axis, $1/[S] = 0$ and therefore $1/v = 1/V_{max}$.

What is the value of $1/[S]$ when the line crosses the x axis, i.e. $1/v = 0$

Answer:

$$\frac{1}{v} = 0 = \frac{K_m}{V_{max}}\left(\frac{1}{[S]}\right) + \frac{1}{V_{max}}$$

$$\therefore \frac{K_m}{V_{max}}\left(\frac{1}{[S]}\right) = -\frac{1}{V_{max}}$$

$$\therefore K_m\left(\frac{1}{[S]}\right) = -\frac{V_{max}}{V_{max}}$$

$$\therefore \frac{1}{[S]} = -\frac{1}{K_m}$$

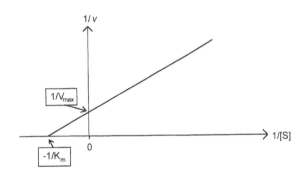

Major point 27: In a Lineweaver Burk plot 1/v is plotted against 1/[S] and where it crosses the Y axis is 1/Vmax and where it crosses the x axis is -1/[K_m]

One of the uses for determining K_m and V_{max} is to compare enzymes. For example, a group in the USA may have extracted a phosphodiesterase from brain and a group in Australia a phosphodiesterase from lung tissue. A quick indication as to whether or not these are exactly the same enzyme, is to compare their K_m and V_{max}. If these are the same then it is highly likely that they are the same enzyme. If different, then they are probably isoenzymes (isozymes). These are enzymes that catalyse the same reaction but differ slightly in structure hence giving them different K_m and V_{max}.

Another use for determining K_m and V_{max} is to determine how enzyme inhibitors function. An inhibitor may affect enzyme activity by binding to the same site on the enzyme as the normal substrate (competitive inhibitor) or may bind to a different site on the enzyme from the substrate and change the shape of the enzyme such that the conversion to substrate to product is much slower. To determine which type of inhibition is occurring, a family of Lineweaver Burk graphs are constructed from experiments where the velocity of the enzyme at different substrate concentrations is compared with the velocities of reactions in the presence of different concentrations of inhibitor.

Recall that K_m is equal to the concentration of the substrate required to reach ½ V_{max}. If very little substrate is required to reach ½V_{max} then the substrate reacts readily with the enzyme, and K_m is small. If the substrate does not react readily with the enzyme then a large amount of substrate would be required to reach ½ V_{max} i.e. a large K_m.

Remembering that V_{max} is a measure of the fastest possible rate, and K_m is an indication of how the binding site is interacting with the substrate, would you expect

V_{max} to change if a competitive inhibitor were added to the reaction mixture?

K_m to change if a competitive inhibitor were added to the reaction mixture?

V_{max} to change if a non-competitive inhibitor were added to the reaction mixture?

K_m to change if a non-competitive inhibitor were added to the reaction mixture?

Answer: *With a competitive inhibitor V_{max} would not change because one could imagine an extreme situation where there was so much substrate compared with inhibitor that the enzyme would always bind to substrate and not inhibitor and hence it would be just the same as having substrate and no inhibitor. K_m would be increased because the substrate reacts less with the binding site because it must compete with the inhibitor for this site. Results from such an experiment would look like this*

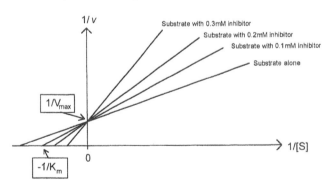

If the inhibitor were non-competitive, then it would be expected that the binding of the substrate to the enzyme would not change because the inhibitor is binding to a different site hence K_m is unchanged. On the other hand, the shape of the enzyme is different with the inhibitor bound and hence it does not matter how much substrate is present the enzyme

will always be less effective due to this shape change, hence V_{max} is decreased (note $1/V_{max}$ is increased)

Q&A 12: If the following results were obtained, is the inhibitor competitive or non-competitive?

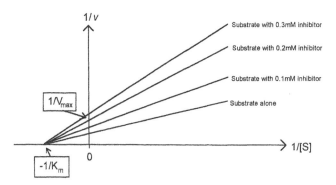

Nomenclature: The activity of an enzyme is expressed in standard units U = the amount of activity of an enzyme which catalyses the transformation of 1 mmol/min.

The specific activity of an enzyme is the number of units per mg of protein. The reason for needing this is that often the enzyme is not pure and there is contaminating protein in the sample

The catalytic constant is units of enzyme activity per mol of protein (mmol/min/mol enzyme).

Katals (Kat) are the conversion of 1 mol/sec (international units)

Important things to remember about enzyme kinetics are

- The Michaelis Menten equation
- Understand that this equation refers to steady state reactions at the beginning of a reaction.
- The Lineweaver Burk plot of the Michaelis Menten equation and what the intercepts are.
- The effects of competitive and non-competitive inhibitors on the Lineweaver Burk plot, K_m and V_{max}.

Sometimes, the experimental data from an enzymatic reaction does not form a straight line on a Lineweaver Burk plot (generally sigmoid). This simply means that it does not follow Michaelis Menten kinetics. Quite often this is due to allosteric effects in much the same way as described for the haemoglobin binding of oxygen. The activity of enzymes is sometimes allosterically affected by the product. The product of the reaction non-covalently binds to the enzyme at a different site from the substrate and inhibits its activity. This would slow the reaction down when the product was in high concentration and the reaction would speed up when the product was in low concentrations.

Control of enzyme/protein activity

We all realise that for the most part, our blood does not clot. However, put a hole in ourselves and let some of the blood out, then it clots. Proteins (enzymes) cause blood to clot and these proteins are in the blood all of the time. Therefore, why doesn't blood clot until we want it to, or in other words, how do we control the activity of these proteins? There are several mechanisms for controlling the activity of proteins.

1. Gene regulation

The genes for synthesising proteins can be regulated. The sex hormones oestrogen and testosterone are two very important regulators of genes. You probably noticed at puberty.

2. Converting proteins into active and inactive states.

(a) The charge on proteins can be regulated. Particularly important in this case are amino acids whose R groups carry an hydroxyl group, serine, threonine and tyrosine. These can be phosphorylated by esterifying a phosphate group to them. This puts a large negative charge in this region of the protein, which causes it to change its shape and its activity. Other proteins can bind Ca^{2+} (calcium binding proteins) which puts a positive charge on the protein. Still others bind cyclic nucleotides (cyclic nucleotide binding proteins). Importantly, the changes to these proteins is reversible and hence their activity can be switched on and off.

(b) Zymogens. Several proteins are synthesised in inactive forms. These are called zymogens. e.g. protein digesting enzymes and blood clotting proteins. To activate zymogens, a small amount of the protein is cleaved from one end. This causes the protein to change shape and activates it. These changes are not reversible.

References and further reading

Aldhous, P. (1993) Managing the genome data deluge. *Science* **262**:502-503

Ang, D., Liberek, K., Skowyra, D. Zylicz, M. and Georgopoulos, C. (1991) Biological role and regulation of the universally conserved heat shock proteins. *J. Biol. Chem.* **266**:24233-24236

Bohen, S.P., Kralli, A. and Yamamoto, K.R. (1995) Hold'em and fold'em: chaperones and signal transduction.. *Science* **268**:1303-1304

Brandon, C. and Tooze, J. (1991) *Introduction to Protein structure.* New York: Garland Publishing

Chothia, C. (1984) Principles that determine the structure of proteins. *Ann. Rev. Biochem.* **53**:537-572

Creighton, T.E. (1993) *Proteins: Structures and Molecular Properties,* 2nd Ed. New York: W.H. Freeman

Devlin, T.M. (1992) *Textbook of Biochemistry with Clinical Correlations,* 3rd ed. New York: Wiley-Liss

Dickerson, R.E. and Geis, I. (1983) *Hemoglobin: Structure, function, evolution, and pathology.* Menlo Park, California: Benjamin/Cummings Publishing Company

Dickerson R.E. and Geis, I. (1969) *The Structure and Action of Proteins.* New York: Harper and Row

Ellis, R.J. and van der Vies, S.M. (1991) Molecular chaperones. *Annu. Rev. Biochem.* **60**:321-347

Eyre, D.R. (1980) Collagen: Molecular diversity in the body's protein scaffold. *Science* **207**:1315-1322

Fridovich, I. (1995) Superoxide radical and superoxide dismutases. *Annu. Rev. Biochem.* **64**:97-112

Gelb, M.H., Jain, M.K., Hanel, A.M. and Berg, O.G. (1995) Interfacial enzymology of glycerolipid hydrolases. Lessons from secreted phospholipase A2. *Annu. Rev. Biochem.* **64**:653-688

Goodsell, D.S. and Olsen, A.J. (1993) Soluble proteins: size shape and function. *Trends Biochem. Sci.* **18**:65-68

Greenstein, B. and Greenstein, A. (1996) *Medical Biochemistry at a Glance.* Oxford: Blackwell Science

Hall, S.S. (1995) Protein images update natural history. *Science* **267**:620-624

Hendrick, J.P. and Hartl, F.-U. (1993) Molecular chaperone functions of heat-shock proteins. *Annu. Rev. Biochem.* **62**:349-384

Hill, R.L. (1965) Hydrolysis of proteins. *Adv. Protein Chem.* **20**:37-107

Jennings, M.L. (1989) Topography of membrane proteins *Ann. Rev. Biochem.* **58**:999-1027

Kamtekar, S, Schiffer, J.M., Xiong, H., Babik, J.M. and Hecht, M.H. (1993) Protein design by binary patterning of polar and nonpolar amino acids. *Science* **262**:1680-1685

Klotz, I.M., Langerman, N.R. and Darnell, D.W. (1970) Quaternary structure of proteins. *Ann. Rev. Biochem.* **39**:25-62.

Kroaut, J. (1988) How do enzymes work?. *Science* **242**:533-540

Matthews, C.R. (1993) Pathways of protein folding. *Annu. Rev. Biochem.* **62**:653-683

Monod, J., Wyman, J., and Changeux, J.-P. (1965) On the nature of allosteric transitions: A plausible model. *J. Molec. Biol.* **12**:88-118.

Perutz, M.F. (1979) Stereochemistry of co-operative effects in haemoglobin. *Nature* **228**:726-734

Rizo, J. and Gierasch, L.M. (1992) Constrained peptides: Models of bioactive peptides and protein substructures. *Annu. Rev. Biochem.* **61**:387-418

Robyt, J.F. and White, B.J. (1987) *Biochemical techniques: Theory and Practice.* Prospect Heights, IL: Waveland Press

Rossman, M.G. and Argos, P. (1981) Protein folding. *Ann. Rev. Biochem.* **50**:497-532.

Q&A Answers

1   Proteins are ubiquitous in the body and have a wide variety of functions: catalytic proteins – enzymes; structural proteins - collagen; contractile proteins – actin, myosin; natural defence proteins – antibodies; digestive proteins – gastrointestinal tract enzymes; transport proteins – haemoglobin; blood proteins – fibrinogen; hormonal proteins – insulin; respiratory proteins – cytochromes; represser proteins – regulate expression of genes on chromosomes, receptor proteins – transport information to cell interior after interacting with proteins on the outside; ribosomal proteins – associated with proteins synthesis; toxin proteins – venoms; vision proteins – rhodopsin.

2   It is necessary to specifically define the carboxyl and amine groups forming the amide bond as α groups (attached to the αC) because there are amino acids with carboxyl and amine groups in their R groups: aspartic acid (**D**), glutamic acid (**E**), arginine (**R**), asparagine (**N**) and glutamine (**Q**).

3   No, a monomer would not have quaternary structure because quaternary structure by definition is the relationship of protein subunits to form a complete physiologically active protein. It must be a dimer or greater.

4   Samples are run from the cathode to the anode. Remember that the protein has been saturated in negative charge. Larger proteins cannot move as easily through the gel matrix as smaller proteins and hence are retarded.

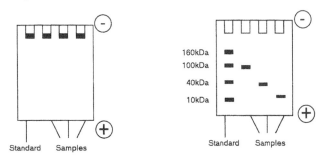

5   The sequence of the protein is:
    GVEFMKLHWIRAYKDGSFTQRGPF

6   Haemoglobin is a tetramer with each subunit containing a haeme group and hence it has 4 haeme groups.

7   It serves no purpose to shift the Hb binding curve to the left (i.e. increase the affinity for O_2) at high altitudes because although it increases the binding of arterial O_2, it decreases the release into tissues more. Instead, people living at high altitude increase the amount of red blood cells in their blood.

8   The M2-M3 loop is extracellular.

9   If the protein had 4 membrane spanning regions instead of 5, then the protein would cross the membrane one less time and hence the C terminus would be extracellular.

10 The name for the enzyme that catalyses the oxidation of alcohol by removing hydrogen is alcohol dehydrogenase.

11 The name be for the enzyme that catalyses the breaking of the ester bond in acetylcholine is acetylcholinesterase.

12 If the inhibitor were non-competitive, then the binding of the substrate to the enzyme would not change because the inhibitor would be binding to a different site, hence K_m is unchanged. On the other hand, the shape of the enzyme would be different with the inhibitor bound and hence it does not matter how much substrate is present; the enzyme will always be less effective due to this shape change, hence V_{max} is decreased (note $1/V_{max}$ is increased).

6 Complex Sugars and their Functions

In this chapter you will:

- learn about homopolysaccharides and hetero-polysaccharides

- study in detail the structure and functions of the polysaccharides – cellulose, chitin, starch and glycogen

- study the structure of common glycosaminoglycans (GAGs): hyaluronate, chondroitin sulphate, dermatin sulphate, heparan sulphate, keratan sulphate

- learn about the acid substituted sugars, particularly glucuronic and iduronic acids, and sulphated sugars.

- learn about the structure of bacterial cell walls

- investigate the nature and functions of proteoglycans, oligosaccharides and glycoproteins: laminin, fibronectin, integrin

- study the functions of plant, animal and bacterial lectins

- study the role of sugars in the formation of cerebrocides

- learn the importance of sugar residues for determining blood groups

Homopolysaccharides

> **Nomenclature:** Homopolysaccharides are sugar polymers made from one particular sugar unit. Heteropolysaccharides are polymers made from different sugar units. In both cases these may be linear or branched polymers.

Celluloses

Celluloses are the major components of plants comprising 20%–45% of their cell wall mass, and as such, the most abundant organic compound on Earth. It enables trees to attain their great heights and yet it is, unbelievably, just a polymer of glucose. Its basic unit is a chain of glucose linked by $\beta(1\rightarrow4)$ glycosidic bonds (cellobiose). These chains are in parallel alignment and held in place by hydrogen bonding between the chains. Major variations are in the degree of polymerisation which appears to be biphasic: either being less than 500 glucose units or between 2500 and 4500 glucose units per chain.

These hydrogen bonded chains are further strengthened by the presence of other polysaccharides, such as hemicellulose, pectin, and lignin, which function as cementing materials.

Higher level

The hemicellulose of dicotyledons is different from the hemicellulose in monocotyledons.

The hemicellulose of dicotolydons is mainly *xyloglucan*, which comprises about 20% of cell wall structure, whereas in monocotolydons it comprises about 2% of cell wall structure. Chemically, it has a backbone of β-1,4-linked D-glucosyl residues with D-xylosyl side chains α-linked to O-6 of some of the glucosyl residues. All xyloglucans are structurally similar in different species and differ only in some of the side chains. Xyloglucan has both a structural and regulatory role. It cross-links with cellobiose by hydrogen bonding to strengthen cellulose. It has a regulatory function controlling cell elongation. Small amounts of xyloglucan are released from the cell wall during auxin-stimulated growth and auxin causes accumulation of endo-β-1,4-gluconase, which cleaves xyloglucan. By breaking down xyloglucan the cellobiose is no longer constrained by the crosslinking, and hence the cell can elongate. Xyloglucan inhibits 2,4-D-stimulated cell wall elongation.

Draw the structure of the pentose, D-xylose

Answer:

D-xylose

Xylans constitute the major hemicellulose in the primary cell walls of monocots and are only found in small amounts in dicots. All xylans consist of a primary backbone of β-1,3-linked xyloxyl residues and this forms hydrogen bonds to cellulobiose. No biological activity has yet been associated with these molecules.

Functions of the polysaccharides in cell walls

Insect attack is obviously a problem for plants because they cannot escape, and so they must have a different mechanism of defence. Mechanical injury to plants causes the plant to produce large amounts of proteins that inhibit insect and microbial proteinases (enzymes produced by plant invaders to break down the plant structure). Fragments of pectin polysaccharides, which are released during the insect invasion, induce the synthesis of these proteinase inhibitors. Another cell wall fragment causes death of the plant cells. This sacrifice of a few cells protects the whole plant against microbial invasion by slowing the microbial invasion while other defence mechanisms come into play. Unfortunately, plants are yet to evolve a mechanism for protecting themselves against chainsaws.

Glucose glucose everywhere but not a drop to eat

The nutritional value of cellulose is virtually nil for higher animals except for ruminants because we do not have cellulase. Cellulases are found in snails, bacteria, fungi and insects. Ruminants have bacteria in their stomachs (rumens) which digest the cellulose for the ruminant. Even with cellulases from bacteria, cows have to chew their cud which means that they vomit up the half-digested cellulose and eat it again. Rabbits are even worse. They pass soft pellets at night and then eat them to produce the hard pellets one normally sees.

Despite obtaining little nutritional value from cellulose, it is important in our digestive processes because it provides roughage which is believed to be important for keeping the contents of our intestines mobile and in reducing bowel cancer.

Chitin

A subtle variation of cellulose is found in chitin (*kite-in*) which is a linear homopolysaccharide consisting of *N*-acetylglucosamine residues linked by $\beta(1\rightarrow4)$ bonds. This is the major component of the exoskeleton of invertebrates. It is also found in most fungi, algae and yeasts as a cell wall component. Chains are bundled together with hydrogen bonding.

Starches

Comprise a group of polyglucose molecules of varying size and shape which occur exclusively in plants. They contain only D-glucose residues. Starches are usually a mixture of 2 distinct polysaccharides. One component is called amylose and the other amylopectin. Both are poly α-D-glucose molecules. Amylose is linear and linked by $\alpha(1\rightarrow4)$ bonds. Amylopectin, on the other hand, is branched as a result of a small number of $\alpha(1\rightarrow6)$ linkages in a chain consisting of mainly $\alpha(1\rightarrow4)$ linkages. Amylose appears to prefer a helical coiled conformation. Its biological role is the storage of carbon and energy in plants.

Most plants have 2 distinct hydrolysing enzymes α-amylase and β-amylase. Cleavage with α-amylase is random and

yields a mixture of glucose and maltose. Cleavage by β-amylase, which is much more common, is characterised by successive removal of maltose units beginning at a non-reducing terminus (the reducing end is the end with the free C1, i.e. the C1 is not involved with the formation of a glycosidic bond). Neither enzyme is capable of hydrolysing the $\alpha(1\rightarrow6)$ linkage. This is done by other enzymes. Starch is digestible by humans due to salivary amylase and pancreatic amylase.

> **Nomenclature:** In a branched polysaccharide, there is only 1 reducing end and multiple non-reducing ends.

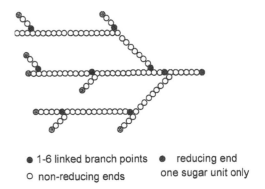

● 1-6 linked branch points ● reducing end
○ non-reducing ends   one sugar unit only

Glycogen

This is the storage form of glucose in animals and may comprise 10% of the mass of the liver. It is structurally the same as amylopectin, but $\alpha(1\rightarrow6)$ glycosidic bonds occur every 12–18 units whereas in amylopectin they occur about every 25. Glycogen may have several hundred thousand glucose residues whereas amylopectin has only 300–6000 glucose residues.

> **Q&A 1:** Why is it a distinct advantage to store glucose as a polymer (glycogen) rather than simply glucose?

Heteropolysaccharides

Heteropolysaccharides are usually composed of repeating disaccharide units rather than a large number of different sugars. They exist in the free state or conjugated (joined) to lipids, peptides or proteins. This is one of the most exciting regions of new knowledge in biochemistry and is generating immense interest. These heteropolysaccharides are involved with viral and bacterial attachment to cells, the immune system, the migration of normal cells, the migration of cancerous cells, the fertilisation of ova, and the changes of cells in development. The excitement is similar to that surrounding genetics some 20 years ago.

Glycosaminoglycans

One important group of heteropolysaccharides comprises those formed from repeating disaccharide units, either an

N-acetylglucosamine (GluNAc), or an *N*-acetylgalactosamine (GalNAc) (glycosamine) and another sugar (glycan). These are major components in the extracellular space of tissues and the matrix of connective tissues including bone and cartilage.

What would be a suitable name for such a heteropolysaccharide?

Answer: *Glycosaminoglycan or GAG.*

Major point 28: Glycosaminoglycans are unbranched polysaccharides which consist of a repeating disaccharide units comprising a sugar linked to either *N*-acetylglucosamine or *N*-acetylgalactosamine.

The most important aspect of these polysaccharides is that they carry a negative charge. This can be carried by the glycan, which is commonly glucuronic acid (remember that glucuronic acid (GlcUA) is glucose with a carboxyl (-COOH) substitution at C6, Chapter 3), or L-iduronic acid.

D-glucuronate   L-iduronate

The negative charge may also be carried by the glycosamine (*N*-acetylglucosamine or *N*-acetylgalactosamine), which is often sulphated via a sulphate ester bond to 1 or more of the free hydroxyl groups.

Why are the negative charges on GAGs so important?

Answer: *These are negative charges because they are attached to the sugar polymers and are not free to move. This means that positive charges, such as sodium and calcium, are attracted to them from the extracellular fluid. The increase in ions in the sugar matrix causes an osmotic difference between the matrix and the extracellular fluid and water moves into the matrix. This gives the sugar matrix a gel-like quality which can be mineralised as in the case of bone. This distribution of fixed negative charge also tends to make these molecules linear rather than folded.*

Common GAGs

Hyaluronate is a polymer of a repeating disaccharide GlcUA-β(1→3)-GluNAc connected by β(1→4) linkages.

Q&A 2: Draw the structure of hyaluronate.

Hyaluronates are enormous molecules having molecular weights of over a million. They form the viscous lubricants of

joints and the gel like substance inside the eyes – the vitreous humour.

- **Formation of a complex proteoglycan with hyaluronate as a base.**

Quite complex molecules called proteoglycans may use hyaluronate as the basic framework or core structure. Proteins stick to this core by hydrogen bonding and to these proteins other polysaccharides such as keratan sulphate and chondroitin sulphate (see below) are covalently attached. This type of structure is seen in cartilage, for example.

Note that simpler proteoglycans can form by covalent attachment of GAGs such as chondroitin sulphate or heparan sulphate to a core protein without hyaluronate.

| Nomenclature: Chondro means cartilage. Dermis means skin.

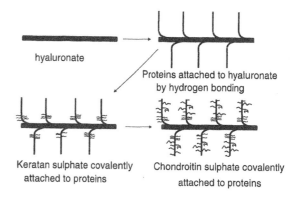

hyaluronate

Proteins attached to hyaluronate by hydrogen bonding

Keratan sulphate covalently attached to proteins

Chondroitin sulphate covalently attached to proteins

Other GAGs

- Chondroitin sulphate and dermatan sulphate are galactosaminoglycans.
- Heparan sulphate, heparin, and keratan sulphate are glucosaminoglycans.

Heparin, heparan sulphate and dermatan sulphate contain both iduronic acid and glucuronic acid units whereas chondroitin sulphate has only glucuronic acid as the hexouronic acid component. Keratan sulphate has galactose as the hexose component.

What is the structure of chondroitin sulphate?

Answer:

GlcUA   GalNAcSO₄⁻

What is the structure keratan sulphate?

Answer:

Gal   GlcNAcSO₄⁻

Due to variability in sulphate substitutions, all GAGs, particularly those with glucuronic acid and iduronic acid, have considerable sequence heterogeneity within and between GAG chains.

Higher level

The biosynthesis of heparin or heparan sulphate

Biosynthesis is initiated by the formation of a polysaccharide chain with the structure [GlcUA-β(1→4)-GlcNAc-α(1→4)-]ₙ. This polymer is modified by *N*-deacylation and *N*-sulphation. It subsequently undergoes C5 epimerisation of some of the D-GluA to L-IdoA units. Just to remind you, this is the reversal of the COO⁻ group attached to C5 from above the ring to below the ring. This is further modified by 2-*O*-sulphation of IdoUA, and 6-*O*-sulphation of GlcN units. More rarely, *O*-sulphate substituents are incorporated at C3 of GlcN units and at C2 of GlcA units. The end result is enormous complexity because there are 4 variations of the HexUA units and 6 variations of the GlcN which results in 16 or 17 different HexUA→GlcNAc units and 10-12 different GlcN→HexUA disaccharide sequences. On an individual molecule there tends to be blocks of heavily sulphated regions interdispersed with non-sulphated regions. Hence heparins from different species are different, and heparan sulphates from different tissues are different.

The difference between heparin and heparan sulphate was initially thought to be in their anticoagulant activity. It was first thought that heparin had anticoagulant activity and heparan sulphate did not. It is now known that both forms may or may not have anticoagulant activity. Structurally, it was initially believed that heparan sulphate was fully *N*-acetylated and heparin was *N*-sulphated, but this is now known to be also untrue. In general, heparin contains more *N*- and *O*-sulphated groups and a higher proportion of IdoUA units than heparan sulphate which tends to contain more acetylated GlcN units and more GlcUA units. This is counter-intuitive to what their names suggest.

Other GAGs are synthesised in a similar way.

Nomenclature: What is the difference between heparin and heparan sulphate? In terms of structure virtually none, and hence the term heparin is reserved for the GAG component of proteoglycans released by mast cells and all other structurally related GAGs are heparan sulphate. Note that the 'an' designation (hepar*an*) relates to the adjective and the 'in' designation (hepar*in*) to the noun.

Proteoglycans

The roles of proteoglycans are extremely varied, and include cell adhesion, motility, proliferation, differentiation and tissue morphogenesis. Covalent bonding to the protein core is usually via a serine (**S**) residue of the protein and any protein may have from 1 to 100 glycosaminoglycans attached to it. There is an enormous variety of core proteins and they have no distinctive features. Hybrid proteoglycans, where different GAGs are bound to the same protein core, are common. Linkage to the core protein is essentially the same for all proteoglycans. GAG→GlcAβ(1→3)Galβ(1→3) Galβ(1→3)Xylβ→O-serine bridge. In heparan sulphate and chondroitin sulphate the xylose is phosphorylated at C2. Corneal keratan sulphate is *N*-linked through asparagine (**N**) units via mannose and GlcUA (glucuronic acid) units.

Bacterial cell walls

Bacterial cell walls are an interesting variation of glycosaminoglycans. To remind you, the cell wall of bacteria is outside the cell membrane. Unlike our cells, the normal osmotic pressure inside a bacterium is much higher, and the wall stops the bacterium from bursting. Some antibiotics, such as the penicillins, prevent the normal synthesis of cell walls and the bacterium bursts.

Essentially bacterial cell walls are linear chains of a β(1→4) linked disaccharide unit made of *N*-acetylglucosamine connected via a β(1→4) linkage to a 9-carbon sugar (nonose) called *N*-acetylmuramic acid which is a variation of *N*-acetylglucosamine. These linear chains are cross-linked by a small peptide bridge. The peptide bridge is interesting because it contains D-amino acids, which are not normally found in proteins, alternating with L-amino acids.

Higher level

The disaccharide units are in β(1→4) linkage. Bacterial cell walls are formed by a polymer of a disaccharide. The disaccharide is *N*-acetylglucosamine β(1→4) *N*-acetylmuramic acid.

N-acetylmuramic acid is a nonose (9-carbon sugar) formed from *N*-acetylglucosamine. The extra 3 carbons come from lactic acid which forms an *O*-glycosidic bond with C3 of *N*-acetylglucosamine.

These polysaccharide chains are cross-linked by a peptide comprising L and D amino acids and a chain of 5 glycines (see below). This polysaccharide structure, which forms the cell wall, is technically extracellular and anchored to the bacterial cell membrane by lipoteichoic acid. Note that lysine (**K**) is an amino acid with an amine group in its R group. This means that it can make 2 peptide bonds - 1 with glycine (**G**) through its R group and 1 with isoglutamate through its αC amine group. This allows the linkage between chains.

Oligosaccharides

These polysaccharides differ from GAGs in that they are polysaccharides other than GAGs which are attached to proteins. They can vary from disaccharides to complex branched structures (GAGs are unbranched), but they have some features in common.

The carbohydrate residue bound to a protein is an acylated osamine, frequently *N*-acetyl-D-glucosamine. The attachment to the protein usually occurs at a serine (**S**) or threonine (**T**) residue (*O*-linked), but there are also attachments to the amide nitrogen of asparagine (**N**) residues (*N*-linked). The remainder of the chain is formed by sequential addition of other constituent residues. However, residues of L-fucose (see below) and *N*-acetylglycosamines are at the ends of the chains.

L-fucose (6-deoxy-L-galactose)

| **Nomenclature:** A glycoprotein is a protein that has oligosaccharide attachments.

Glycoproteins have an enormous variety of functions including lubrication, many being secreted, and another large group is involved with cell adhesion.

Glycoproteins and lubrication

Mucous

This is a glycoprotein found in various secretions. It contains a single polypeptide protein core comprising some 5000 amino acid residues, of which 1300 are seryl and threonyl residues. About 800 disaccharide units are attached to these residues. Each disaccharide is an α-*N*-acetyl-D-neuraminyl-(2→6)-β-N-acetylglucosamine, accounting for about 60% of the molecular weight.

> **Q&A 3:** Given the following information, attach α-*N*-acetyl-D-neuraminyl-(2→6)-β-*N*-acetylglucosamine to a protein core at a seryl group.

α-*N*-acetylneuraminic acid 9-phosphate
(sialic acid 9-phosphate)   β-*N*-acetylglucosamine   seryl group on protein

Mucin is a very long molecule densely covered with short carbohydrate bristles. The polar character of the carbohydrate attracts water, so there is little hydrophobic interaction between the large surfaces. The many negative charges carried by the terminal *N*-acetylneuraminyl residues repel each other, serving to keep the molecule extended and overcoming any tendency to stack up. The sheath of water and the tendency to keep apart enable the molecules to lubricate. There is a whole family of mucins with slightly different characteristics and these are called Muc 1, Muc 2, etc.

Glycoproteins and secretion

The oligosaccharide attachments appear to be an important component of secretory proteins. These proteins need their carbohydrates for the actual secretion process, because in the Golgi apparatus the oligosaccharides serve as a signal to package these proteins for secretion.

The following table has examples of glycoproteins which are secreted.

Protein	Function	Secreted by
Immunoglobulin, e.g. IgG	antibody	lymphocytes, plasma cells
Fibrinogen	clot formation	liver
Transferrin	iron transport	liver
Thyrotropin	hormone	anterior pituitary
Ribonuclease B	enzyme	pancreas
Ferro-oxidase	enzyme	liver

Glycoproteins and cell adhesion

Something that is obvious, but rarely thought about, is that cells must adhere to a substratum. If they didn't, they would fall off. As simple as it may seem, the adhesion of cells to substrata is a very intricate interaction. At times the cells need to actually move across the substratum such as occurs in wound repair. This requires that the cells partially lose their adhesion to the substratum and then reattach. In this whole process, glycoproteins forming a layer beneath cells (the basement membrane) are key elements. Important glycoproteins of the basement membrane are laminin and fibronectin.

Laminin This is a glycoprotein only found in the basement membrane. It is confined to the part of the basement membrane closest to the cell (lamina rara). Generally it has been extracted from tumour cell lines which generate a large amount of laminin and collagen type IV. About 12%-15% of its mass is carbohydrates which are mainly oligosaccharides linked to asparagine (*N*-linked). It has a distinctive cruciform shape and one part of the protein attaches to the cell and another part attaches to collagen fibres (type IV) of the basement membrane. This effectively anchors the cell to the basement membrane. It also has an attachment site for heparan sulphate.

Fibronectin It is found in the extracellular matrix of many cells and is also a plasma protein (0.3 g/L; plasma fibronectin) produced by the liver. Besides oligosaccharides, it has the GAGs heparan sulphate and polygalactosaminoglycan attached to it. The carbohydrates protect fibronectin against proteolysis. It mediates cell attachment and spreading of cells on collagen, fibrin and artificial tissue culture substrates. It has numerous effects on cell morphology and intracellular functions. It has binding sites for collagen, fibrin, heparin, hyaluronic acid, and actin. Fibronectin also stimulates cell movement by mechanisms that are still not understood but which may be important to cell migratory events in embryonic development and wound healing.

Its ability to bind to fibrin is extremely important in wound repair. Fibrin is one of the main elements of a blood clot which forms when tissue is damaged. Fibronectin in the basement membrane of cells links to fibrin through a glutamine residue and this enables the cells to migrate into a clot during the repair process.

Glycoproteins and cell signalling

Large insoluble proteins of the extracellular matrix can contribute to signal generation processes. They affect cell growth and differentiation.

Integrins are glycoproteins. They are made from 2 non-covalently linked protein strands called α and β (α/β heterodimers). Integrins have 3 domains, 1 part lies inside the cell, 1 part is in the cell membrane (membrane spanning domain) and 1 part is extracellular.

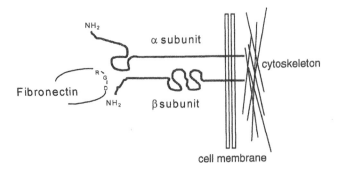

In mammals, there are 12 different α subunits and 8 different β subunits and these can associate in different combinations. The combination determines which ligands they will bind to, e.g. $\alpha5/\beta1$ is a fibronectin receptor and binds specifically to fibronectin, whereas $\alpha3/\beta1$ binds fibronectin, laminin and collagen. Several integrins bind to proteins at the amino acid sequence of Arg-Gly-Asp (RGD). Both the α and β subunits are involved in this process.

Significant for the integrins is the fact that part of their structure is intracelular and part extracellular. Therefore, a ligand (normally water soluble and hence unable to pass through the cell membrane) binding to an integrin on the outside can transmit a signal directly to the inside of the cell. This occurs even though integrins have only a short cytoplasmic domain. Some changes which have been observed in cells when integrins are activated are: changes in cytoplasmic pH; changes in Ca^{2+} levels, phosphorylation of tyrosine in proteins (places a negative charge on the proteins) and changes to cytoskeletal shape. Their activation activates T lymphocytes and can also cause tumours.

Glycoproteins and cell recognition

Cells recognise each other by pairs of complementary structures. Realisation that the sites may be sugars was only determined in the late 60s. Until then, sugars were regarded as storage and structural molecules.

Knowing the answer to these questions it becomes apparent why polysaccharide groups are valuble for cell recognition. All cells carry a sugar coat and it only needs a few sugars in different combinations to give a cell a unique identity on its surface.

Lectins

Proteins that bind to sugars are called lectins. The strict definition of a lectin is a free (non-membrane bound) sugar specific cell agglutinating protein, but this has been broadened to include all non-antibody proteins, both free and membrane bound, which bind specifically to saccharides. Some lectins are specific for particular oligosaccharides, while others are specific for monosaccharides. Lectins were once thought to be exclusive to plants, but are now known to be ubiquitous and are strategically placed on animal cells to combine with carbohydrates on neighbouring cells.

Nomenclature: The broad definition of a lectin is a protein that binds to a sugar group. It is not an antibody and may exist as a free protein or may be membrane bound.

Plant lectins are found primarily in seeds. As a rule, there is 1 sugar binding site per subunit. Therefore, a lectin comprising 4 subunits, such as soybean lectin and lima bean lectin, has 4 binding sites. One of the functions of lectins in plants is to mediate the symbiosis between plants and microorganisms, and another is to protect plants from phytopathogens (organisms which cause disease in plants).

Some examples of plant lectins are: concanavalin A (Con-A), which binds to D-mannose groups; wheat germ agglutinin (WGA), which binds to N-acetyl D-glucosamine or sialic acid; and soybean agglutinin (SBA), which binds to N-acetyl D-galactosamine.

Use of lectins in biochemistry

Lectins are used in the laboratory for affinity chromatography. Suppose a protein, in which you were interested, had a large number of sugar groups on it and you wished to purify the protein. If a particular sugar predominates, a lectin specific for that sugar can be attached to the matrix of a column. The mixture containing the protein of interest is then run through the column and the protein of interest binds to the column and hence separated from the other proteins. For example, dopamine-β-hydroxylase is an enzyme that has a relatively large number of mannose groups. To purify it, adrenal medullary cells, a rich source of the enzyme, are homoginised and the proteins extracted. To separate dopamine-β-hydroxylase from the other proteins, a column with Con-A attached to the solid support (beads) is made and the protein mixture run over it. The Con-A binds to the mannose groups on dopamine-β-hydroxylase and the other proteins are washed away.

To retrieve dopamine-β-hydroxylase from the column, what would the wash solution contain? Remember: Con A on the column has a high affinity for D-mannose.

Answer: *The wash solution would contain high amounts of D-mannose which displaces the dopamine-β-hydroxylase from the Con A and then it is collected in the eluant.*

Nomenclature: Affinity chormatography is a method for separating a particular chemical, usually a protein, by running it through a column which has a substance bound to it which has a high affinity for the particular protein, e.g. an antibody specific for the particular protein could be bound to the column, or a lectin specific for a particular sugar group could be bound to the column.

Vertebrate lectins

Vertebrate lectins are divided into 2 classes - integral membrane lectins, which require membrane solubilisation to extract; and soluble lectins. Animal lectins have a variety of functions among which are control of endocytosis/phagocytosis and regulation of cell migration and adhesion. The ability of cancerous cells to metastasise (migrate into blood vessels or lymphatic vessels to seed another organ with cancer) is believed to be due to lectins. During inflammation, lectins called selectins or LEC CAMS (leucocyte cell adhesion molecules) are responsible for the adhesion of white blood cells to endothelial cells of blood vessels. Once the adhesion occurs, the white blood cells are able to migrate out of the bloodstream into the tissue.

> **Nomenclature:** Lectins found on tissue such as the endothelial cells of blood vessels, which <u>select</u> for particular cell types, are called selectins. The often used acronym CAM as in LEC CAM above stands for cell adhesion molecule and refers to a specific selectin, e.g. N-CAM = neural cell adhesion molecules, which are glycoproteins, and these promote cell-to-cell and cell-to-matrix adhesion. ICAM = intercellular cell adhesion molecule.

Higher level

Adhesion receptors on inflammatory cells

An example of dynamic interaction of cells occurs in the immune system, which requires sophisticated signalling systems and regulation that are very important in the areas of infection, cancer, inflammation, allergy and autoimmune disease.

During an inflammatory response, leucocytes migrate from the circulation into tissues to kill pathogenic microorganisms. The obvious question that arises is how do they actually know to move out of the blood vessel and how do they do it? All of the answers to this problem have not been solved, but some information is available.

Damaged cells and macrophages produce cytokines to signal to other cells that they need help because an invasion is occurring. Two of these cytokines are interleukin 1 (IL-1) and tissue necrosis factor (TNF). Endothelial cells of the blood vessels have receptors for these molecules and respond to them by expressing adhesion molecules to which leucocytes attach.

What would be a suitable acronym for a molecule produced by endothelial cells which causes adhesion of leucocytes?

Answer: Endothelial cell derived leucocyte adhesion molecule-1 (ELAM-1)

ELAM 1 mediates the binding of neutrophils to blood vessels before their extravasation into the tissue. Note that normally the neutrophils are whooshing along in the middle of the blood vessel. In order to crawl out of the blood vessel into the tissue, they firstly need to attach to the blood vessel wall (margination), otherwise it would be like trying to hop off a moving train. The receptor on the neutrophils, which binds to ELAM-1, is a glycoprotein, and its active site is a tetrasaccharide called sialyl-Lewis X.

The adhesion of these cells can be blocked by liposomes coated with sialyl-Lewis X sequences, because these liposomes adhere to the receptors blocking binding of the neutrophils. Such techniques could be used for supplementary therapy in pathological inflammation and in metastasis of carcinoma where ELAM-1 is thought to be involved. Another possible medical use involving the ELAM-1 receptor is to couple drugs to analogues of the sialyl-Lewis X sequence so that they are targeted to the inflammatory site.

Other carbohydrate-based receptors

A distinct pentasaccharide has been shown to inhibit attachment of the blastocyst in mice to the endometrium. This creates new possibilities for contraception! Sugar in your tea?

Bacterial lectins

These often occur in a filamentous form. Indeed, the fimbriae, which often coat bacteria, are lectins. For *E. coli*, it is specific for D-mannose.

Have you ever wondered why certain bacteria or viruses attack particular species or indeed a specific tissue, e.g. *Neisseria gonorrhoeae* exclusively infects human genitalia and oral epithelia. The reason is that bacteria (or viruses) have lectins specific for the sugar groups of their host cells and so only bind and infect these cells. Attachment of microbes to animal cells prevents them from being washed away. An example of an attachment site is the intercellular adhesion molecule (ICAM-1), the receptor for most rhinoviruses, which cause the common cold. For influenza the receptor attachment site is sialic acid. Using low molecular weight analogues to block the binding site might afford protection as an alternative to immunisation.

Attachment of carbohydrates to cell membrane lipids

Sugars can also be attached to lipids (fats). One particular fat structure is a sphingolipid which is one of the lipids found in cell membranes (see Chapter 11 for details on lipid structure).

sphingosine

This is a sphingolipid and is not very different from diacylglycerides. Here the X represents any defined group and can be a sugar group attached by an ether (*O*-glycosidic bond).

If X= H, then this structure is called ceramide.

If X = glucose, then it is called glucosylcerebrocide.

Q&A 7: What would be a suitable name for such a molecule if X were galactose?

Most of us have heard about the ABO blood groups. Blood group A has an antigen on the surface of its red blood cells called A antigen and if this type of blood is given to a person who is blood group type B, they destroy the A group blood cells because they have antibodies against the A antigen.

Q&A 8: Based on the above logic for people with blood group A, what type of antigens would there be on the red blood cells of someone with respectively blood groups: B; AB; and O?

These ABO bloodgroup antigens are different sugars attached to a cerebroside (lipid) in the red blood cell membrane. Knowing just how strong the immune reaction is against the wrong blood group, how many sugars would be involved and how many sugars would each of these differ by.

a) 4–6 sugar groups involved and differ by 1 sugar group

b) 15–20 sugar groups involved and differ by 2–5 sugar groups

c) 20–40 sugar groups involved and differ by 5–10 sugar groups

d) 50–60 sugar groups involved and differ by 10–15 sugar groups

Answer: *Unbelievably, 'a' is the correct answer. Only 4 to 6 sugar groups are involved and they differ by 1 sugar group. Even more surprising is that the sugar group they differ by is such a small change, galactose compared with N-acetylgalactosamine.*

Note that the antigens are not confined to red blood cells, but are also present in normal cells. In normal cells, the antigens are carried by a protein, not a lipid. In this case the attachment sugars differ, but the terminal sugars are exactly the same. Note that 'sial' is sialic acid, a 9-carbon sugar (nonose). This serves to illustrate

- the importance of sugars
- the sensitivity of our immune system
- sugars can be linked to lipids in cell membranes or to proteins, in this case in a cell membrane, but other proteins not in cell membranes may also be glycosylated.

Details of the sugar linkages are illustrated below. It is not necessary to commit this to memory.

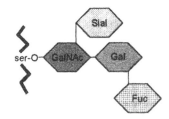

GalNAcα(1→3)Galβ(1→3)GalNAc-O-ser-R   Galα(1→3)Galβ(1→3)GalNAc-O-ser-R

A-antigen   B-antigen

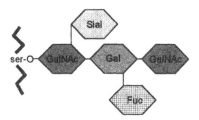

O antigen

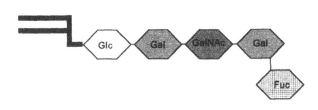

A antigen

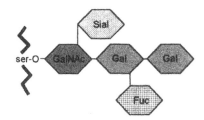

B antigen

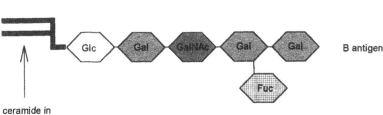

ceramide in outer envelope of cell membrane

In red blood cells sugar groups attached to a lipid (ceramide) in cell membrane

In other cells sugar groups attached to an intramembranous protein via a serine linkage

---

Q&A 9: What would be the gene product that leads to the production of the A antigen and the B antigen?

Anecdote: In the early days of heart transplant, patients died from failure of the new heart, or a massive immune response and rejection of the new heart. These days in Australia, a major cause of death following heart transplants is skin cancer. Following the heart transplant, the patient is put on immunosuppressive drugs. As mentioned above cancerous cells typically have different sugar groups on their surfaces and the body's immune system is constantly recognising them as foreign and destroying them. Once the immune system is suppressed, the cancerous cells are not destroyed and hence take over and kill the patient.

References and further reading

Abeijon, C. and Hirschberg, C.B. (1992) Topography of glycosylation reactions in the endoplasmic reticulum. *Trends Biochem. Sci.* **17**:32-36.

Aspinall, G.O. (1982) *The Polysaccharides* Vols. 1 and 2. New York: Academic Press.

Brennen, P.J. and Nikaido, H. (1995) The envelope of mycobacteria. *Annu. Rev. Biochem.* **64**:29-63.

Cerami, A., Viassara, H. and Brownlee, M. (1987) Glucose and aging. *Sci. Amer.* **265(5)**:90-97.

Collins, P.M. (1987) *Carbohydrates*. London: Chapman and Hall.

Devlin, T.M. (1992) *Textbook of Biochemistry with Clinical Correlations*, 3rd ed. New York: Wiley-Liss.

Dwek, R.A., Edge, C.J., Harvey, D.J., Wormald, M.R. and Parekh, R.B. (1993) Analysis of glycoprotein-associated oligosaccharides. *Annu. Rev. Biochem.* **62**:65-100.

Feeney, R.E., Burcham, T.S. and Yeh, Y (1986) Antifreeze glycoproteins from plar fish blood. *Annu. Rev. Biophys. Chem.* **15**:59-78.

Greenstein, B. and Greenstein, A. (1996) *Medical Biochemistry at a Glance*. Oxford: Blackwell Science.

Kjellen, L. and Lindahl, U. (1991) Proteoglycans: structures and interactions. *Annu. Rev. Biochem.* **60**:443-475.

Lasky, L.A. (1992) Selectins: interpreters of cell-specific carbohydrate information during inflammation. *Science* **258**:964-969.

Lodish, H.F. (1991) Recognition of complex oligosaccharides by the multisubunit asialoglycoprotein receptor. *Trends Biochem. Sci.* **16**:374-377.

McNeil, M., Darvill, A.G., Fry, S.C. and Albersheim, P. (1984) Structure and function of the primary cell walls of plants. *Annu. Rev. Biochem.* **53**:625-664.

Rademacher, T.W., Parekh, R.B. and Dwek, R. (1988) Glycobiology. *Annu. Rev. Biochem.* **57**:785-838.

Roehrig, K.L. (1984) *Carbohydrate Biochemistry and Metabolism*. Westport, Conneticut: AVI Publishing Company.

Rothman, J.E. (1992) The compartmental organization of the Golgi apparatus. *Sci. Amer.* **253(3)**:74-89.

Ruoslahti, E. (1989) Proteoglycans in cell regulation. *J. Biol. Chem.* **264**:13369-13372.

Weis, W.I. and Drickaner, K. (1996) Structural basis of lectin-carbohydrate recognition. *Annu. Rev. Biochem.* **65**:441-474.

Q&A Answers

1 If glucose were stored as glucose rather than a polysaccharide such as glycogen or starch, it would cause an enormous osmotic force within the cell. This would cause water to move into the cell and the cell would burst.

2 The structure of hyaluronate is a repeating polysaccharide made from repeating $\beta1{\rightarrow}4$ disaccharide units comprising glucuronate in $\alpha(1{\rightarrow}3)$ linkage with N-acetylglucosamine.

GluUA   GluNAc

3 See next page.

4 The membrane spanning domain of integrins would be lipid soluble and hence hydrophobic. This is true for all proteins that have intramembranous portions.

5 Four different nucleotides can combine to form 24 tetra nucleotides, whereas 2 identical monosaccharides can bind to form 11 different disaccharides.

6 Amazingly, 4 different monosaccharides (hexoses) can make 35 560 unique tetrasaccharides. Therefore, carbohydrates can carry enormous amounts of information.

7 Galactosylcerebrocide is the membrane lipid ceramide with galactose attached to the hydroxyl group via an O-glycosidic bond.

8 A person with blood group B would have B antigen on their red blood cells; a person with blood group AB would have both A and B antigens on their red blood cells; and a person with blood group O would have O antigen on their red blood cells, but there are no natural antibodies against O antigens and hence people with O group are regarded as universal donors.

9 The gene-product for the A-antigen is α-D-N-acetylgalactosamine transferase which is the enzyme which catalyses the transfer of α-D-N-acetylgalactosamine to the growing sugar polymer. The gene-product for the B-antigen is α-D-galctosyl transferase which is the enzyme which catalyses the transfer of α-D-galactose to the growing sugar polymer. Of course, someone with AB blood type produces both enzymes.

3   α-*N*-acetyl-D-neuraminyl-(2→6)-β-*N*-acetylglucosamine is formed by *O*-glycosidic bonds between *N*-acetyl-D-neuramic acid, *N*-acetylglucosamine and serine. This is the basic oligosaccharide connection formed in mucous.

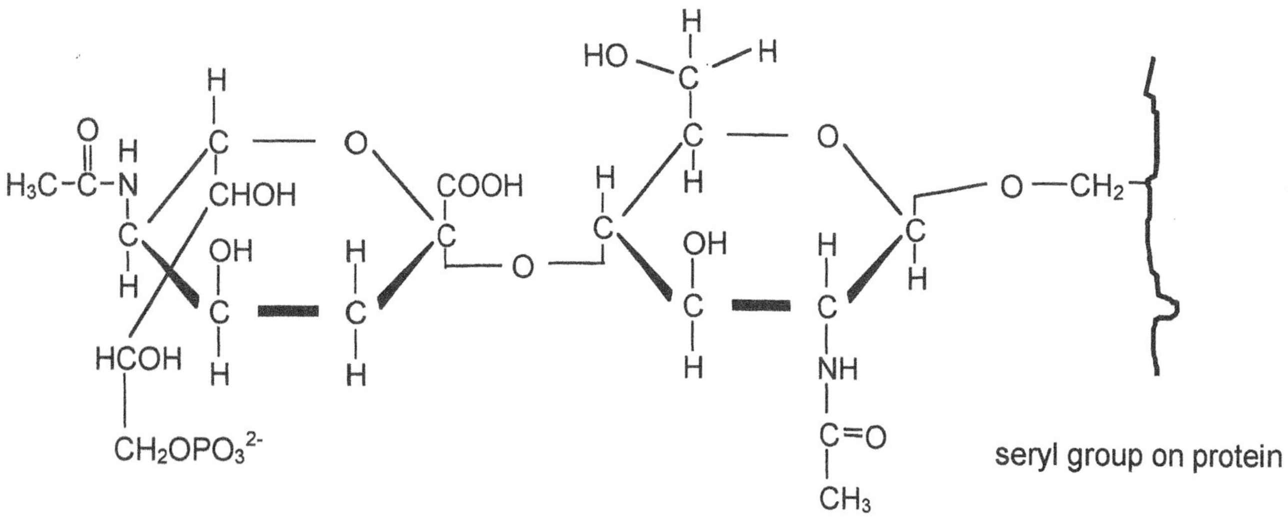

α-*N*-acetylneuraminic acid 9-phosphate
(sialic acid 9-phosphate)

β-*N*-acetylglucosamine

seryl group on protein

7 Metabolism of Sugars

In this chapter you will learn

- about glycolysis
- how sugars are activated by converting them to UDP-glycose or ADP-glycose so that they can be attached to other molecules, e.g. to synthesise glycogen
- how sugar amines are synthesised
- about the synthesis of muramic and sialic acid
- how sugars are attached to proteins and how this leads to the intracellular transport of proteins.

Sugars do more than just provide energy. In the previous chapter we saw that they were the basis of chitin, cellulose, bones, cartilage, lubricants such as mucus, cell recognition and some membrane lipids (sphingosines). They are also the structural basis for DNA and RNA, a source of carbon skeletons for some amino acids, the basis of some intracellular messenger systems. As an energy source, they are not only broken down, but when there is a surplus, they are stored as starch and glycogen. Therefore, it is important to realise that sugars are involved in a host of metabolic pathways, and not just the provision of energy. However, we shall firstly examine the use of sugars as an energy source starting with glucose.

Glycolysis

Given that energy is locked up in chemical bonds, energy can be extracted from glucose by breaking some of its chemical bonds. Surprisingly, this can be achieved without the use of oxygen. Essentially this is achieved by splitting glucose into 2 pyruvic acids which are molecules containing 3 carbons. This process is called glycolysis.

> **MAJOR POINT 29:** Glycolysis is an anaerobic process for extracting energy from glucose by breaking it down into the 3-carbon molecule pyruvic acid. Two pyruvic acids are obtained from 1 molecule of glucose.

Nomenclature: Anaerobic literally means without air, but in practice means without oxygen.

Draw the structures of both the beginning and end products of glycolysis – glucose and 2 pyruvic acids.

Answer:

β-D-glucopyranose   pyruvic acid

Look closely at these molecules and decide which atoms are lost when 1 molecule of glucose is converted to 2 molecules of pyruvic acid.

Answer: *Both have 6 carbon atoms, and 6 oxygen atoms, but the 2 pyruvic acids have a total of 8 hydrogen atoms, which is 4 less than in 1 molecule of glucose. Therefore 2 events have occurred during glycolysis: glucose has been split into 2; and there has been a loss of 2 hydrogen molecules or 4 hydrogen atoms. Loss of a hydrogen molecule is a form of oxidation and therefore during glycolysis, glucose is oxidised by loss of 2 hydrogen molecules to form 2 pyruvates. This oxidation is an anaerobic process.*

> **Q&A 1:** Pyruvic acid is a 3-carbon ketoacid. Draw the chemical structure of pyruvic acid and indicate the ketone group and the acid group.

The sequence of events in glycolysis will now be examined. Firstly, the cell takes up glucose. In order for the cell to prevent glucose simply diffusing back out of the it, the glucose is made much more water-soluble by placing a large negative charge on it. This is achieved by esterifying a phosphate to C6. This is a condensation and the phosphate is obtained from ATP, which is converted to ADP in the process, which means that energy is lost in this step. However, this is better than losing glucose out of the cell.

Esterify a phosphate group to glucose on C6. Remember this is a condensation reaction and water is formed.

Answer:

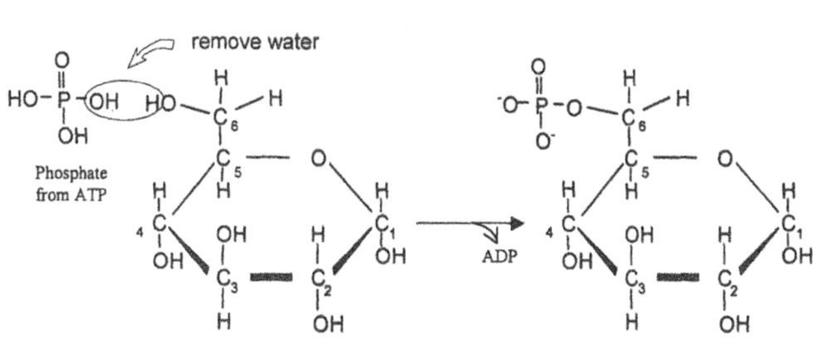

Note that phosphoric acid loses its H ions at neutral pH, hence it is drawn as a phosphate group carrying negative charges.

What would be a suitable name for the molecule formed in this reaction and the name of the enzyme involved, remembering that an enzyme which catalyses a phosphorylation and uses ATP as its source of phosphate is called a kinase.

Answer: *Glucose has been converted to glucose 6-phosphate (the phosphate has been condensed to C6) and there are 2 possible enzymes depending upon the tissue. One is **glucokinase** (found in the liver), which is highly specific for glucose and the other is **hexokinase** (muscle and fat), which will catalyse the phosphorylation of most hexoses, including glucose.*

Clearly to split anything in 2 it is much easier if it is bilaterally symmetrical. Glucose 6-phosphate is not a very symmetrical molecule and would be easier to split if it were. To make it more symmetrical it is firstly converted to fructose 6-phosphate.

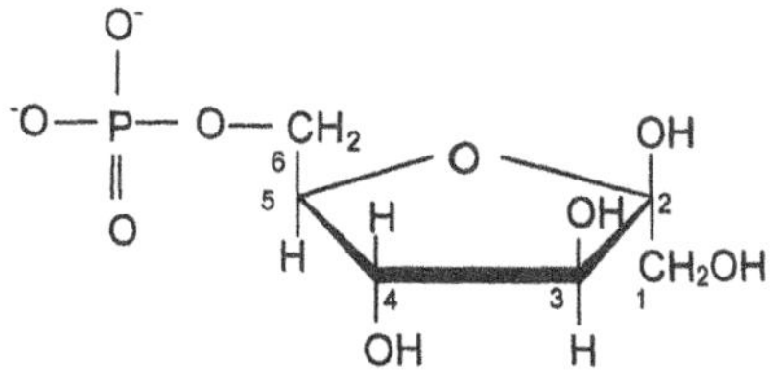

fructose-6-phosphate

This is a little more symmetrical because carbons 1, 2, and 3 are now similar to carbons 6, 5, and 4, with carbons 1 and 6 now both being outside the ring. Fructose 6-phosphate is an isomer of glucose 6-phosphate.

What would be a suitable name for the enzyme which catalyses the conversion of glucose 6-phosphate to its isomer, fructose 6-phosphate?

Answer: *Glucose 6-phosphate isomerase.*

In order to make the molecule symmetrical it is necessary to esterify a phosphate to C1

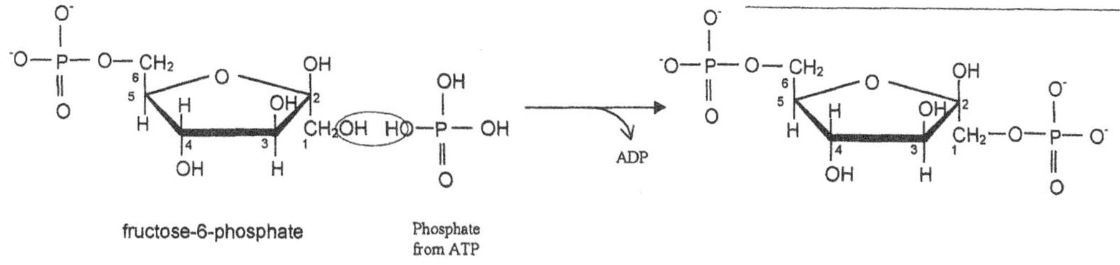

fructose-6-phosphate   Phosphate from ATP

We now have a molecule that is almost symmetrical and again the phosphate group came from ATP. At this stage of glycolysis, a large amount of energy, 2 ATPs has been lost.

What would be a suitable name for the enzyme involved in the addition of a phosphate to phosphorylated fructose (fructose 6-phosphate)? Remember from above that an enzyme that adds a phosphate using ATP as its source is called a kinase.

Answer: *Phosphofructokinase (often abbreviated to PFK).*

MAJOR POINT 30: Phosphofructokinase is the main control point in the glycolytic pathway. When activated, glycolysis increases so more energy is obtained from glucose. It is inactivated when energy is not needed from glucose.

To understand what happens in the next reaction it is necessary to draw fructose 1,6-bisphosphate in its linear form, and then split it in 2. The enzyme involved is an aldolase. This is an enzyme which splits an aldose (sugar) and it is not specific for fructose 1,6-bisphosphate, hence its general name.

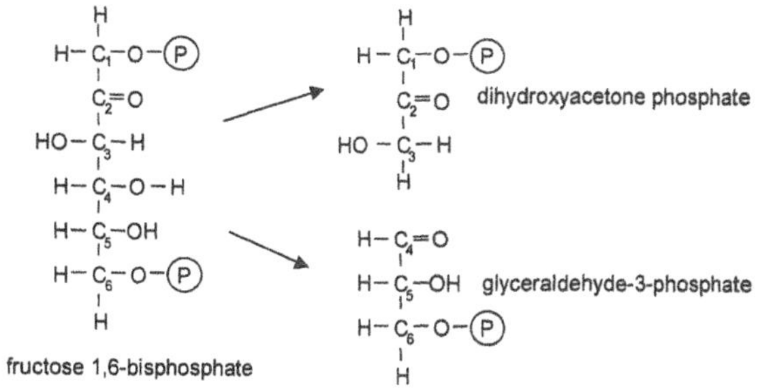

fructose 1,6-bisphosphate

Note that P is a shorthand notation for phosphate.

To comprehend what is happening in this reaction, it is necessary to concentrate on the H of the hydroxyl group (OH) of C4.

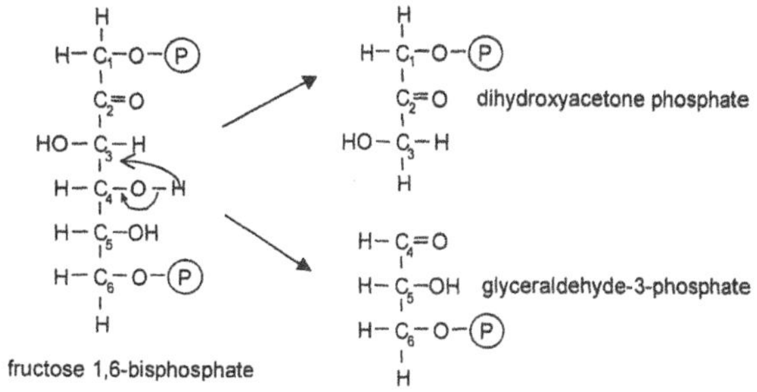

fructose 1,6-bisphosphate

Effectively, the H of the OH group on C4 is transferred to C3 and the remaining oxygen forms a double bond with C4.

This produces a phosphorylated aldehyde (glyceraldehyde 3-phosphate) and a phosphorylated ketone (dihydroxyacetone phosphate).

> **Q&A 2:** Draw glycerol, glyceraldehyde and glyceraldehyde 3-phosphate.

How is the name dihydroxyacetone derived? It is simple if you know the structure of acetone.

acetone   dihydroxyacetone   dihydroxyacetone phosphate

Dihydroxyacetone is the keto form of glycerol. Remember that dihydroxyacetone is the simplest keto-sugar, having only 3 carbons, and glyceraldehyde is the simplest aldo-sugar. Dihydroxyacetone phosphate is formed by esterifying phosphate to one of the hydroxyl groups of dihydroxyacetone.

The next step is the conversion of dihydroxyacetone phosphate to its isomer, glyceraldehyde 3-phosphate. This results in 2 glyceraldehyde 3-phosphates having been formed from the original glucose; 1 directly from the cleavage of fructose 1,6-bisphosphate and the other from dihydroxyacetone phosphate.

What would be an appropriate name for the enzyme which catalyses the conversion of one triose phosphate, dihydroxyacetone phosphate, to its triose isomer, glyceraldehyde 3-phosphate?

Answer: *Triose phosphate isomerase.*

In order to understand the next steps in the pathway it is useful to compare the structures of the original molecule, glyceraldehyde 3-phosphate, with the end-product pyruvate. It is also important to remember that there are 2 glyceraldehyde 3-phosphates from the original glucose molecule entering this part of the pathway, and therefore all of the following steps occur twice, once for each glyceraldehyde 3-phosphate.

Amazingly, each of the carbons undergoes oxidation. Energy is produced by addition of oxygen at C1, loss of hydrogen at C2, and by loss of phosphate (C3). In other words, we obtain a large amount of energy from these reactions. Hopefully, enough to replace the energy lost so far (2 ATPs) plus some more.

Oxidation of C1 occurs firstly and an enormous amount of energy is released in the process. It takes 2 reactions for this oxidation to take place. In the first of these, the H atom is removed and replaced with a phosphate group. In this reaction glyceraldehyde 3-phosphate loses hydrogen.

1,3-bisphosphoglycerate

$NAD^+ \longrightarrow NADH + H^+$

What would be a suitable name for the enzyme that catalyses the dehydrogenation of glyceraldehyde 3-phosphate to 1,3-bisphosphoglycerate?

Answer: *Glyceraldehyde 3-phosphate dehydrogenase.*

There are 2 remarkable things about this reaction:

• Firstly, it is an oxidation because there is a loss of hydrogen. This hydrogen is transferred to nicatinomide adenine dinucleotide (NAD^+) and hence NAD^+ is reduced. In this reduced form, NADH stores energy and when it eventually gives up this energy (loses its hydrogen), the energy is used to do work.

• Secondly, the phosphate group added to glyceraldehyde 3-phosphate is organic phosphate. This means that energy is <u>not</u> lost by the addition of phosphate as it would have been if it had come from ATP, but in fact energy is gained. The energy is stored in a usable form in the next step of the oxidation when 1,3-bisphosphoglycerate transfers its phosphate from C1 to ADP which is converted to ATP.

What would be the name of the product after the removal of a phosphate from the C1 of 1,3-bisphosphoglycerate?

Answer: *3-phosphoglycerate.*

> **Q&A 3:** Illustrate the conversion of 1,3-bisphosphoglycerate to 3-phosphoglycerate. Do not forget that the phosphate group is transferred to ADP.

In this reaction 1 ATP is regained for each 1,3-bisphosphoglycerate (2 altogether). The enzyme which catalyses the loss of phosphate from 1,3-bisphosphoglycerate is phosphoglycerate kinase. This is not logical because the name suggests that the enzyme is catalysing the addition of phosphate (kinase) to phosphoglycerate. In other words, it is describing the reaction for the reverse direction. Indeed, this reaction can go in the reverse direction - if the levels of 3-phosphoglycerate are high there is enough ATP to supply the phosphate groups. I am sure it was named this way simply to confuse biochemistry students.

What happens next in the formation of pyruvate is that phosphate is transferred from C3 of 3-phosphoglycerate to

ADP to produce an ATP, and an hydrogen molecule (H_2) is removed from C2 and a keto group is formed at C2.

3-phosphoglycerate

This is a good way to remember what happens <u>but it is not what happens</u>! This process does not actually occur because the phosphate cannot be removed and replaced with a hydrogen group. Phosphate groups are removed by hydrolysis and hence water is added to the bond leaving an hydroxyl group. To overcome this problem, 3-phosphoglycerate undergoes a couple of rearrangements. Firstly, the phosphate is transferred from C3 to C2. Logically, the new molecule formed is 2-phosphoglycerate and the enzyme involved is phosphoglycerate mutase.

Water is then removed across C2 and C3 and a double bond is formed between C2 and C3 forming phosphoenolpyruvate. This reaction is catalysed by an enolase.

Nomenclature: A double bond in a carbon chain is called an 'ene', hence the term ene in phospho**eno**lpyruvate.

The double bond between C2 and C3 gives easy access to the phosphate group at C2 which can be transferred to ADP to form another ATP, i.e. 2 ATPs for each glucose molecule

The conversion of phosphoenolpyruvate to pyruvate is similar to the conversion of 1,3-bisphosphoglycerate to 3-phosphoglycerate in that it can also go in the reverse direction. The enzyme is named for the reverse pathway, pyruvate kinase. In other words the enzyme catalyses the addition of a phosphate to pyruvate.

It is essential that you remember how glycolysis occurs. The following describes how I think about this pathway, but I will not be sitting the exam. You need to write it out in your own words and build up the details of the chemical reactions based on the skeleton formed by this logic process. Practise it, step by step until you really know it, and not just remember it for an exam.

> ✴ **Logic process: Glycolysis is an anaerobic process that involves the conversion of glucose to 2 pyruvate molecules. Firstly, glucose is phosphorylated to prevent it from leaving the cell. The phosphate comes from ATP and hence represents an energy loss. It is then balanced in preparation for cleavage by converting it to fructose and phosphorylating it again. Once again, the phosphate is obtained from ATP and hence an energy loss to the cell. It is then cleaved to produce indirectly 2 phosphoglyceraldehydes. Each of the carbons of the 2 phosphoglyceraldehydes is oxidised to produce energy. Oxidation of the aldehyde group to an acid involves 2 major processes. An H is removed and is captured by NAD+ to produce NADH, which is a high-energy compound, and a phosphate is added. The addition of the phosphate is a gain of energy because it comes from organic phosphate. It is quickly removed by transferring it to ADP to form ATP and a phosphoglycerate is formed from the aldehyde. The remaining phosphate is removed from the phosphoglycerate and transferred to ADP to form ATP. In the process, which involves some rearrangements, the phosphoglycerate is converted to pyruvate. The net energy gain is 2 ATPs and 1 NADH for each phosphoglyceraldehyde, but 2 ATPs are lost in the preparation of glucose for cleavage into the 2 glyceraldehydes, and hence the total gain in glycolysis is 2NADH + 2ATP.**

Using sugars to form components of other molecules

Activation of sugars

For a sugar to be used as a building block for other molecules such as glycogen or as glycosaminoglycans, it must be firstly activated by attaching it to a nucleotide (nucleotides are ATP, UTP, GTP, TTP etc). This attachment commonly occurs at C1.

MAJOR POINT 31: A sugar must be activated by attaching it to a nucleotide before it can be used as a building block for attaching to other molecules.

Let us examine this process as part of the synthesis of glycogen from glucose. The first step is to protect glucose from diffusing out of the cell. Therefore, it is made more water-soluble by phosphorylating it to form glucose 6-phosphate.

Q&A 4: What is the name of the enzyme that catalyses the phosphorylation of glucose?

The phosphate group is then transferred from C6 to C1 as that is where the attachment of the nucleotide will occur. The enzyme that catalyses this reaction is phosphoglucomutase.

Nomenclature: Mutases are enzymes which catalyse the transposition of functional groups.

glucose-6-phosphate   glucose-1-phosphate

The next step activates the sugar by attaching a nucleotide to it. This places the sugar (glucose) in a high-energy state ready to be transferred to other molecules. The most commonly used nucleotide is uridine triphosphate (UTP).

It is worthwhile reviewing the structure of nucleotides at this stage because learning their structures is relatively easy if you know the structure of ribose. Indeed, the foundation to nucleotide structure is ribose. The following is a key to learning the structure of nucleotides using UTP as an example, and is not the biochemical pathways for their formation (see Chapter 12).

Draw β-D-ribose in its furanose form. Remember that it is a 5C sugar.

Answer:

Condense a phosphate to C5 to form a phosphoester bond.

Answer:

Attach another phosphate to the phosphate by removing water (an acid anhydride bond meaning an acid without water).

Answer:

Attach another phosphate to the phosphate that was just added. What is removed in the process?

Answer:

uridine 5'-triphosphate

Next add a base to the ribose at C1. In this case, the pyrimidine base uracil will be added to form uridine triphosphate. When bases are added, the C1 of ribose is attached to an amine group of the base. What is removed to form the *N*-glycosidic bond with the base?

Q&A 5: Name 3 other bases. (Hint: these are used to make DNA.)

Nomenclature: A **tri**phosphate differs from a **tris**phosphate and a **di**phosphate from a **bis**phosphate in that, for a **tri**phosphate or diphosphate the phosphates are joined to each other, whereas in a **tris**phosphate or **bis**phosphate (e.g. fructose 1,6-**bis**phosphate), the phosphate groups are attached at different places on the sugar.

The 5' (five prime) shows that the triphosphate is attached to C5 of ribose. The prime designation occurs because the atoms of the ring in the base (in this case uracil) are given numbers without primes. This distinguishes between the atoms of the uracil ring and the ribose ring.

Returning to the synthesis of glycogen and the activation of glucose, uridine triphosphate (UTP) is attached to glucose 1-phosphate to form uridine diphosphoglucose (UDPGlc). The energy needed to form this bond comes from cleaving a diphospho group (pyrophosphate: PP_i) from the UTP molecule. The enzyme which catalyses this reaction is UDP-glucose pyrophosphorylase.

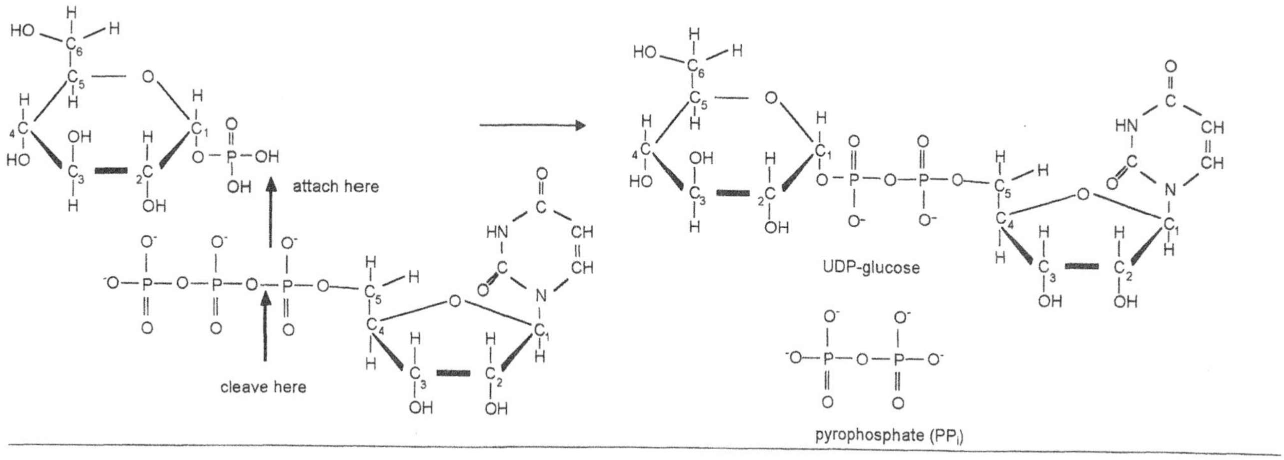

UDP-glucose

pyrophosphate (PP$_i$)

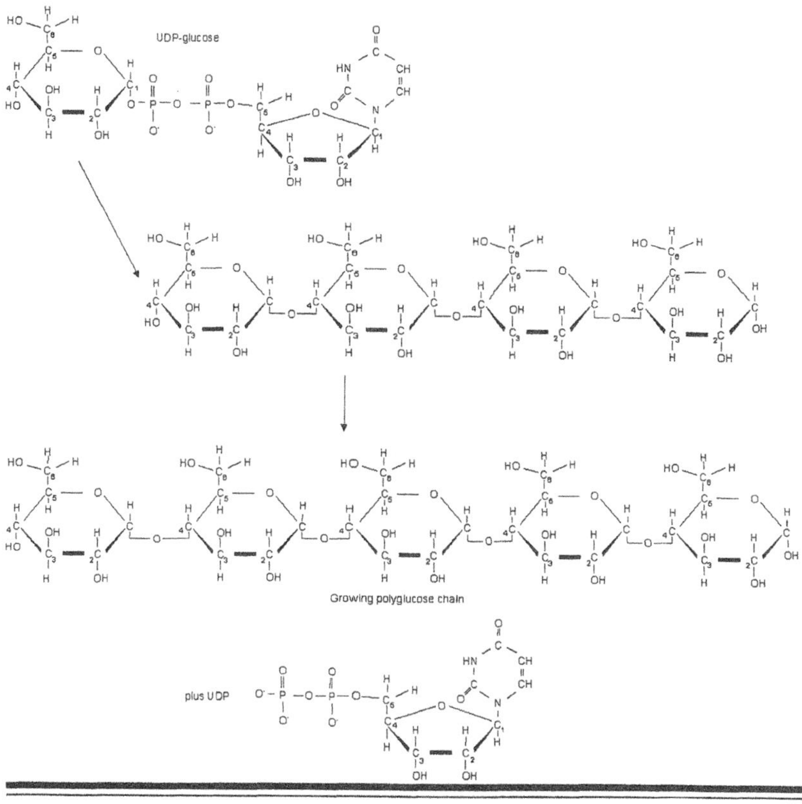

attach here

cleave here

UDP-glucose

Growing polyglucose chain

plus UDP

UDP-glucose can then be used to attach glucose to another sugar molecule. In the formation of glycogen, UDP-glucose is attached to the C4 of a growing polyglucose chain. The enzyme catalysing this process is glycogen synthase. In plants, for the formation of amylose and starch, ADP-glucose is used instead of UDP-glucose.

Higher level

Synthesis of glycogen

Glycogen is not simply a straight chain of $\alpha1\rightarrow4$ linked glucose units (amylose), but has branches that are formed by $\alpha1\rightarrow6$ linkages. Branching chains are formed by another enzyme which cleaves the $\alpha1\rightarrow4$ linkage and forms an $\alpha1\rightarrow6$ linkage (amylo-$(1,4\rightarrow1,6)$transglycosylase). A block of usually 7 residues is transferred to a more interior site. Usually the block that is cleaved comes from a chain more than 11 residues long and is transferred to a place at least 4 residues from a branch point.

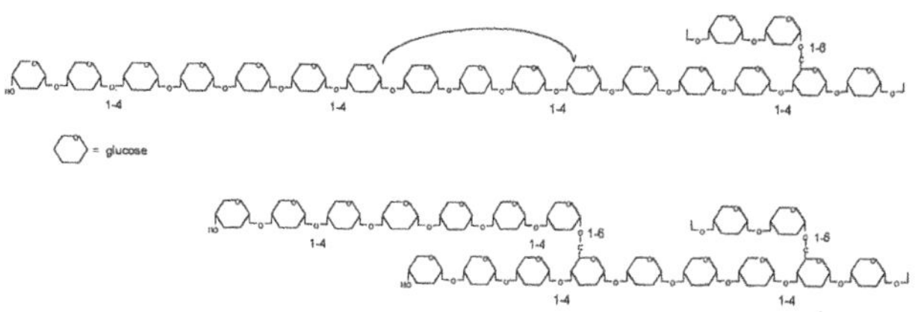

= glucose

Branching is important because it increases the solubility of glycogen and provides a large number of non-reducing sugar terminals which are sites of activity for glycogen phosphatase, the enzyme that breaks down glycogen.

Sugar amine synthesis

Another occurrence of nucleotide activation of sugars is found in the synthesis of sugar amines. Sugar amines or glycosamines are a very important component of polysaccharides known as glycosaminoglycans (GAGS) (see Chapter 6). The first steps in this pathway are identical to those above with the conversion of glucose to glucose 6-phosphate. Glucose 6-phosphate is then converted to fructose 6-phosphate in a reaction catalysed by glucose 6-phosphate isomerase. Importantly, fructose 6-phosphate has a ketone group at C2, which is the site where an amine group can be transferred (c.f. the conversion of glutamate (E) to α-ketoglutarate in Chapter 8). In this case the amine group is donated by glutamine (Q) to form 2-amino-D-glucose 6-phosphate which has the common name of glucosamine 6-phosphate.

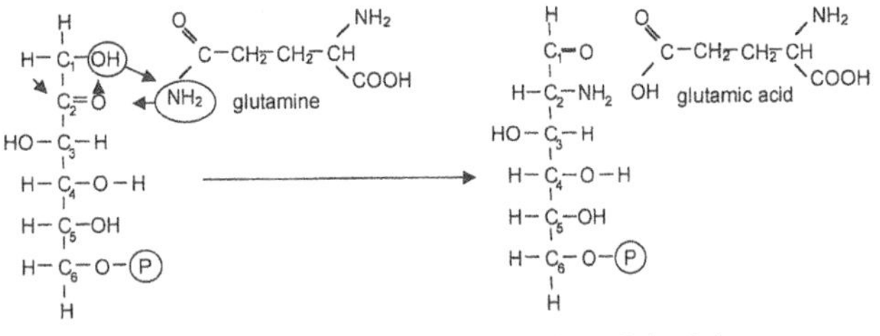

fructose-6- phosphate

2-amino-D-glucose-6-phosphate
(glucosamine-6-phosphate)

> **Q&A 6:** Draw glucosamine 6-phosphate in its pyranose form and number the carbon atoms.

The phosphate group on C6 is then transferred to C1 to form glucosamine 1-phosphate. The enzyme involved is glucosamine 6-phosphate mutase.

In the next step, the amine on C2 is acetylated (acetic acid is transferred to it). This occurs by forming an amide bond. In biochemical reactions, the acetyl group is carried by coenzyme A as acetyl-coenzyme A (acetyl CoA).

glucosamine-1-phosphate   N-acetylglucosamine-1-phosphate

$CH_3C - SCoA$ ⟶ CoASH

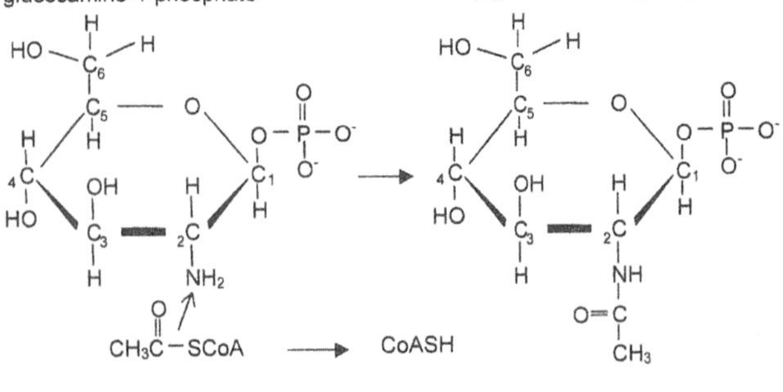

> **MAJOR POINT 32:** The source of acetyl groups in biochemical reactions is acetyl-CoA.

Biochemically, N-acetylglucosamine is used for building polymers such as the GAG components of proteoglycans (see Chapter 6). As shown above, to use a sugar molecule for building a polymer, it must be raised to a high energy state by at-

taching it to a nucleotide (UDP is most common). Therefore the next step is nucleotidyl transfer.

To which carbon of N-acetylglucosamine 1-phosphate would UTP be added?

Answer: UTP is added to the phosphorylated carbon of N-acetylglucosamine, C1 and pyrophosphate is released.

N-acetylglucosamine-1-phosphate   UDP-N-acetylglucosamine

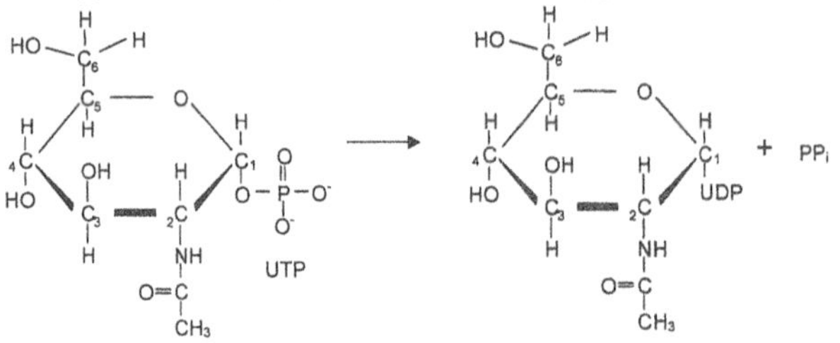

Synthesis of muramic acid

UDP-N-acetylglucosamine is formed and this can be used for making glycosaminoglycans. It is also the precursor for muramic acid, which is used in the synthesis of bacterial cell walls.

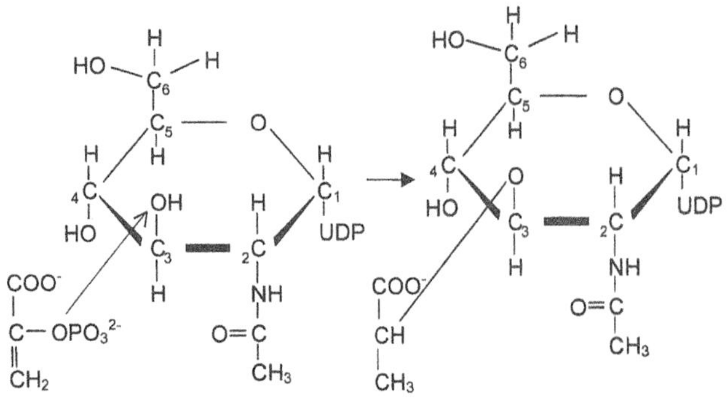

phosphoenolpyruvate
(or lactic acid)

UDP-muramic acid

In order to form the other major sugar used in GAGs, N-acetylgalactosamine, there is a simple rearrangement of the hydroxyl group associated with C4 of UDP-N-acetylglucosamine to form UDP-N-acetylgalactosamine.

> **Q&A 7:** Rearrange UDP-N-acetylglucosamine to form UDP-N-acetylgalactosamine.

UDP-N-acetylgalactosamine is now in its active form to produce GAGs or another sugar, sialic acid. Sialic acid is one of the sugars associated with the ABO blood groups and has 9 carbons.

> **Q&A 8:** What is the general name of a sugar with 9 carbons. Remember that the general name for a sugar with 6 carbons is a hexose.

Higher level

Formation of sialic acid

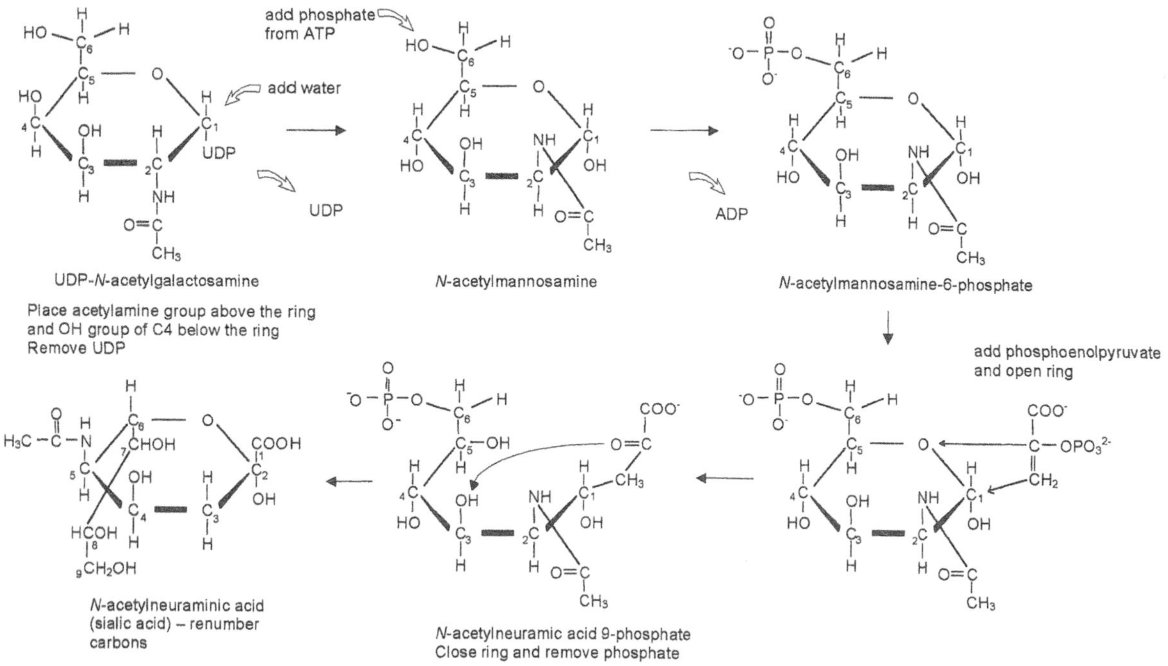

UDP-*N*-acetylgalactosamine
Place acetylamine group above the ring and OH group of C4 below the ring
Remove UDP

N-acetylmannosamine

N-acetylmannosamine-6-phosphate

add phosphoenolpyruvate and open ring

N-acetylneuraminic acid (sialic acid) – renumber carbons

N-acetylneuramic acid 9-phosphate
Close ring and remove phosphate

Detailed knowledge of the pathway for the formation of UDP-*N*-acetylglucosamine and UDP-*N*-acetylgalactosamine is not an essential component of a budding biochemist's armoury. However, it is desirable that the pathway and the principles behind it are understood because these support other understanding processes in biochemistry. The logic of this pathway is outlined below. You should write it out in your own words and practise it.

> **★ Logic process: The process begins with phosphorylation of glucose to glucose 6-phosphate, which is then converted to its isomer, fructose 6-phosphate. The significance of fructose 6-phosphate is that it is a ketone and amine groups can be readily transferred to ketones. An amine group is transferred to the keto (C=O) group to form glucosamine 6-phosphate from glutamine (Q), and by donating its amine group, glutamine (Q) becomes glutamate (E). The phosphate group is transferred from C6 to C1 to form glucosamine 1-phosphate. The amine group is then acetylated to form *N*-acetylglucosamine 1-phosphate. It is important to remember that for acetylation, the acetyl group is carried by acetyl-CoA. Before the cell can use N-acetylglucosamine as a building block for polysaccharides, it must be**

> **activated by adding a nucleotide (UTP). This occurs at the phosphorylated carbon, C1. Once formed, UDP-*N*-acetylglucosamine can either be incorporated into glycosaminoglycans or transformed into other sugar derivatives, UDP *N*-acetylgalactosamine, UDP-muramic acid or sialic acid.**

Attachment of sugars to proteins

One of the less obvious roles for sugars in cells is for sorting proteins. Once proteins are formed, the cell must determine whether to excrete or to store the protein in a particular subcellular compartment. The process of moving proteins about in a cell and placing them in specific compartments is called protein translocation. One of the main signals for protein translocation is the sugar groups attached to the proteins.

Sugars can be attached to proteins at the amino acids serine (**S**) or threonine (**T**) (*O*-glycosylation) or to asparagine (**N**) (*N*-glycosylation). Serine (**S**) and threonine (**T**) have an hydroxyl group to which sugars can attach (*O*-glycosidic bond) and asparagine (**N**) has an amine group to which sugars can attach (*N*-glycosidic bond) (see Chapter 3 for details).

There are 2 important features of glycosylation of proteins:

- Nucleoside activated sugars are used for this process either directly or indirectly.
- The proteins to be glycosylated are shunted into the endoplasmic reticulum or Golgi apparatus where glycosylation takes place.

MAJOR POINT 33: Proteins are glycosylated in the endoplasmic reticulum or the Golgi apparatus.

Translocation of the protein from the cytoplasm into the endoplasmic reticulum

As a new protein is synthesised, the initial part of translation of the RNA message occurs in the cytoplasm via ribosomes associated with the granular endoplasmic reticulum. Proteins destined to be glycosylated are transferred to the inside of the endoplasmic reticulum. For this to occur, they must cross a membrane, the membrane of the endoplasmic reticulum.

Would you expect the amino acids of the proteins destined for transfer across the endoplasmic reticulum to be hydrophobic or hydrophilic?

Answer: *Hydrophobic – lipid soluble.*

Transfer of the protein occurs well before translation has been completed. Twenty hydrophobic amino acids at the NH_2 terminus are a signal that the protein is destined for the inside of the endoplasmic reticulum. These 20 hydrophobic amino acids are recognised by a signal recognition particle, which is a protein.

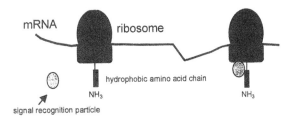

When the signal recognition particle binds to the 20 amino acids, translation stops. The signal recognition particle bound to the 20 amino acids is recognised by a receptor on the membrane of the endoplasmic reticulum (the signal recognition particle receptor) and a ribosomal receptor protein.

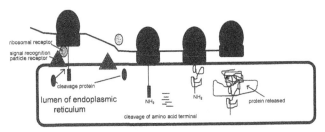

The amino acid chain is transferred across the membrane of the endoplasmic reticulum, and the signal recognition particle and its receptor are cleaved from the ribosome. The growing chain continues to extend in the lumen of the endoplasmic reticulum. Signal peptidase cleaves off the 20 basic amino acids from the NH_2 terminal, creating a new NH_2 terminus. When protein synthesis is complete, soluble ER proteins are released into the lumen of the endoplasmic reticulum and membrane spanning proteins remain attached to the membrane.

O-linked glycosylation

Proteins destined to have O-linked sugars are incorporated into vesicles which are budded off the endoplasmic reticulum membrane and transported to the Golgi membranes. Here, an O-glycosidic linkage is formed with N-acetylgalactosamine. UDP-N-acetylgalactosamine carries the sugar group and a transferase catalyses the reaction. Usually, a galactose is added next and then specific sugars follow. Each is carried by UDP and each requires a specific glycosyltransferase.

N-linked glycosylation

This is considerably more complex than O-linked glycosylation. Initially, an oligosaccharide is formed on the outside of the endoplasmic reticulum and then transferred to the inside of the endoplasmic reticulum. More sugar groups are added and then it is transferred to a protein via an N-glycosidic bond. Depending upon the final fate of the glycoprotein, some of the sugar groups are trimmed from the oligosaccharide and others are added. During this process, the sugars are transferred from the endoplasmic reticulum to the Golgi apparatus by budding and from there to other organelles, such as lysosomes, depending upon the sugar signal.

The exact details are outlined below as higher level. It is important that the principles are remembered.

Higher level

Initially, the oligosaccharide is formed on dolichol. This is a special lipid alcohol embedded in the membrane of the smooth endoplasmic reticulum. This lipid has a fat-soluble component (in the membrane) and a water-soluble component (the hydroxyl group).

$$CH_3 \qquad\qquad CH_3 \qquad\qquad CH_3$$
$$H_3C - C = CH - CH_2 - (CH_2 - C = CH - CH_2)_{17} - CH_2 - CH - CH_2 - CH_2 - OH$$

dolichol

Q&A 9: How many carbons in dolichol?

Q&A 10: Which part of dolichol is: a) lipid soluble (inserted in the membrane of the endoplasmic reticulum); b) water soluble; and c) to which part could a sugar could be attached?

In its active form, dolichol exists as dolicholphosphate. Its phosphate is esterified to the hydroxyl group which makes

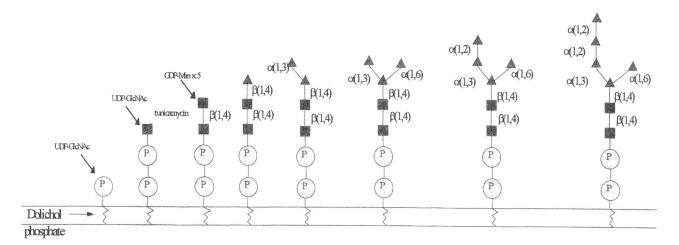

this part of the molecule even more water soluble. It also provides an attachment site for sugars. The lipid component anchors the growing sugar chain to the membrane of the endoplasmic reticulum.

Sugar groups are transferred to dolichol from nucleoside diphosphosugars. The sequence and bonds formed are illustrated above. Each transfer is catalysed by a specific glycosyltransferase.

Once the fifth mannose group has been attached, the oligosaccharide, still anchored to the membrane of the endoplasmic reticulum by dolichol, is translocated to the lumen of the endoplasmic reticulum.

Although more sugars are added to the oligosaccharide, there is a major difference - the sugars to be transferred are carried by dolichol rather than a nucleoside. The nucleoside transfers its sugar to dolichol on the outside of the endoplasmic reticulum. Amphomycin inhibits this particular transfer. The sugar is then carried into the lumen of the endoplasmic reticulum and transferred to the growing oligosaccharide. Mannose is transferred from GDP mannose to dolichol and then to the growing chain. Four mannose groups are added in this manner.

After mannose, glucose is attached to the growing chain. Once again dolichol phosphate is the carrier.

Once 3 glucose groups have been attached, the oligosaccharide is transferred to a protein and becomes *N*-linked to an asparagine (**N**). Castanospermine inhibits this process. Note that the protein is often still being synthesised when the oligosaccharide is attached.

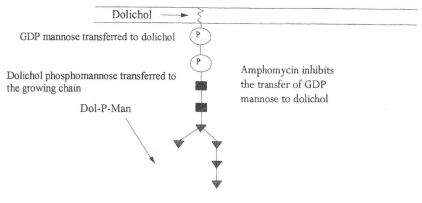

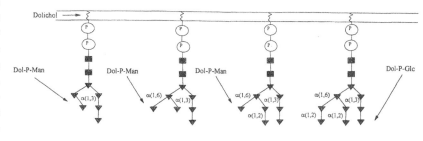

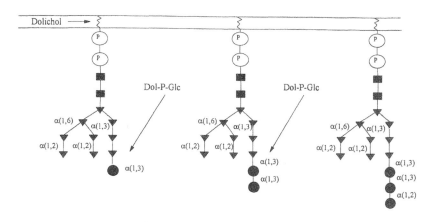

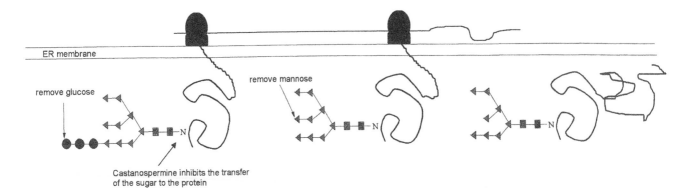

ER membrane

remove glucose

remove mannose

Castanospermine inhibits the transfer
of the sugar to the protein

Once the linkage has occurred, the oligosaccharide is trimmed. Firstly the glucoses are removed and then a mannose. At this stage, the glycoprotein is budded off the endoplasmic reticulum and transferred to the Golgi. Here, further sugars are added or removed. In the Golgi, sugars are transferred to the oligosaccharide chain from nucleosides. Fucose, galactose and sialic acid are added in the trans Golgi. The final complement of sugars creates the signal that de-termines the eventual

ifate of the protein - whether it will bencorporated in a lysosome, cell membrane or be excreted. For example, lysosome-destined proteins have phospho-*N*-acetylglucosamine added to mannose and glucose is removed. The membranes in the Golgi destined to become a lysosome have a mannose 6-phosphate receptor, which binds to the oligosaccharide and these proteins become incorporated in the lysosome when it buds from the Golgi membranes.

References and further reading

Abeijon, C. and Hirschberg, C.B. (1992) Topography of glycosylation reactions in the endoplasmic reticulum. *Trends Biochem. Sci.* **17**:32-36.

Alonso, M.D., Lomako, J., Lomako, W.M. and Whelan, W.J. (1995) A new look at the biogenesis of glycogen. *FASEB J.* **9**:1126-1137.

Devlin, T.M. (1992) Textbook of Biochemistry with Clinical Correlations, 3rd ed. New York: Wiley-Liss.

Fothergill-Gilmore, L. (1986) The evolution of the glycolytic pathway. *Trends Biochem. Sci.* **11**:47-51

Greenstein, B. and Greenstein, A. (1996) Medical Biochemistry at a Glance. Oxford: Blackwell Science.

Q&A Answers

1   Pyruvic acid has both a ketone (circled) and an acid group (arrowed) and hence it is a ketoacid.

ketone

$$CH_3 - C - C \begin{matrix} O \\ \\ OH \end{matrix}$$

acid

2   Notice the relationship between these 3 molecules. The aldehyde and phosphate groups are circled.

glycerol   glyceraldehyde   glyceraldehyde-3-phosphate

3

1,3-bisphosphoglycerate

$$(P) - O - C_1 = O$$
$$H - C_2 - OH$$
$$H - C_3 - O - (P)$$
$$H$$

3-phosphoglycerate

$$HO - C_1 = O$$
$$H - C_2 - OH$$
$$H - C_3 - O - (P)$$
$$H$$

ADP $\longrightarrow$ ATP

4   The enzyme that catalyses the phosphorylation of glucose is glucokinase (liver) or hexokinase (muscle or fat cells).

5   The other bases besides uracil are adenine, thymine, guanine and cytosine.

6   The structure of glucosamine 6-phosphate in its pyranose form.

glucosamine-6-phosphate

7 Conversion of UDP-*N*-acetylglucosamine to UDP-*N*-acetylgalactosamine involves a simple transposition of the hydroxyl group on C4 from below the ring to above the ring.

8 The general name of a sugar with 9 carbons is a nonose.

9 Dolichol has 95 carbon atoms. It is made from farnesyl pyrophosphate and 19 isoprene units are added. The farnesyl units are important in the synthesis of other lipids such as cholesterol (see Chapter 11).

10 a) The lipid-soluble part of dolichol is the uncharged long carbon chain.

b) the water-soluble part is the hydroxyl (OH) group because of hydrogen bonding.

c) a nucleotide sugar is transferred to the hydroxyl group by forming an *O*-glycosidic bond. The nucleotide is removed in the process.

8 Metabolism of Amino Acids

In this chapter you will learn:

- that the catabolism of the amine group and carbon skeletons of amino acids are by quite different pathways
- how the amine group of amino acids is processed to form urea
- the structure of urea
- the relationship between urea and uric acid
- the urea cycle
- how α-keto acids are formed from of glutamate, aspartate and alanine
- transamination and deamination of amino acids
- the importance of the B vitamins in amino acid metabolism
- how the breakdown pathways of the amino acids lead to the formation of elements of the TCA cycle or ketones
- strategies for learning about amino acid breakdown
- about the synthesis of amino acids
- assimilation of ammonia
- formation of the carbon skeletons of amino acids
- the essential and non-essential amino acids.

Amino acid degradation

Amino acids are the building blocks of proteins (Chapter 4). In our bodies, proteins are continually being broken down into amino acids, and some of these amino acids are used for making new proteins, but the remainder are broken down further. In this chapter, we will examine the pathways for the breakdown of amino acids.

The degradation of amino acids is traditionally regarded as a difficult topic to study. Although it is not simple, it is impossible if you do not know the structures of the 20 amino acids that make up proteins. You must <u>know</u> their structures and to almost know their structures is not good enough for learning about their metabolism. Every time you read about an amino acid, look up its structure if you do not know it and spend a minute trying to commit it to memory. Without this foundation you miss out on one of the most exhilarating pieces of learning – the degradation of amino acids. It really is a thrill to see how these amino acids relate to each other and fit into other biochemical pathways. Above all, it gives you a different handle on amino acid structures, which assists you to understand their structures and not be burdened with the unpalatable task of just learning them. As the degradation of the various amino

acids is described, really focus on the molecular groups that are being removed or altered and the **patterns** of these changes. These are the handles that will enable you to understand and learn the intricate details of amino acids. It is recommended that you complete Chapter 4 before proceeding.

Processing the amine group

What is the main problem for our bodies that is associated with amino acid degradation?

Answer: *The amine group ($-NH^+_3$) is the problem. When cleaved from an amino acid, it forms ammonia (NH_3), which is toxic at high concentrations. You are probably already aware of this because many household cleansers contain ammonia to kill bacteria.*

Anecdote

Some people who have cats find that their cats have a tendency to urinate on their kitchen benches or in the kitchen sink. The problem is that the kitchen has been cleaned with a product containing ammonia. A breakdown product of urine is ammonia and the cat thinks that the benches smell like a lavatory and treat it as such. Changing the cleaning agent should solve the problem.

How to remove ammonia was one of the most important evolutionary advancements that was required for animals to move from an aqueous environment to a land environment. In an aqueous environment, excess ammonia can be simply excreted into the surrounding water as soon as it is produced. Marine invertebrates remove ammonia by excreting it into the sea directly and fish excrete ammonia into the water via their gills or by their swim bladders. Once animals moved onto the land, they had to develop a mechanism for storing their waste products. Since ammonia is toxic, its storage is not possible and hence different strategies had to be developed for dealing with nitrogenous wastes. Mammals evolved urea synthesis as a mechanism for removing nitrogenous wastes, whereas many reptiles and birds evolved to excrete nitrogenous wastes in the form of uric acid.

Urea   Uric acid

Q&A 1: How many ureas does uric acid contain?

The formation of uric acid and how to draw its structure is dealt with in Chapter 12. Just take note at this stage that it is a purine, which is the same underlying structure of the bases guanine and adenine. Here, we will concentrate on the synthesis of urea and as a first step it is important to know the origin of the amine (-NH$_2$) groups. One comes from free ammonium ion and the other from aspartic acid (**D**).

> **Major point 34:** One of the 2 amine groups of urea comes from aspartic acid (D) and the other from free ammonia.

Most of the free ammonium ion is formed from the following reaction. Note that the chemicals are protonated/deprotonated as they would be at physiological pH.

The focus for learning is the relationship between α-ketoglutarate and the amino acid glutamate (**E**). Note that glutamate (**E**) is oxidised in its conversion to α-ketoglutarate and the energy released is transferred to NAD$^+$ or NADP in the form of H (NADH or NADPH). The name of the enzyme which catalyses the reaction is glutamate dehydrogenase.

Q&A 2: Explain how the name α-ketoglutarate is formed.

This reaction is the key to the removal of the amine group from all amino acids. Just a little bit of lateral thinking shows why. The amine group from amino acids can be transferred to α-ketoglutarate. This forms glutamate (**E**) and the particular α-keto-acid. Glutamate (**E**) can then be converted to α-ketoglutarate and free ammonia as described above. The transfer of an amine group to α-ketoglutarate is called transamination.

The co-factor, pyridoxal-5'-phosphate, is required for this reaction to take place. This is because the amine group is firstly transferred to pyridoxal-5'-phosphate to form pyridoxamine-5'-phosphate and it is transferred from pyridoxamine-5'-phos-

phate to α-ketoglutarate. The important point here is that pyridoxal-5'-phosphate is derived from pyridoxine which is vitamin B6.

> **Major point 35:** The amine group is removed from amino acids by transferring it to α-ketoglutarate to form glutamate (E).

Remember that the other amine source for urea is aspartate (**D**). Although aspartate is directly involved in transferring its amine to form urea, the amine may come indirectly from other amino acids by a reaction that parallels the reaction above. The keto-acid of aspartate is oxaloacetate.

Q&A 3: Using the information below, draw the transfer of an amine group from a general amino acid to oxaloacetate to form aspartate (D), include the structure of oxaloacetate (hint: examine the structural relationship between α-ketoglutarate and glutamate (E)).

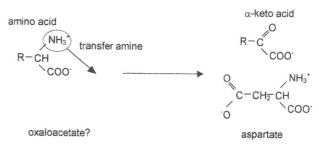

Most amino acids prefer to transfer their amine group to α-ketoglutarate to form glutamate (**E**) (see above). Glutamate (**E**) can readily transfer its amine group to oxaloacetate.

Q&A 4: What is the overall reaction for the transfer of the amine group from glutamate (E) to oxaloacetate? (Write out the reaction, do not draw structures).

In this process, the amine group from the original amino acid is transferred to aspartate (**D**) by firstly being transferred to α-ketoglutarate to form glutamate (**E**) and then it is transferred from glutamate (**E**) to oxaloacetate. The enzyme which catalyses this reaction is glutamate-aspartate amino transferase.

Therefore, glutamate (**E**) becomes the key molecule for supplying the amine groups for urea: free ammonium ion by hydrolysing glutamate (**E**) to α-ketoglutarate and free ammonium; and aspartate (**D**) by transfer of the amine group from glutamate (**E**) to oxaloacetate.

The liver is the main organ where these transaminations and urea synthesis occur. Consequently, amino acids, which are to be broken down, must be transported from the other organs to the liver. The kidneys also form some urea.

> **Major point 36:** Urea is formed in the liver.

Muscles have a special problem, particularly in anaerobic conditions (heavy workloads). They can use glucose and the carbon skeletons of amino acids as fuels. This presents them with 2 problems: what to do with the amine group from the amino acids; and what to do with pyruvic acid which is the end-product of anaerobic breakdown of glucose.

Given the structure of pyruvic acid (below), could the amine groups of amino acids be transferred to it, and if so, what would be formed from pyruvate?

Answer: Yes, the carbonyl (ketone) group is a site for amine group transfer, and alanine (A) would be formed.

The strategy behind this reaction is very clever. The alanine (A) formed is transported to the liver thus removing the waste products, pyruvate and nitrogen, at the same time from the muscle. The liver converts the alanine (A) back to pyruvate by a transamination to α-ketoglutarate. It then synthesises glucose from the pyruvate (gluconeogenesis) and transports it back to the muscle. It is also extraordinarily clever because if pyruvic acid were transported back to the liver as pyruvic acid rather than alanine (A), it would alter the pH of the blood. This cycle between the muscle and the liver is called the glucose-alanine cycle.

Other tissues transfer the amine from amino acids to glutamate (E) to form glutamine (Q) and this travels in the bloodstream to the liver where it is hydrolysed to form glutamate (E).

> **Q&A 5:** Draw the transfer of the amine group of an amino acid to glutamate (E) to form the α-keto-acid of the amino acid and glutamine (Q). Name the enzyme which catalyses this reaction.

> **Q&A 6:** Draw the hydrolysis of glutamine to glutamate (E) and free ammonia and name the enzyme which catalyses this reaction. This reaction occurs in the liver.

To summarise how different tissues in the body deal with ammonia (think about the reactions outlined here, do not just commit them to memory).

- In the liver, glutamate (E) is cleaved to α-ketoglutarate and free ammonia or the amine group is transferred to oxaloacetate to form aspartate (D).

- In muscle the amine group of amino acids is transferred to pyruvate to form alanine (A). Alanine (A) is transferred to the liver where the amine group is transferred to α-ketoglutarate to form glutamate (E) and the pyruvate is used to form glucose.
- In peripheral tissues the amine group of amino acids is transferred to glutamate (E) to form glutamine (Q) and this is transported to the liver where it is cleaved into free ammonia and glutamate (E).
- Urea is formed in the liver from the free ammonia and aspartate (D) formed by these processes.

The urea cycle

Now that we know how free ammonia and aspartate (D) are formed, it is necessary to examine how urea is formed. The formation of urea occurs in a pathway that is a cycle and hence is called the urea cycle. To understand cycles, I find it is important to start into the cycle with a molecule that you are already familiar with. In this case, start with the amino acid arginine (R).

Draw the structures of arginine (R) and urea.

Answer:

Note that arginine (R) contains a 'urea group'. What could be added to arginine (R) to form urea?

Answer: *Water*

The enzyme that catalyses the breakdown (hydrolysis) of arginine (R) into ornithine and urea is argininase. This takes place in the cytosol of the cell and then ornithine is transferred into the mitochondria. Here, ornithine begins to be cycled back to form arginine (R). In this process it gains 2 amine groups. One from free ammonium (in the form of carbamoyl phosphate) and the other from aspartate (D). It also gains a carbon and this comes from carbon dioxide (also in the form of carbamoyl phosphate).

In the mitochondria, free ammonium ion is combined with carbon dioxide and phosphate to form carbamoyl phosphate which is then transferred to the δC amine group of ornithine.

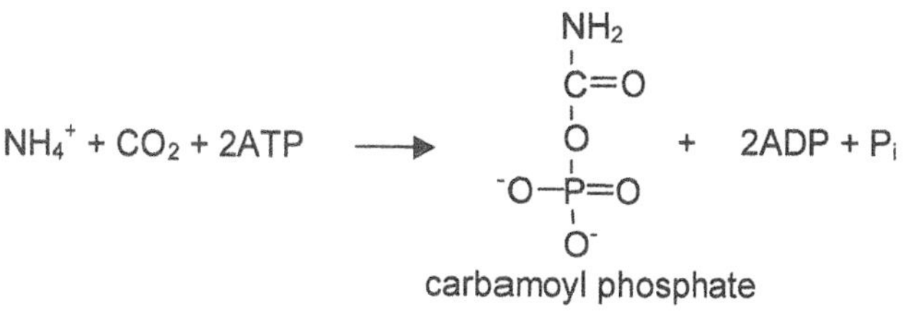

$$NH_4^+ + CO_2 + 2ATP \longrightarrow \text{carbamoyl phosphate} + 2ADP + P_i$$

carbamoyl phosphate

The enzyme that catalyses the synthesis of carbamoyl phosphate in mitochondria is carbamoyl-phosphate synthase type 1. This distinguishes it from carbamoyl phosphate synthase type 2, which is located in the cytoplasm and is used in pyrimidine synthesis. This compartmentalisation allows control over the 2 pathways. This certainly makes sense when you realise that one is a degradative pathway to remove nitrogen from the body and the other a synthetic pathway to build the bases for DNA or RNA. Also note that the reaction consumes 2 high-energy bonds, i.e. 2ATPs.

> **Major point 37:** The formation of carbamoyl phosphate is the rate limiting step of urea synthesis.

Removal of the phosphate group from carbamoyl phosphate allows it to form an amide bond with ornithine. This forms citrulline.

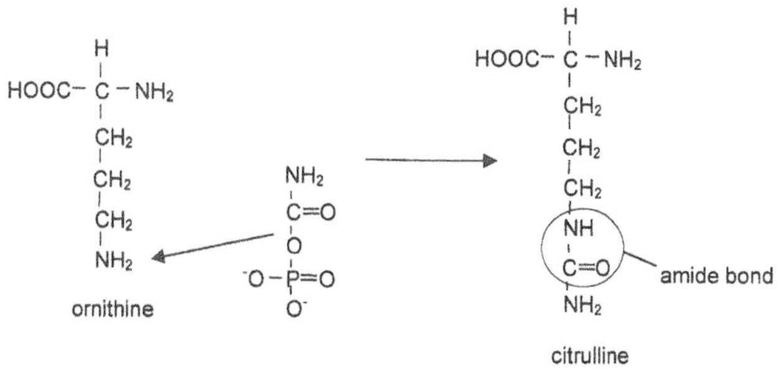

The enzyme that catalyses the transfer of carbamoyl group to ornithine is ornithine transcarbamoylase. This reaction takes place in the mitochondria. Citrulline is then transported out of the mitochondria into the cytoplasm where the next reactions take place.

If you examine the structures of citrulline and arginine (**R**), you will notice that the difference is an amine and therefore an amine has to be transferred to citrulline to form arginine (**R**). This is not a simple process and the amine comes from aspartate (**D**).

Aspartate (**D**) is actually attached to citrulline via its amine group to form argininosuccinate. This reaction requires a great deal of energy and ATP is converted to AMP.

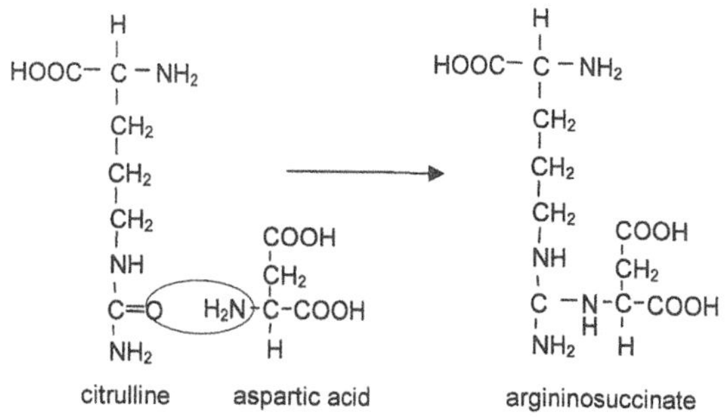

Nomenclature: To understand the name argininosuccinate, 'arginine' can be seen within the structure. 'Succinate' is also part of the structure. Succinic acid is illustrated below.

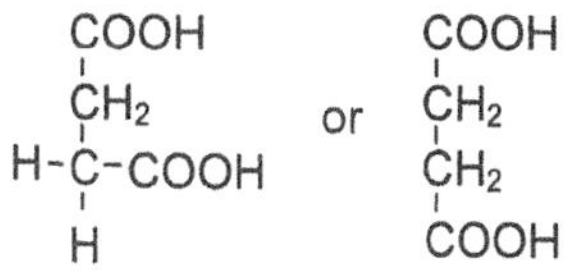

What would be a suitable name for the enzyme which catalyses the synthesis of argininosuccinate? Do not be too imaginative.

Answer: *Argininosuccinate synthase.*

*Draw argininosuccinate and indicate where it has to be cleaved to form arginine (**R**).*

Answer:

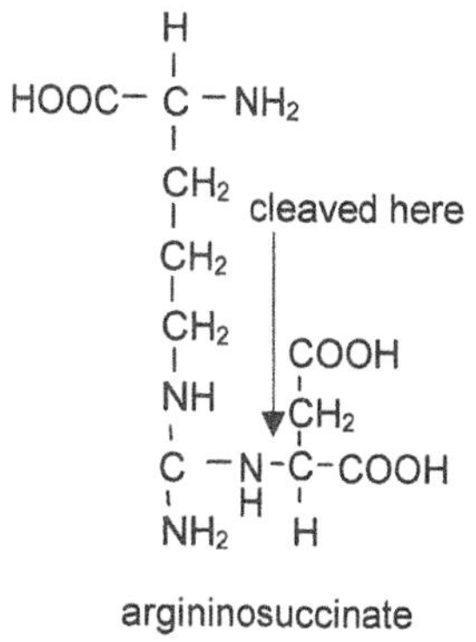

argininosuccinate

Simple, isn't it? This is a simple cleavage and not an hydrolysis. Therefore, nothing is added to the molecule. To balance the final equation, a double bond has to form between the 2 middle carbons of the cleaved carbon skeleton. This forms fumarate rather than succinate.

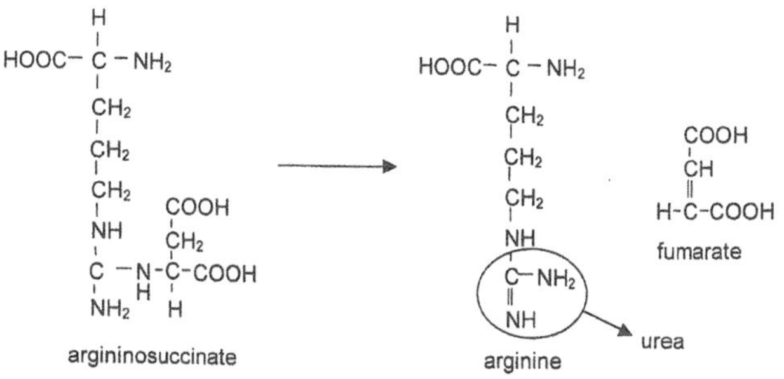

The name of the enzyme which catalyses the reaction is argininosuccinase. The cycle is now complete, with arginine (**R**) ready to be cleaved to form urea and ornithine.

Q&A 7: Draw the whole urea cycle and indicate which parts occur in the mitochondria and which parts occur in the cytoplasm. Include the names of the enzymes and where energy is used. To help, start with arginine (**R**) which is cleaved to urea and ornithine. Ornithine is then transferred to the mitochondria.

Q&A 8: To assist the learning process, become familiar with the similarities in structure between aspartate (**D**), oxaloacetate, succinate and fumarate. Draw their structures.

Degradation of the carbon skeleton of amino acids

Once the amine group is removed, the carbon skeleton of most amino acids can be metabolised to generate energy or form glucose. The carbon skeletons of other amino acids are metabolised to form either acetyl-CoA or acetoacetyl-CoA (we will learn about these in detail later) which are transformed into ketone bodies.

Energy is readily extracted if the carbon skeletons can be formed into molecules of the tricarboxylic acid (TCA) cycle. This cycle takes place in mitochondria and is dealt with in detail in the next chapter.

Important to remember is that:

- α-ketoglutarate, succinyl-CoA (succinate attached to coenzyme A), fumarate and oxaloacetate are all elements of the TCA cycle
- amino acids whose carbon skeletons can be transformed into any of these can readily enter the TCA cycle and be used for energy production or the synthesis of glucose.

The following cartoon shows the general flow in the tricarboxylic acid cycle. (Details are in the next chapter).

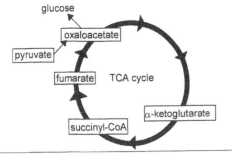

Mnemonic for the order: A cat (α-ketoglutarate) sucks continually at (succinyl-CoA) funny (fumarate) oxen (oxaloacetate)

Breakdown of glutamate (E), glutamine (Q), proline (P), arginine (R) and histidine (H)

These amino acids all enter the tricarboxylic acid cycle as α-ketoglutarate. Clearly, the carbon skeletons of glutamate (**E**) and glutamine (**Q**) enter the cycle here. Glutamine (**Q**) is converted to glutamate (**E**) by deamination (glutaminase) and glutamate (**E**) is converted to α-ketoglutarate by deamination. These were dealt with above. Remember that in this case, the enzyme is glutamate dehydrogenase because a hydrogen molecule is removed from water and transferred to NAD and the oxygen is transferred to form the α-ketoglutarate.

Not so obvious are proline (**P**) and arginine (**R**) and histidine (**H**). These are firstly converted to glutamate (**E**) and then to α-ketoglutarate.

To understand the breakdown of proline (**P**) we need to look at the similarities between the structure of proline (**P**) and glutamate (**E**). In essence, it is a simple unfolding of the proline ring and oxidation of the aldehyde group.

Initially the ring is primed for unfolding by removing hydrogen. Remember that loss of hydrogen is an oxidation. In this case, oxygen captures the hydrogen and water is formed. The product Δ^1-pyrroline-5-carboxylate is a pyrroline ring with the first bond (between N1 and C2) being a double bond (Δ^1) and a carboxyl group coming from C5 of the ring. The ring spontaneously unfolds to form glutamate γ-semialdehyde which is then converted to glutamic acid (**E**). This reaction is catalysed by glutamate semialdehyde dehydrogenase. This name seems strange. However, the hydrogen molecule is removed from water and transferred to NAD$^+$ or NADP and the remaining oxygen atom is added to form glutamate (**E**).

For the breakdown of arginine (**R**), it is first converted to ornithine.

Q&A 9: Recall the conversion of arginine (**R**) to ornithine in the urea cycle and draw the reaction.

Ornithine has the right number of carbons to form glutamate (**E**), but it has an extra amine group. Remember that to remove unwanted amines, the general procedure is to either deaminate or transaminate the compound to form

---

H | remove hydrogen | H | ring opens spontaneously

$^-$OOC$-$C$-$NH$_2^+$ $^-$OOC$-$C$-$NH$^+$

CH$_2$ CH$_2$ CH$_2$ CH$_2$

CH$_2$ CH$_2$

proline Δ^1-pyrroline-5-carboxylate   glutamate γ-semialdehyde   glutamate

omithine → glutamate γ-semialdehyde → glutamate

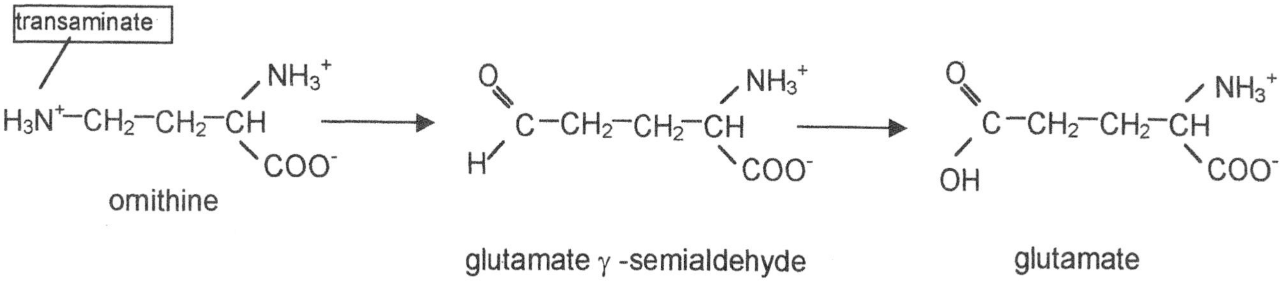

a ketone group. The amine group in ornithine is transaminated to α-ketoglutarate (catalysed by ornithine-δ-aminotransferase) and because the amine in this case was attached to the terminal carbon, an aldehyde rather than a ketone is formed. Ornithine is thus converted to glutamate γ-semialdehyde by a transamination to α-ketoglutarate, which becomes glutamic acid. The aldehyde group of glutamate γ-semialdehyde is oxidised to the acid, glutamic acid. The 2 glutamic acids (**E**) formed are then deaminated to form α-ketoglutarate (as described above).

Histidine (**H**) is also converted to glutamate by a complex series of reactions which you need not commit to memory. However, some aspects of the conversion are interesting. The amine of the αC histidine (**H**) is removed and an amine in the ring of histidine (**H**) eventually becomes the amine of glutamate (**E**). One of the carbons in the ring becomes the carbon of the carboxyl group of the αC and hence the molecule becomes 'turned around' to become glutamate (**E**). Also important is that a derivative of the vitamin, folic acid, is a cofactor in one of the reactions.

Major point 38: Folic acid is required for amino acid degradation.

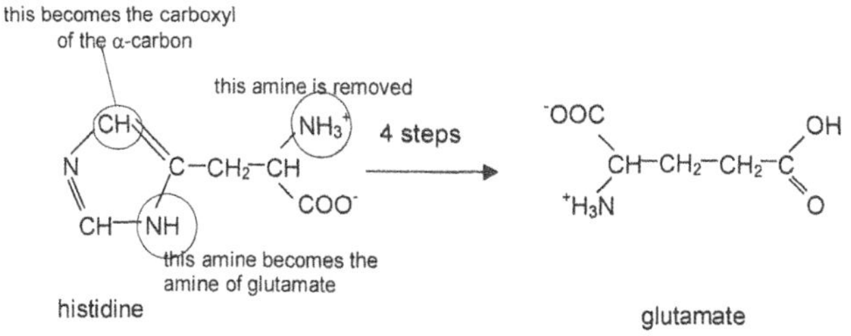

---

histidine → (4 steps) → glutamate

This is an exciting series of reactions because the structural relationship between histidine (**H**) and glutamate (**E**) is not at all obvious until you become aware of this conversion of histidine (**H**) into glutamate (**E**).

Breakdown of isoleucine (I), valine (V), threonine (T) and methionine (M)

The branched amino acids isoleucine (**I**) and valine (**V**), the sulphur containing amino acid, methionine (**M**) and the hydroxylated amino acid threonine (**T**) enter the TCA cycle at succinyl-CoA which is converted to succinate by removal of CoA.

Before entering this section it is necessary to have a little background about coenzyme A. Coenzyme A is an important carrier of carboxylic acid groups. It has a reactive sulphydryl group (-SH) and when coenzyme A is in its native form it is usually abbreviated to CoA-SH to emphasise this reactive group. A carboxylic acid can be linked to CoA-SH by removing water, the OH from the acid and an H from the sulphydryl group.

Q&A 10: For practice, link succinic acid to CoA-SH by removing water to form succinyl-CoA.

The branched methyl group in isoleucine (**I**) and valine (**V**) is difficult for our bodies to deal with. There are 9 steps for conversion of isoleucine (**I**), and 10 steps for the conversion of valine (**V**) to succinyl-CoA. Once again it is unnecessary to remember these pathways unless you are specialising in this area. Reference material can be consulted if details of these pathways are required.

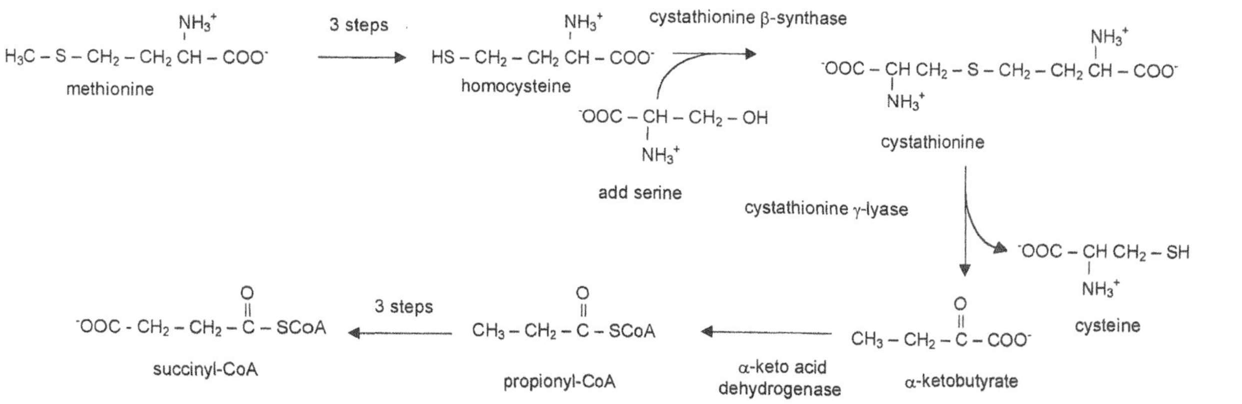

Similarly, the internal hydroxyl group of threonine (**T**) is difficult to deal with and it requires 5 steps to convert it to succinyl-CoA (see higher level below).

The breakdown of methionine (**M**) is interesting because

- it demonstrates an unexpected relationship with serine (**S**)
- cysteine (**C**) is synthesised as a by-product

Methionine (**M**) is demethylated in a complex series of 3 reactions to form homocysteine. Serine (**S**) is condensed to homocysteine (remove water) to form cystathionine. Cleaving cystathionine produces the α-ketobutyrate and the by-product cysteine (**C**). This cleavage includes a deamination, which makes it a little unusual. α-Ketobutyrate is then attached to coenzyme A in a reaction that involves release of CO_2 and the conversion of NAD^+ to NADH. Propionyl-CoA is then converted to succinyl-CoA in a further 3 steps.

> **Major point 39:** In the breakdown of methionine (M), serine (S) is converted to cysteine (C) by transfer of the sulphydryl group.

Breakdown of phenylalanine (F) and tyrosine (Y)

Phenylalanine (**F**) and the closely related tyrosine (**Y**) are converted to fumarate.

This is a complex series of reactions that for the most part would be regarded as higher level, particularly the details of the reactions. However, the general principles are not so difficult. The main problem is to breakdown the ring. It has been shown above for other amino acids with ring structures in their R groups that the ring has to be set-up for unfolding. Initially phenylalanine (**F**) is converted to tyrosine (**Y**).

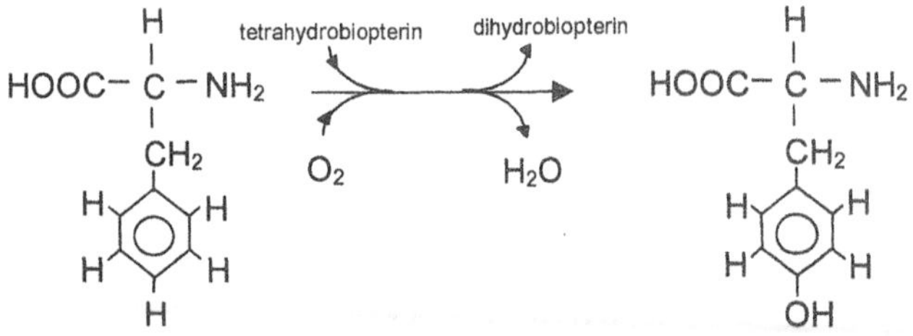

> **Q&A 11:** What is the name of the enzyme that catalyses the conversion of phenylalanine to tyrosine?

The next part of the reaction is to remove the amine group by transamination α-ketoglutarate.

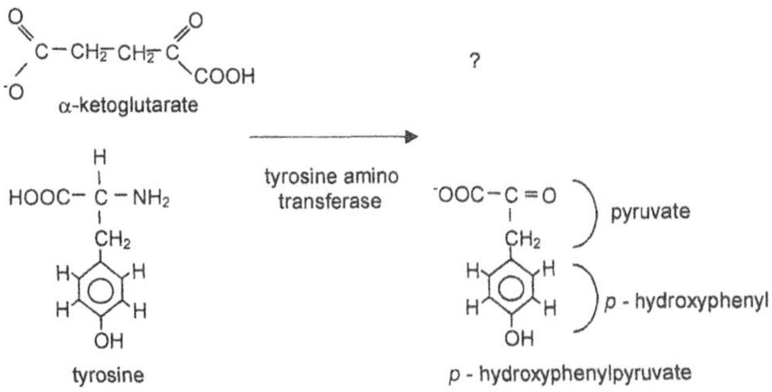

> **Q&A 12:** What forms from α-ketoglutarate in this reaction?

Higher level

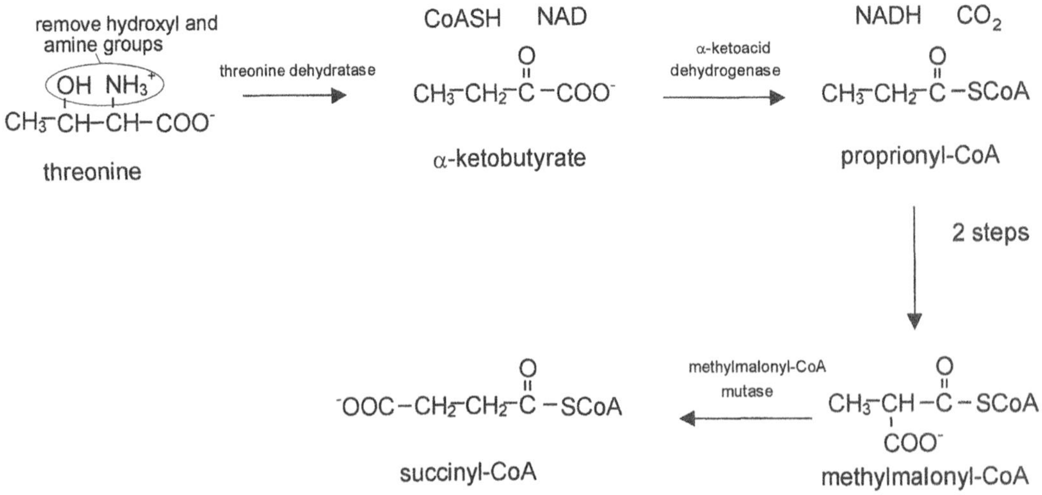

Note that the attachment of α-ketobutyrate to acetyl-coenzyme A (CoA-SH) is unusual as it involves a decarboxylation. This must occur because of the keto group adjacent to the carboxyl group.

The next reaction is very complicated and it sets up the ring for cleavage. It changes the position of the hydroxyl group, adds another hydroxyl group to the ring and removes the terminal carboxyl group to form a new carboxyl group from the ketone. This forms homogentisate – a satiated gay gent. Because 2 oxygen atoms are added to the structure, the enzyme that catalyses the reaction is called *p*-hydroxyphenylpyruvate dioxygenase.

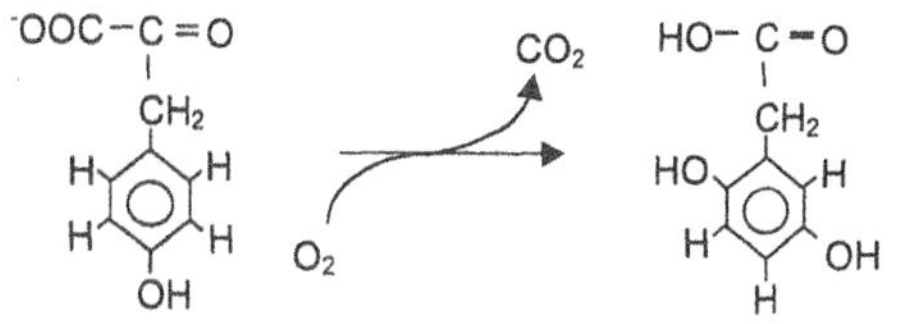

p-hydroxyphenylpyruvate

The ring is then opened (homogentisate 1,2-dioxygenase) and then there is a slight rearrangement of the hydrogen atoms across the double bond, *cis* to *trans* (malylacetoacetate isomerase) and then it is cleaved into fumarate and acetoacetate. This reaction is catalysed by fumarylacetoacetase.

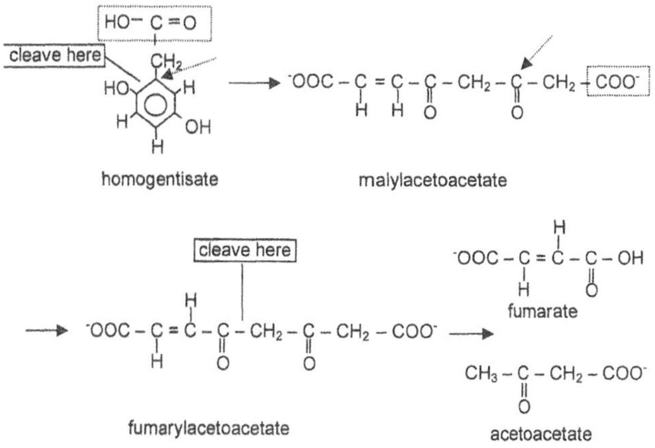

homogentisate   malylacetoacetate

fumarylacetoacetate   acetoacetate

fumarate

The fumarate can then be transferred to the tricarboxylic acid cycle. Note that acetoacetate is formed. This part of the carbon skeleton is used in a different pathway, which will be dealt with later.

Breakdown of asparagine (N) and aspartate (D)

It is relatively obvious which amino acids enter the TCA cycle as oxaloacetate. The deamination of asparagine (**N**) results in the formation of aspartate (**D**). This reaction releases free ammonium ion and is catalysed by asparaginase. Aspartate (**D**) is converted to oxaloacetate by transferring its amine to α-ketoglutarate. This reaction is catalysed by aspartate aminotransferase and of course, glutamate is also formed in this reaction from α-ketoglutarate.

> **Q&A 13:** Draw the reaction sequence showing the conversion of asparagine (**N**) to aspartate (**D**) to oxaloacetate.

Breakdown of alanine (A), cysteine (C), glycine (G), serine (S) and tryptophan (W)

In addition to conversion to elements of the tricarboxylic acid cycle, some amino acids are converted to pyruvate. Pyruvate can be channeled into 2 pathways depending upon the organ or the body's needs.

- It can be channeled into the synthesis of new glucose (gluconeogenesis), which is mainly carried out in the liver.
- It can be linked to coenzyme A to form acetyl-CoA, which is then channeled into the TCA cycle for energy production.

Note that these processes are opposite in intent. The formation of glucose is a mechanism for forming a fuel source (potential energy), whereas the channeling of acetyl-CoA into the TCA cycle is part of a mechanism for extracting energy from the carbon skeleton.

Alanine (**A**) is converted to pyruvate.

Examine the structures of alanine (**A**) and pyruvate. The relationship between alanine (**A**) and pyruvate is homologous to the relationship between glutamate (**E**) and α-ketoglutarate, and between aspartate (**D**) and oxaloacetate.

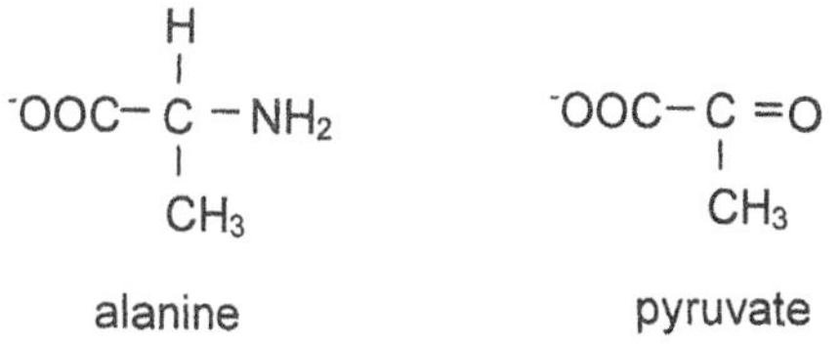

alanine   pyruvate

*How do you think alanine (**A**) could be converted into pyruvate?*

Answer: *Alanine (**A**) is deaminated to form pyruvate. This actually occurs by a transamination.*

As above, α-ketoglutarate is the amine acceptor and when the amine is transferred from alanine (**A**) α-ketoglutarate is converted to glutamate. The enzyme that catalyses this reaction is alanine aminotransferase.

Glycine (**G**) and serine (**S**) are also converted to pyruvate.

If the structure of glycine (**G**) is examined, it might be expected that a methyl group is added to glycine to form alanine (**A**) and then pyruvate is formed. **This is not the case!** The unexpected occurs. Glycine (**G**) is formed into serine (**S**) by a reaction catalysed by hydroxymethyltransferase. For the vitamin conscious, it is important to note that the methyl group is supplied by a derivative of the vitamin folic acid, tetrahydrofolate-N^5,N^{10}-methylene-tetrahydrofolate

> **Q&A 14:** Given the name of the enzyme (hydroxymethyl transferase) draw the conversion of glycine (**G**) to serine (**S**).

Serine (**S**) is then converted in a one-step reaction to pyruvate and this reaction is catalysed by serine dehydratase. Free ammonia is released.

Important to remember is that reactions for the conversion of glycine (**G**) to serine (**S**) and from serine (**S**) to pyruvate, are double whammy reactions. In the first reaction, both an hydroxyl and a methyl group are transferred and in the second, both the hydroxyl and the amine groups are removed.

Tryptophan (**W**) is converted to alanine (**A**) by removal of the ring. This requires 4 separate reactions. Alanine (**A**) is then converted to pyruvate as described above.

The ring of tryptophan (**W**) is processed by a different series of 14 reactions to form acetoacetyl-CoA. Ring structures are difficult to dispose of.

There are a variety of reactions for the breakdown of cysteine (**C**), and all of them end in the formation of pyruvate. In mammals this is a 3-step reaction.

Breakdown of leucine (L), lysine (K), phenylalanine (F) and tyrosine (Y)

Some amino acids have their carbon skeleton, or part thereof, broken down to acetoacetyl-CoA. These are ketogenic.

These are all multistep reactions and quite complicated. Acetoacetyl-CoA is converted to acetyl-CoA, which can enter the TCA cycle. The net effect of these reactions is the formation of ketones. Hence these amino acids are called ketogenic. Lysine (**K**) and leucine (**L**) are purely ketogenic, whereas some parts of the carbon skeletons of tyrosine (**Y**), phenylalanine (**F**) and tryptophan (**W**) enter other pathways. In the formation of ketones the net reaction is such that there is no net production of glucose and hence glucose is not synthesised from the breakdown of these amino acids.

What to learn and likely exam questions

Clearly, it is necessary to learn the urea cycle. This needs practice, and you also need to know which parts of the cycle take place in the mitochondria and which parts in the cytoplasm.

For amino acid degradation, the following priority of learning is recommended.

- Focus on the amino acids that are simply deaminated or transaminated to form important keto-acids. These of course are glutamate (**E**) aspartate (**D**) and alanine (**A**).
- Next think about glutamine (**Q**) and asparagine (**N**) which, with a simple deamination or transamination, can become glutamate (**E**) and aspartate (**D**)
- Once these are under control, focus on amino acids that can be converted to glutamate (**E**) or pyruvate.
- Focus on the structural relationship between glutamate (**E**) and proline (**P**), arginine (**R**) and histidine (**H**), remembering that histidine (**H**) becomes 'turned around' with the nitrogen of the ring becoming the nitrogen of the amine of glutamate (**E**).
- For glycine (**G**) and serine (**S**) remember there are double whammy reactions to convert them to pyruvate. Glycine (**G**) is converted to serine (**S**) by adding 2 groups to it (a methyl and hydroxyl) and serine (**S**) is converted directly to pyruvate by deamination and removal of the hydroxyl group. The non-ring part of tryptophan (**W**) is also converted to pyruvate.
- Next focus on the amino acids which are converted to succinyl-CoA. Remember the difficult amino acids are converted to succinyl-CoA. These are more difficult to learn, but remember that they are the amino acids with a branched R group, isoleucine (**I**) and valine (**V**), the amino acid with the non-terminal hydroxyl group, threonine (**T**), and the amino acid with the methylated sulphur group, methionine (**M**). Of these, the breakdown of methionine is worthwhile learning as it is interesting and it doubles-up because it also teaches you the synthesis of cysteine (**C**). Also remember that serine (**S**) is consumed and hence it is a second pathway for the breakdown of serine (**S**), albeit very indirect. Whether or not you will be tested on details of pathways for the breakdown of the amino acids in this group will depend upon the time dedicated to these in the lectures. If you are unsure, ask the lecturer if the details of these pathways are necessary for the exam.
- The amino acids that are converted to fumarate are the closely related phenylalanine (**F**) and tyrosine (**Y**). Remember that phenylalanine (**F**) is converted to tyrosine (**Y**) and then deaminated. The ring is then set up for opening. The final chain produced contains fumarate and the ketone, acetoacetyl-CoA.
- The final group comprises the ketogenic amino acids. These are lysine (**K**), and the amino acids with rings and it is their ring structures which form the ketones. Tyrosine (**Y**) and phenylalanine (**F**) have already been dealt with and the third one is tryptophan (**W**).
- It is important to be able to list the ketogenic amino acids. (Mnemonic **Wings FLY**).

Q&A 15: Draw the thumbnail sketch of the TCA cycle (see above) and indicate where each of the amino acids may enter this cycle. The ketogenic amino acids may enter the cycle also. They enter it as acetyl-CoA between oxaloacetate and α-ketoglutarate.

Amino acid synthesis

Synthesis of the amine group, glutamic acid (E) and glutamine (Q)

To form amino acids, a very important process is the formation of the amine group (assimilation of ammonia). Many bacteria and some plants can do this by attaching NH_4^+ to α-ketoglutarate.

In this reaction, not only is glutamate (**E**) synthesised, but water is as well. To synthesise water, 2 hydrogen atoms are required. These come from NADPH or NADH and H^+, which means that the reaction consumes energy. The name of the enzyme that catalyses this reaction, glutamate dehydrogenase, is not obvious because it describes the reverse process.

The most common method for assimilating ammonia is to transfer it to glutamate (**E**) to form glutamine (**Q**). This reaction requires energy that is obtained from converting ATP to ADP.

Q&A 16: Considering that glutamine is synthesized, what would be an appropriate name for the enzyme?

Major point 40: In assimilation of ammonia, glutamic acid (E) or glutamine (Q) are formed.

In mammals, the assimilation of ammonia to form glutamine (**Q**) from glutamate (**E**) mainly occurs in muscles and then glutamine (**Q**) is transferred to the liver and kidneys where its ammonium group is transferred to other compounds to form amino acids. Commonly, the amine group of glutamine (**Q**) is transferred to α-ketoglutarate to form glutamate (**E**) and the amine of glutamate (**E**) is transferred to α-keto-acids to form the corresponding

amino acid and α-ketoglutarate. A good example is the synthesis of aspartate (**D**). Oxaloacetate is a keto-acid found in the tricarboxylic acid cycle

Transfer an amine group from glutamic acid (E) to the α-keto group of oxaloacetate and name the products.

Answer:

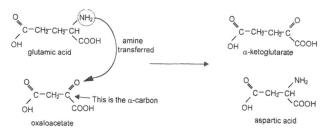

The enzyme which catalyses this reaction is aspartate transaminase. Also required is a co-enzyme, which is derived from vitamin B6 (pyridoxine). A person lacking vitamin B6 will suffer from protein deficiency due to lack of amino acid synthesis.

Two amino acids are different. Lysine (**K**) and threonine (**T**) are not formed by the transamination of their keto-acids because these keto-acids are unstable.

Major point 41: Glutamic acid (E) is a key to the formation of amino acids. Its amine group is transferred to an α-keto-acid to form the corresponding amino acid and α-ketoglutarate.

The carbon skeletons of amino acids

The carbon skeleton of amino acids is derived from intermediates of glycolysis, the tricarboxylic acid cycle or the pentose phosphate pathway. Some amino acids, such as the aromatic amino acids, are quite difficult to synthesise. As a result, plants and bacteria can synthesize all 20 amino acids, but mammals can only synthesise about half.

Major point 42: The amino acids a particular mammal can synthesise are called non-essential amino acids and those which it cannot synthesise are called essential amino acids and must be obtained from foods.

Synthesis of alanine (A), glutamate (E), glutamine (Q), aspartate (D) and asparagine (N)

The synthesis of all of these is simple transaminations or aminations. Learn these, as you will be sure to be tested on some of them. We have already seen how glutamate (**E**) and glutamine (**Q**) are formed. These form by the assimilation of ammonia.

Glutamate (**E**) may also be formed by the transamination of an amine group from an amino acid to α-ketoglutarate or by the deamination of glutamine (**Q**). Therefore, there are 3 ways glutamate may be formed.

Q&A 17: To what is ammonia added to form a) glutamate (E) and b) glutamine (Q)?

Q&A 18: Draw the keto-acid form of alanine (A) and name it and then transaminate it to form alanine (A).

remove hydrogen → transfer amine from glutamate → dephosphorylate

3-phosphoglycerate → 3-phosphohydroxypyruvate → 3-phosphoserine → serine

Alanine (A) and aspartate (D) are formed by the transamination of the amino group from an amino acid to their keto-acids

Q&A 19: Draw the keto-acid form of aspartate (D) and name it and then transaminate it to form aspartate (D).

Asparagine (N) is formed by the transfer of the amine group from glutamine (Q) to aspartate (D). In this process ATP is converted to AMP. The enzyme that catalyses this reaction is asparagine synthase.

Q&A 20: Draw the transamination of aspartate (D) to form asparagine (N). Include the amine source in the reaction and the energy source.

Synthesis of serine (S), glycine (G) and cysteine (C)

These are derived from 3-phosphoglycerate, which is a reaction product in glycolysis.

Nomenclature: To draw 3-phosphoglycerate remember that it is glycerol, except that one of the end carbons forms a carboxylic acid (COOH) instead of an alcohol (OH), and the carbon at the other end has a phosphate group attached to it via an ester bond as illustrated below.

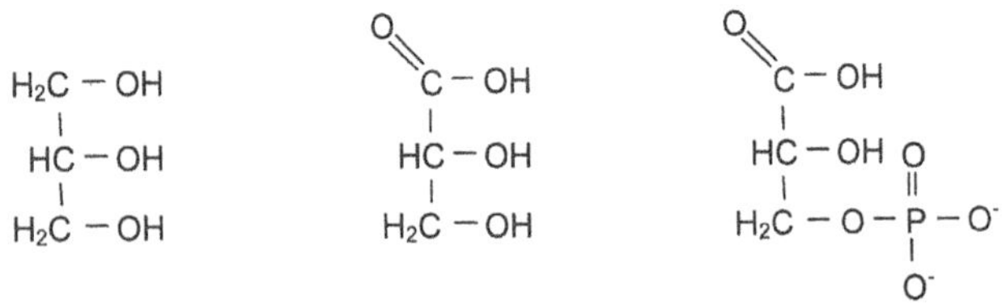

glycerol   glycerate   3-phosphoglycerate

Compare the structure of 3-phosphoglycerate with serine (S) to determine what must be changed to form serine (S).

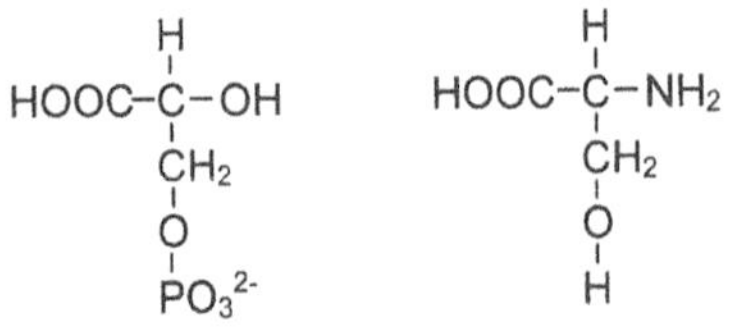

3-phosphoglycerate   serine

An amine has to be added to C2 and the phosphate group has to be removed. Remember that an amine group is normally added by a transamination to a ketone. Therefore it makes sense to change 3-phosphoglycerate into a ketone before transferring the amine. This is what happens.

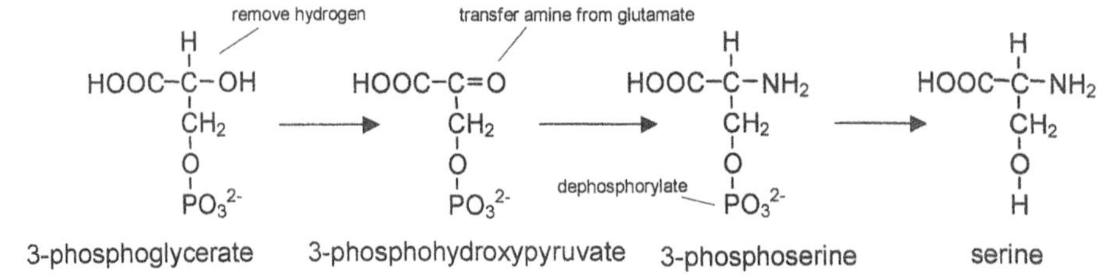

The dehydrogenation of 3-phosphoglycerate is catalysed by 3-phosphoglycerate dehydrogenase and the hydrogen is transferred to NAD^+ to form $NADH + H^+$. This loss of hydrogen is an oxidation and the energy obtained from it is stored as NADH. The transamination is catalysed by phosphoserine transaminase and the dephosphorylation to form serine (S) is catalysed by 3-phosphoserine phosphatase.

Glycine (G) is formed simply by removal of the hydroxymethyl group from serine (S) (the reverse of degradation). This reaction is catalysed by serine hydroxymethyltransferase. An important aspect of this reaction is that the methyl group is transferred to the co-factor tetrahydrofolate, which becomes 5,10-methylene-tetrahydrofolate. This, once again, emphasises the importance of folic acid (a vitamin) for proper amino acid, and hence protein, maintenance in the body.

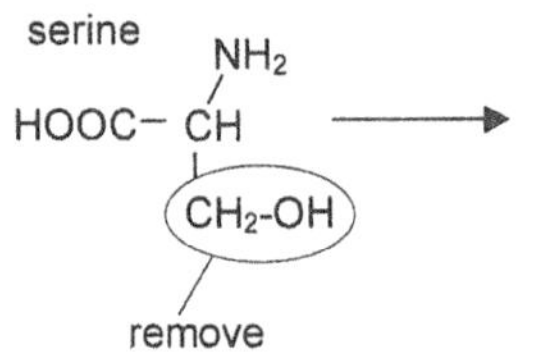

serine → glycine

remove

Cysteine (C) is also formed from serine (S) in mammals. Essentially a sulphydryl group (-SH) is transferred from homocysteine to serine and the OH group removed. This is a 2-step reaction beginning with the condensation of serine (S) and homocysteine to form cystathionine. The enzyme is cystathionine β-synthase.

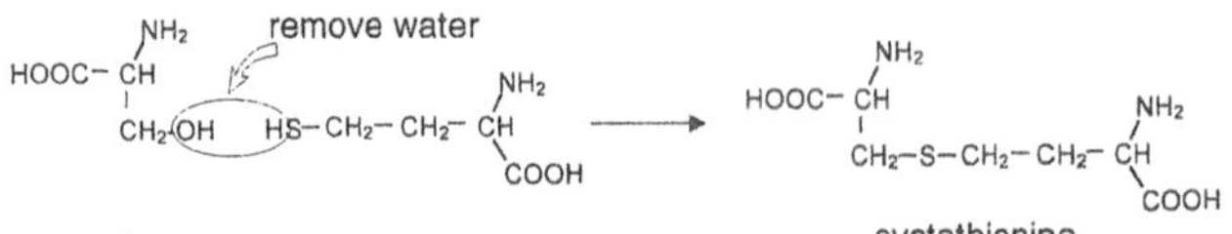

serine + homocysteine → cystathionine (remove water)

The next part of the reaction is not obvious. The cystathionine is cleaved to give cysteine (C) and it is also deaminated to give α-ketobutyrate. This reaction has been covered above in the degradation of methionine (M).

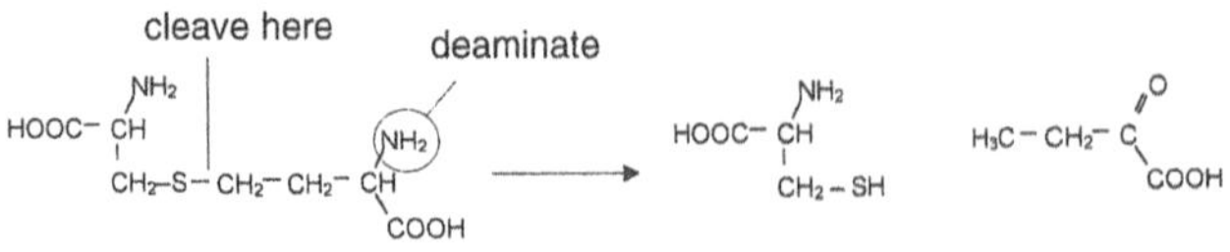

cystathionine → cysteine + α-ketobutyrate (cleave here, deaminate)

Synthesis of proline (P)

Proline (**P**) is formed from glutamate (**E**). Compare their structures. These are not very different and realising this helps to envisage how proline (**P**) forms from glutamate (**E**). Both have 5 carbons and C5 (circled) in glutamic acid (**E**) is linked to the amine group to form proline (**P**).

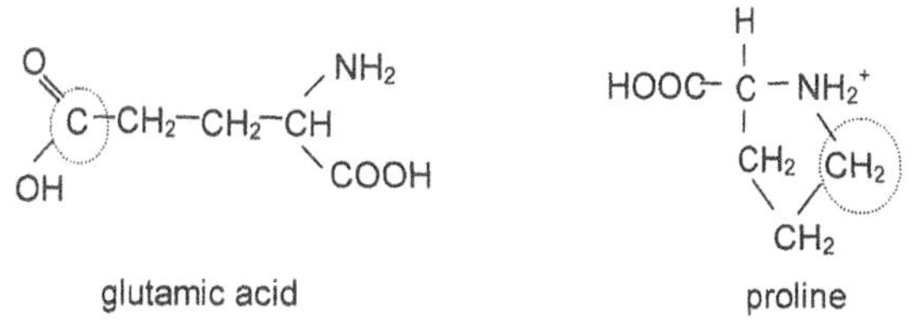

glutamic acid   proline

To synthesise proline, glutamate (**E**) is firstly phosphorylated at the γ-carboxyl group.

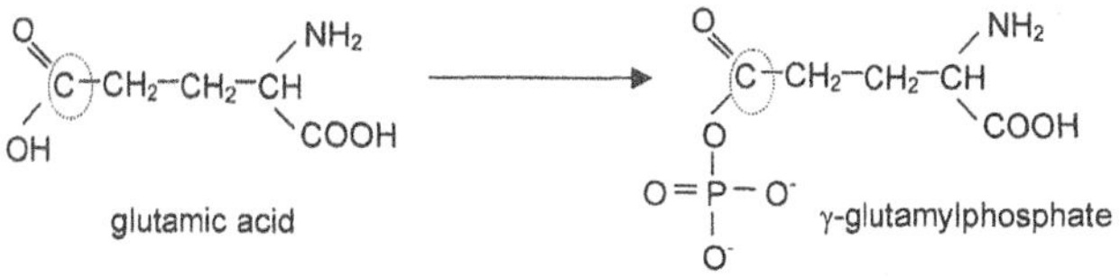

glutamic acid γ-glutamylphosphate

The phosphate comes from ATP and enzymes which catalyse the addition of a phosphate to a molecule from ATP are called kinases. In this case, the carbon atom of glutamate, which is phosphorylated, must be designated and hence the enzyme is called γ-glutamate kinase.

The next reaction is not obvious. The phosphate is removed and replaced by a hydrogen atom. This forms an aldehyde. The hydrogen comes from NADH, which is converted to NAD^+.

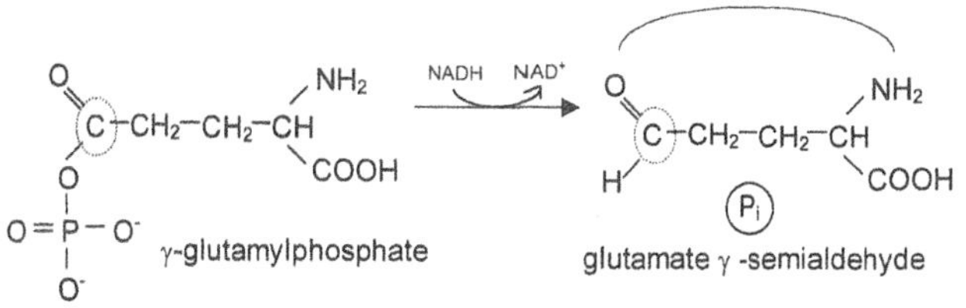

γ-glutamylphosphate   glutamate γ-semialdehyde

The enzyme that catalyses this reaction is glutamate γ-semialdehyde dehydrogenase, which really describes the reverse reaction. The formation of glutamate γ-semialdehyde exactly parallels the synthesis of aspartate β-semialdehyde from aspartic acid which is used for synthesising lysine (**K**) threonine (**T**) and methionine (**M**) – see below.

The oxygen of the aldehyde group then interacts spontaneously with the amine group of the αC and a ring is formed. This involves the removal of water.

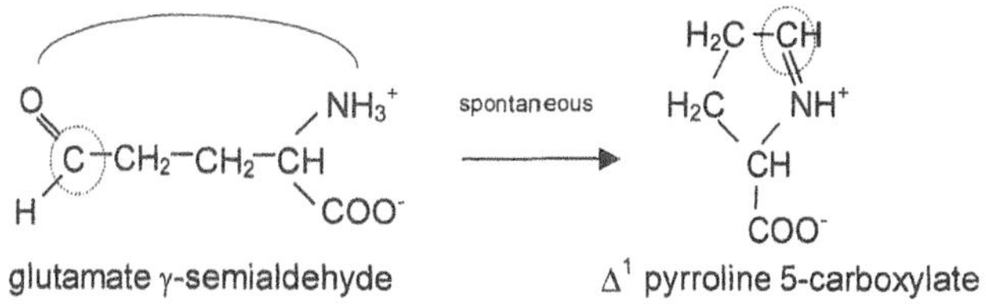

glutamate γ-semialdehyde Δ^1 pyrroline 5-carboxylate

The molecule formed is Δ^1-pyrroline 5-carboxylate.

Nomenclature: The ring structure with a nitrogen and 4 carbons is called a pyrroline. The numbering begins with the nitrogen and goes in the direction of the double bond. The double bond is signified by the Δ symbol and since it is the first bond it is given the superscript 1. The carboxyl group is attached to C5 of the ring and hence the molecule is given the name Δ^1-pyrroline 5-carboxylate.

The final reaction is to add hydrogen across the double bond to form proline (**P**). The hydrogen comes from NADH or $NADPH + H^+$. Addition of an hydrogen to a molecule is a reduction and hence the enzyme that catalyses this reaction is called Δ^1-pyrroline 5-carboxylate reductase.

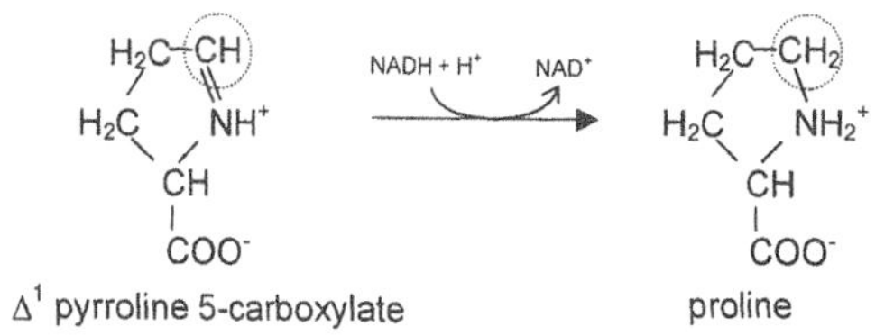

Δ^1 pyrroline 5-carboxylate   proline

Synthesis of arginine (R)

Not surprisingly, this is not so dissimilar from elements of the urea cycle. One of the main differences is that 2 tissues are involved: intestinal epithelial cells and kidney cells. In the intestine, glutamate γ-semialdehyde is converted to ornithine by a transamination to the aldehyde. Carbamoyl phosphate is then attached to the amine of the γC to form citrulline.

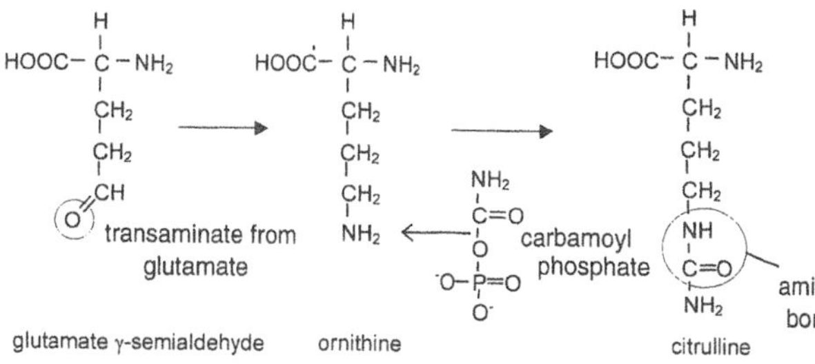

glutamate γ-semialdehyde   ornithine   citrulline

Citrulline then passes into the bloodstream and is taken up by the kidneys where aspartate is added to form argininosuccinate and then fumarate is cleaved from this to form arginine (**R**).

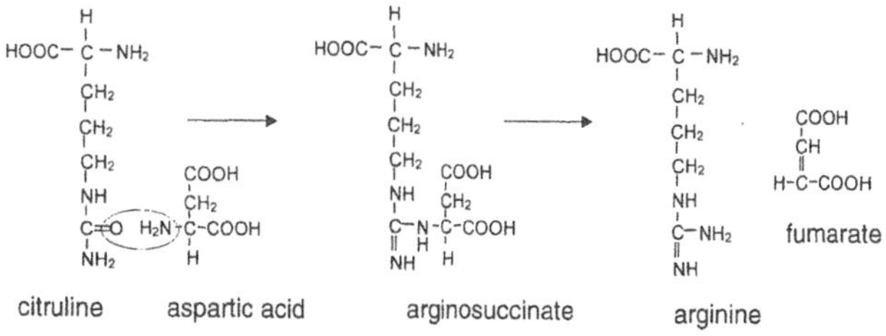

citruline   aspartic acid   arginosuccinate   arginine

Synthesis of tyrosine (Y)

In mammals, tyrosine (**Y**) is synthesised from phenylalanine (**F**). We have already covered this reaction as it is the first part of the reaction for the breakdown of phenylalanine. Important in the reaction is that the oxygen used comes from molecular oxygen and the free oxygen atom is combined with

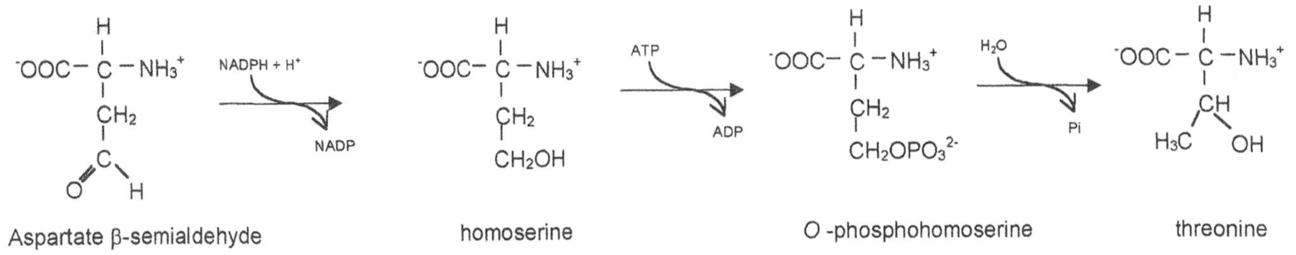

Aspartate β-semialdehyde   homoserine   O-phosphohomoserine   threonine

hydrogen, which comes from tetrahydrobiopterin. Tetrahydrobiopterin is regenerated by obtaining its hydrogen from NADH + H⁺.

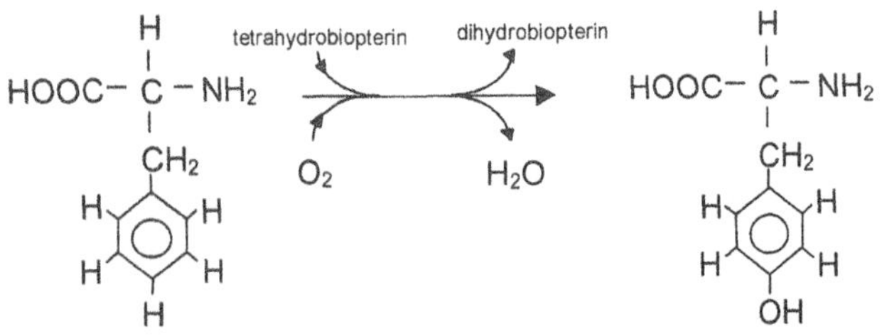

> **Q&A 21:** What is the name of the enzyme that catalyses the hydroxylation of phenylalanine (**F**) and what is the name of the genetic disease caused by a mutation in the gene that codes for this enzyme? (This has been covered earlier.)

Other amino acids are not synthesised by mammals and hence have to be obtained from the diet. These are synthesised by bacteria and plants.

Synthesis of lysine (K), threonine (T) and methionine (M)

The first common step in this pathway is to form aspartate β-semialdehyde. This exactly parallels the formation of glutamate γ-semialdehyde. The β carboxyl group of aspartate (**D**)

is phosphorylated and then dephosphorylated with the addition of hydrogen to form an aldehyde.

Eight reactions are required in bacteria to change aspartate β-semialdehyde into lysine (**K**). In some yeast and algae, lysine (**K**) is formed form α-ketoglutarate in a series of 7 reactions.

For the formation of threonine (**T**) and methionine (**M**), the next step is common. Hydrogen is added to the aldehyde group converting aspartate β-semialdehyde into homoserine. To form threonine the hydroxyl group is phosphorylated and then in a complex single-step reaction dephosphorylated and an hydroxyl group added to the βC. (See double-column diagram above.)

To form methionine (**M**), homoserine is converted to homocysteine in 3 reactions and then homocysteine is methylated with the methyl group being donated by 5-methyltetrahydrofolate.

Synthesis of isoleucine (I), valine (V) and leucine (L)

These branched-chain amino acids are also not synthesised by mammals and are therefore essential amino acids. Their synthesis is quite complex, but the pathways have a similar series of reactions and the initial part of the synthesis of valine (**V**) and leucine (**L**) is the same.

Some important points to note are:

- isoleucine (**I**) is derived from threonine (**T**), and hence initially from aspartate β-semialdehyde
- valine (**V**) and leucine (**L**) are derived from pyruvate

The actual pathways I would regard as higher level.

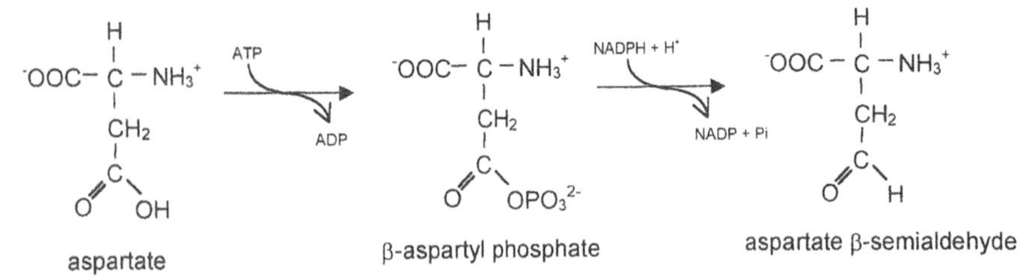

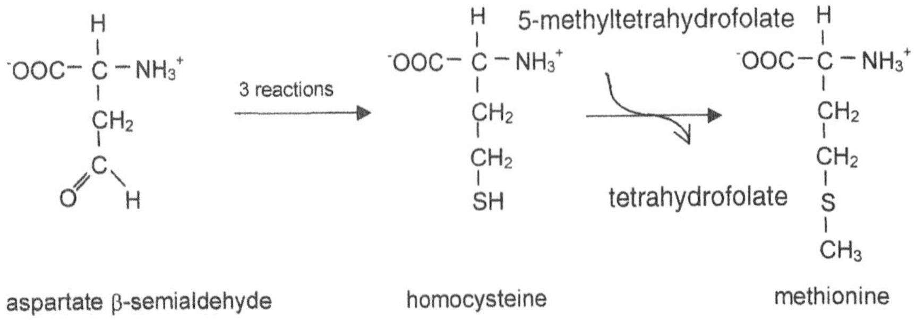

aspartate   β-aspartyl phosphate   aspartate β-semialdehyde

aspartate β-semialdehyde   homocysteine   methionine

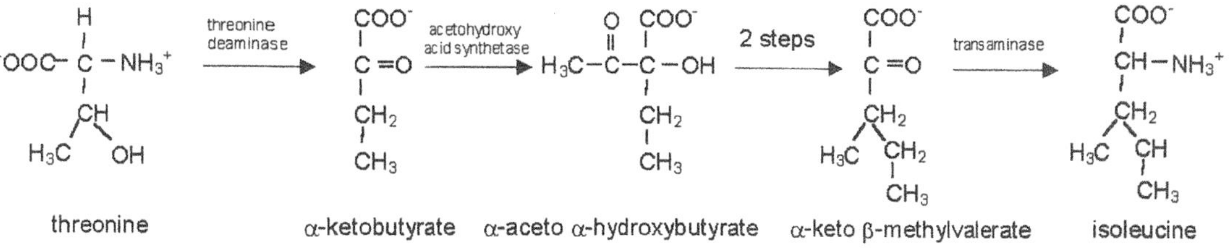
threonine deaminase
acetohydroxy acid synthetase
2 steps
transaminase
threonine
α-ketobutyrate
α-aceto α-hydroxybutyrate
α-keto β-methylvalerate
isoleucine

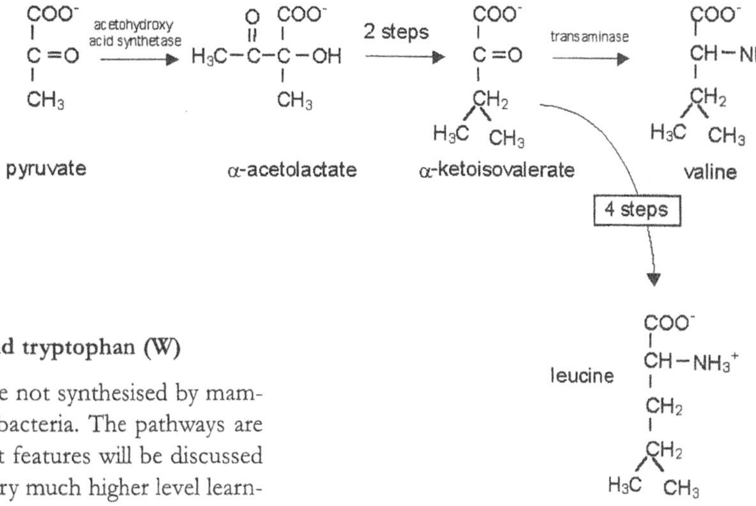
acetohydroxy acid synthetase
2 steps
transaminase
pyruvate
α-acetolactate
α-ketoisovalerate
valine
4 steps
leucine

Synthesis of phenylalanine (F) and tryptophan (W)

These ring-structured molecules are not synthesised by mammals, but are made by plants and bacteria. The pathways are rather complex and only the salient features will be discussed here. Learning these pathways is very much higher level learning and really only necessary if you were to specialise in this area.

The starting molecules for the synthesis of phenylalanine (F) are phosphoenolpyruvate and erythrose-4-phosphate. This is interesting because phosphoenolpyruvate is formed during glycolysis and erythrose 4-phosphate is formed in a pathway that breaks off the glycolysis pathway called the pentose phosphate shunt (see Chapter 10).

Phosphoenolpyruvate and erythrose 4-phosphate are fused together to form a deoxy sugar with a furan ring. This is modified to form a 6-sided ring structure and eventually shikimate. Shikimate is converted to chorismate which is converted to phenylalanine (F).

Chorismate is also the basis of tryptophan (W). In this case it is modified and connected to phosphoribosylpyrophosphate, which eventually leads to the formation of indole. This is then fused to serine (S).

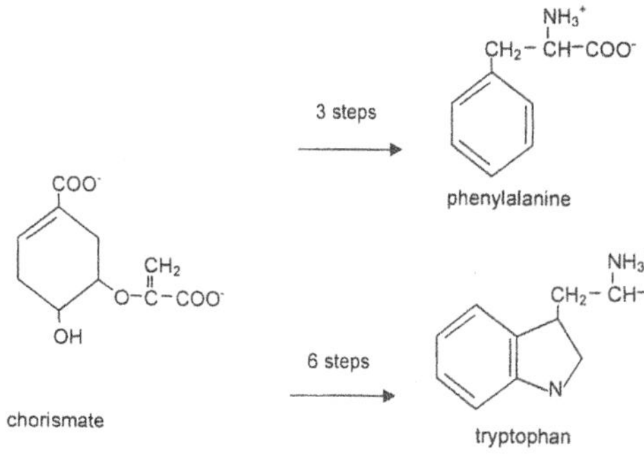
3 steps
phenylalanine
6 steps
chorismate
tryptophan

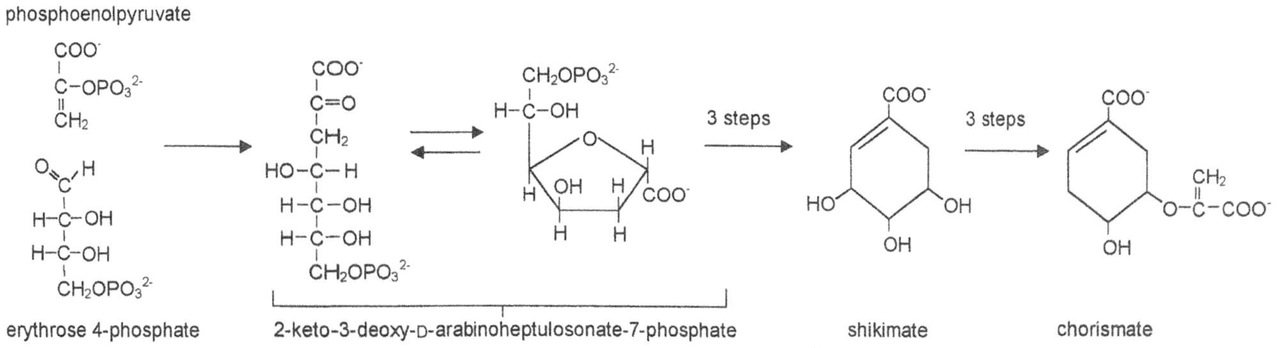
phosphoenolpyruvate
3 steps
3 steps
erythrose 4-phosphate
2-keto-3-deoxy-D-arabinoheptulosonate-7-phosphate
shikimate
chorismate

Synthesis of histidine (H)

Histidine is synthesised in a very complex pathway beginning with the fusion of 2 similar molecules, phosphoribosylpyrophosphate and adenosine triphosphate (ATP). The bridging complex between these 2 molecules is the adenine group of the ATP. Nine more steps take place to form histidine (H).

Key points about amino acid synthesis

- Incorporation of ammonia is called assimilation.
- Ammonia is commonly assimilated by adding it to glutamic acid (E) to form glutamine, but may also be assimilated by adding it to α-ketoglutarate to form glutamic acid.
- Learn the pathways for the synthesis of all the non-essential amino acids.
- Alanine (A), glutamate (E), aspartate (D), asparagine (N) and glutamine (Q) are formed by simple aminations or transaminations.
- Serine (S), glycine (G) and cysteine (C) are derived from 3-phosphoglycerate. These reactions begin with the synthesis of serine (S).
- Remember that proline (P) and glutamate (E) are structurally related; both have a total of 5C. Proline is synthesised from glutamate (E).
- Tyrosine (Y) is formed from a simple hydroxylation of phenylalanine (F).
- The synthesis of arginine (R) involves a similar set of reactions to those found for the synthesis of urea. The initial synthesis of citrulline from glutamate γ-semialdehyde occurs in intestinal epithelial cells. The final conversion of citrulline to arginine (R) occurs in kidney cells.
- The remaining amino acids are non-essential and their synthesis is more complicated.
- Lysine (K), threonine (T) and methionine (M) are synthesised from aspartate (D).
- The branched amino acids, isoleucine (I), valine (V) and leucine (L), are synthesised in complex pathways. Isoleucine (I) is formed from threonine (T) and valine (V) and leucine (L) from pyruvate.
- Isoleucine (I) and valine (V) biosynthesis shares 4 enzymes.
- Biosynthesis of phenylalanine (F), and tryptophan (W) has shikimate and chorismate as key intermediates. The ring structure is formed by condensation of erythrose 4-phosphate and phosphoenolpyruvate.
- Histidine (H) formation involves the transfer of ATP to the ribose phosphate part of phosphoribosyl pyrophosphate.

References and further reading

Atkinson, D.E. and Camien, M.N. (1982) The role of urea synthesis in the removal of metabolic bicarbonate and the regulation of blood pH. *Curr. Topics Cell. Regulation* **26**:261-302.

Bender, D.A. (1985) *Amino Acid Metabolism*, New York, Wiley.

Devlin, T.M. (1992) *Textbook of Biochemistry with Clinical Correlations*, 3rd ed. New York, Wiley-Liss

Greenstein, B. and Greenstein, A. (1996) *Medical Biochemistry at a Glance*. Oxford, Blackwell Science

Q&A Answers

1   Uric acid has the equivalent of 2 ureas in its structure

2   α-Ketoglutarate obtains its name because a ketone forms on the α carbon of glutaric acid. Note that glutaric acid has 5 carbon atoms, which include the carbons of 2 terminal carboxylic acid groups.

3

4   glutamate + oxaloacetate → α-ketoglutarate + aspartate

5   In tissues other than muscle the amine groups of amino acids are often transferred to glutamate (E) to form glutamine (Q). The enzyme which catalyses the reaction is glutamine synthase.

6   Glutamine (Q) is hydrolysed in the liver to form glutamate (E). The enzyme which catalyses this reaction is glutaminase.

7 Note that in the urea cycle, it makes sense that every time a citrulline passes out of the mitochondria ornithine passes in. This is what happens and the protein in the mitochondrial membrane that allows this to happen is called an ornithine citrulline exchanger. Remember to practise the structures of the molecules as well.

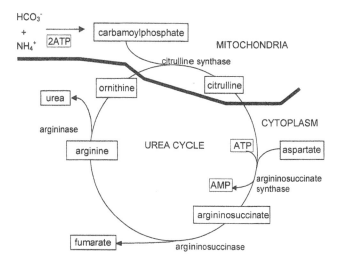

8 Knowing the similarities between these molecules is important

COOH	COOH	COOH	COOH
CH₂	CH₂	CH₂	CH
H₂N-C-COOH	O=C-COOH	H-C-COOH	H-C-COOH
H		H	
aspartate	oxaloacetate	succinate	fumarate

9 The initial step in the breakdown of arginine (**R**) is its conversion to ornithine by adding water.

10 CoA-SH is linked to succinic acid by removing water. The type of bond formed is a thioester bond and the new molecule is succinyl-CoA.

11 The name of the enzyme that converts phenylalanine (**F**) to tyrosine (**Y**) is phenylalanine hydroxylase. Some people lack this enzyme and end up with large amounts of phenylketones in their urine as phenylalanine (**F**) is deaminated. This condition is called phenylketonuria

12 Glutamate (**E**) forms from α-ketoglutarate when an amine group is transferred to it from tyrosine (**Y**).

13 To convert asparagine (**N**) to oxaloacetate, asparagine (**N**) is firstly converted to aspartate (**D**) and free ammonia by deamination. This reaction is catalysed by asparaginase. Aspartate (**D**) is converted to oxaloacetate by transferring its amine to α-ketoglutarate. This reaction is catalysed by aspartate aminotransferase and of course, glutamate (**E**) is also formed in this reaction from α-ketoglutarate.

14 The conversion of glycine (**G**) to serine (**S**) is catalysed by hydroxymethyl transferase. The methyl group comes from N^5,N^{10}-methylene-tetrahydrofolate which is converted to tetrahydrofolate in the process.

15 The relationship between the various amino acids and the tricarboxylic acid cycle is shown in the following thumbnail sketch.

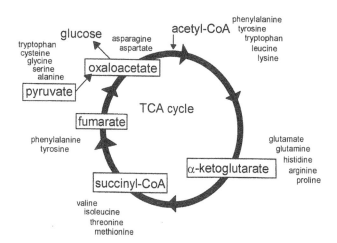

16 The name of the enzyme that catalyses the synthesis of glutamine(**Q**) from glutamate(**E**) and ammonia is glutamine synthase.

17 a) Ammonia is added to α-ketoglutarate to form glutamate (**E**).
b) Ammonia is added to glutamate (**E**) to form glutamine (**Q**).

18 Pyruvic acid is the α-keto-acid of alanine (**A**) and its transamination forms alanine.

Pyruvic acid   Alanine

19 The keto-acid of aspartate (**D**) is oxaloacetate and its transamination forms aspartate (**D**).

oxaloacetate   aspartate

20 Asparagine (**N**) is formed by the transfer of the amine group from glutamine (**Q**) to aspartate (**D**) and in the process ATP is converted to AMP.

21 Phenylalanine hydroxylase catalyses the conversion of phenylalanine (**F**) to tyrosine (**Y**) and lack of phenylalanine hydroxylase leads to a condition known as phenylketonuria.

9 The Tricarboxylic Acid Cycle

In this chapter you will learn:

- where the TCA cycle takes place.
- the relationship of aspartic acid and glutamic acid and the compounds of the TCA cycle oxaloacetate and α-ketoglutarate
- where the acetyl group of acetyl-CoA originates
- how pyruvate is transferred across the mitochondrial membrane and converted to acetyl-CoA
- in which parts of the cycle $FADH_2$, NADH, GTP and CO_2 are produced
- the compounds and enzymes of the cycle
- how the TCA cycle is controlled
- The parts of the TCA cycle used for amino acid synthesis or breakdown, the part used for fat synthesis, and the part used for energy production
- The conversion of oxaloacetate to phosphoenolpyruvate for gluconeogenesis

Location and purpose of the tricarboxylic acid cycle

The tricarboxylic acid (TCA) cycle is one of the most fascinating and central cycles in biochemistry. It is also known as Kreb's cycle, and the citric acid cycle. This cycle takes place in mitochondria. It has 2 main functions: to shuffle carbon skeletons, and to break down some compounds for generation of energy. The energy obtained is captured by reducing NAD and FAD to NADH and $FADH_2$ respectively. To obtain full usage of the energy generated in this cycle, NADH and $FADH_2$ are processed by another pathway (oxidative phosphorylation) where their energy is converted to ATP.

> **MAJOR POINT 43:** The TCA cycle takes place inside mitochondria.

For me, the most fascinating part of this cycle is that different tissues use different parts of the cycle depending upon the primary function of that tissue. Muscle and brain, which require large amounts of energy, use the TCA cycle to produce high energy compounds such as ATP and NADH. On the other hand, the liver, which requires little energy, uses the TCA cycle to shuffle carbon skeletons to form amino acids and fats. Fat cells primarily use the TCA cycle to channel the carbon atoms of glucose to form (unfortunately) fat.

We already know many of the elements of the TCA cycle from our knowledge of amino acids and their denaturation. In particular, the amino acid glutamate can be fed into the cycle by deaminating it to its keto acid form α-ketoglutarate. From there, its skeleton can be cycled to form other amino acids or glucose. Other amino acids are also fed into the cycle at succinyl-CoA, fumarate and oxaloacetate.

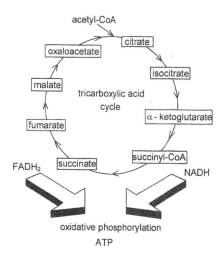

The reactions of the TCA cycle

One of the most daunting jobs for students is to learn the TCA cycle, but it is much easier to learn if you know the structures of the amino acids, aspartate (**D**), glutamate (**E**) and alanine (**A**). Let us start with aspartate (**D**).

*What is the keto acid of aspartate (**D**)?*
Answer: *Oxaloacetate, which is one of the molecules in the TCA cycle.*

The relationship between these 2 molecules is not only important structurally, but functionally as well. This is how the carbon skeleton of aspartic acid is fed into the TCA cycle for either degradation to obtain energy or for transformation into other amino acids.

The next step in the cycle (see above) is the synthesis of citrate. This occurs by the transfer of an acetyl group (-CH$_2$COOH, essentially acetic acid) to the carbon of oxaloacetate with the ketone group (= O). The ketone group becomes an hydroxyl group in the process. Importantly the acetyl group (-CH$_2$COOH) comes from acetyl-CoA.

For this reaction the thioester bond of acetyl-CoA is hydrolysed (add water) and the methyl group transferred to the carbon of oxaloacetate that contains the ketone. In the process, the ketone group gains an hydrogen and becomes an hydroxyl group.

The obvious follow-up question is where does the acetyl-CoA come from? The acetyl part of acetyl-CoA comes from the breakdown (oxidation) of fatty acids, from ketone bodies, which are formed by the breakdown of many amino acids (threonine (**T**), valine (**V**), isoleucine (**I**), leucine (**L**), methionine (**M**), tryptophan (**W**): Mnemonic – method for trying to isolate 3 valiant letchers) or from pyruvate.

Q&A 1: What are the 2 main sources of pyruvate?

For these substances to enter the TCA cycle, they must be transferred from the cytoplasm into the mitochondria. They cannot just diffuse through the mitochondrial membranes, which form a barrier. The solution is to have proteins embedded in the membrane of the mitochondria which carry or transport these substances across the mitochondrial membrane so that they can be formed into acetyl-CoA. These special proteins are called transporters. The advantage of such a system is that it can be controlled. For example there is a 'pyruvate/H$^+$ symport', which carries both pyruvate and H$^+$ into the mitochondria from the cytoplasm. The driving force for this reaction is the difference of H$^+$ concentration across the mitochondrial membrane.

MAJOR POINT 44: Molecules going from the cytoplasm into mitochondria or vice versa must be transported across mitochondrial membranes by special carrier proteins.

Once inside mitochondria, they are converted to acetyl-CoA, and are then effectively trapped inside the mitochondria.

MAJOR POINT 45: Neither acetyl-CoA, nor co-enzyme A can cross the mitochondrial membrane.

The fact that coenzyme A cannot cross the mitochondrial membrane means that there are 2 separate pools of co-enzyme A, 1 in the mitochondria and 1 in the cytoplasm. Hence, the acetyl-CoA found in mitochondria must be formed in the mitochondria.

Firstly the formation of acetyl-CoA from pyruvate will be examined. Pyruvate is derived from

- the amino acid alanine (**A**) by a deaminination. Glycine (**G**) and serine (**S**) are also converted to alanine (**A**)
- the breakdown of glucose (glycolysis).

*Draw pyruvate. Remember it is the keto acid of alanine (**A**) (alanine deaminated).*

Answer:

To form acetyl-CoA, the carboxyl group of pyruvate must be removed from the acetyl group.

Q&A 2: What is formed from the carboxyl group when it is removed from pyruvate?

It is actually quite difficult to remove the carboxyl group and then add the remaining acetyl group to CoA-SH. Firstly, pyruvate is carried into mitochondria where it is processed by a cluster of enzymes anchored in the inner mitochondrial membrane. These are called the pyruvate dehydrogenase complex. By grouping the enzymes together, the overall reaction can occur more efficiently because the substrate from one reaction does not diffuse away from the next enzyme. This is typical of many other reaction sequences. Details of the reaction sequence are outlined in the higher level section below.

The main features of this process are

- Carbon dioxide is removed from pyruvate and the remaining acetyl group is added to CoA-SH to form acetyl-CoA.

- The high energy compound NADH is formed from NAD$^+$.

- Three different enzymes are involved in the process. These form the complex known as pyruvate dehydrogenase complex.

- Four cofactors or co-enzymes necessary in this reaction sequence are derived from vitamins. The vitamins are thiamine (vitamin B$_1$), riboflavin (vitamin B$_2$), pantothenic acid, and niacin.

- The reaction is irreversible - it is not possible to go from acetyl-CoA to pyruvate.

MAJOR POINT 46: In obtaining energy from glucose, 2 pyruvates are formed in the cytoplasm. As each of these is converted to acetyl-CoA in the mitochondria, the high energy compound NADH is formed (2 NADH for each glucose molecule).

Higher level

The first enzyme encountered catalyses the decarboxylation of pyruvate and is called either pyruvate decarboxylase (in yeast) or pyruvate dehydrogenase (E_1). The active enzyme is a complexed with the vitamin B_1 derivative thiamine pyrophosphate, which is a common reactant in decarboxylation reactions.

The next enzyme in the complex, dihydrolipoamide acetyltransferase (E_2) has a long lipid like arm with a disulphide group at the terminal end. This arm picks up acetyl group from the thiamine pyrophosphate complex and transfers it to co-enzyme A to form acetyl-CoA (co-enzyme A is made from the vitamin, pantothenic acid). In the process, the disulphide group of dihydrolipoamide becomes reduced. This creates a problem in that if it remains in this reduced state, it cannot repeat the process and the conversion of pyruvate to acetyl-CoA would stop. Therefore, the oxidised form of dihydrolipoamide acetyltransferase must be regenerated. This occurs by removing hydrogen.

What would be an appropriate name for an enzyme which oxidises dihydrolipoamide?

Answer: *Dihyrolipoamide dehydrogenase (E_3) regenerates the oxidised form of dihydrolipoamide acetyltransferase*

The hydrogen is from the reduced form of dihydrolipoamide acetyltransferase is transferred to flavin adenine dinucleotide (FAD) which becomes $FADH_2$. FAD is derived from riboflavin, vitamin B_2. Once again, FAD must be regenerated or the reaction stops. Here nicotinamide adenine dinucleotide NAD^+ becomes reduced to NADH by accepting hydrogen from $FADH_2$. NADH is then used to generate ATP in a process called oxidative phosphorylation. NAD^+ is derived from the vitamin niacin.

It is worthwhile noting that there are multiple copies of each enzyme in the complex, and the exact numbers vary between species. There are about 60 molecules of the transacetylase and 20 to 30 each of the other 2 molecules forming 1 complex in mammals.

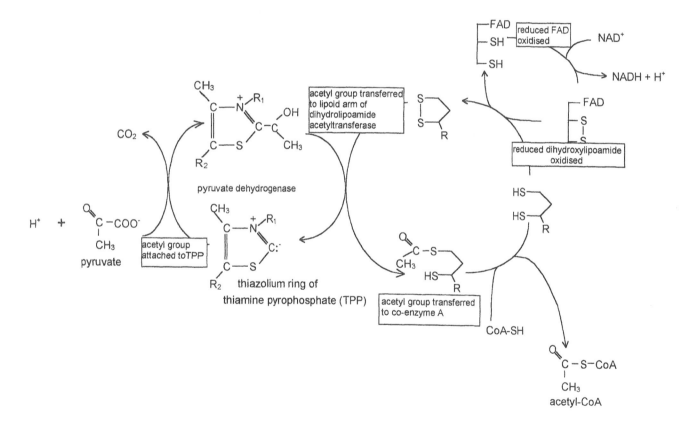

Once acetyl-CoA is formed in the mitochondria, the acetyl group (2 carbons) is transferred to oxaloacetate (4 carbons) to form citrate (6 carbons, see overpage). The cycle progresses to form oxaloacetate again. Therefore, it is logical that 2 carbons must be lost along the way. The first part of this cycle is the rearrangement of citrate to form isocitrate. The enzyme that catalyses this reaction, aconitase, obtains its name from an intermediate in the reaction, *cis*-aconitate.

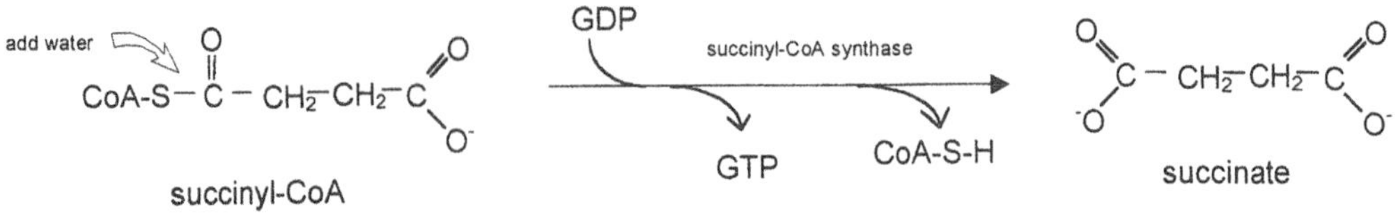

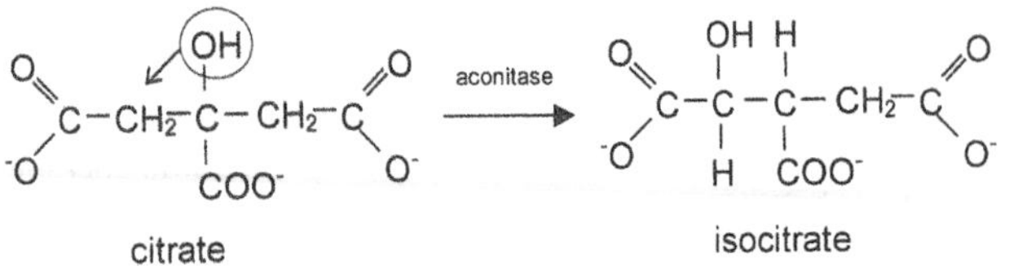

succinyl-CoA

GDP → GTP   succinyl-CoA synthase   CoA-S-H → succinate

citrate   aconitase   isocitrate

This sets up the molecule to be changed into α-ketoglutarate by removal of the carboxyl group (-COO) on the central carbon and removal of hydrogen from the carbon with the hydroxyl group (-OH). Remember that removal of hydrogen is an oxidation and generates energy in the form of NADH. The carboxyl group is released as CO_2. The whole process is called an oxidative decarboxylation and the enzyme that catalyses the reaction is called isocitrate dehydrogenase.

isocitrate   isocitrate dehydrogenase NAD^+ NADH + H^+ CO_2 α-ketoglutarate

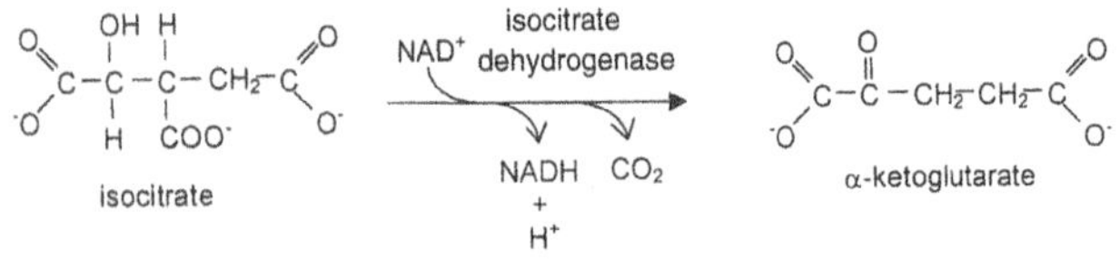

You already know the structure of α-ketoglutarate because it is the keto form of an amino acid. This means that besides being obtained from isocitrate, α-ketoglutarate can be obtained from the skeletons of some amino acids, and hence this part of the tricarboxylic acid cycle is also an entry point for the carbon skeleton of some amino acids.

> **Q&A 4:** Which amino acid may be deaminated to enter the TCA cycle at this point?

Co-enzyme A is again involved in the next step in the cycle. Essentially, it forms a bond (a thioester bond) with the α-C and in so doing, the carboxyl group of the α-C is released as carbon dioxide. This reaction gains further importance because it is where a second high energy compound, NADH, is formed. It is important to note that the carbon skeletons of several branched chain amino acids enter the TCA cycle at succinyl-CoA.

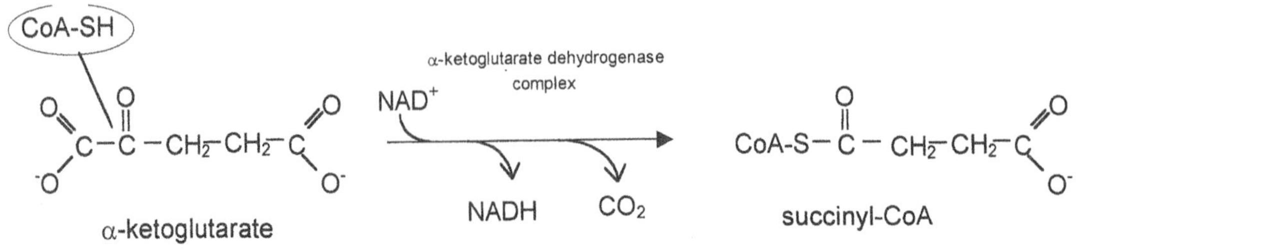

In the next step, succinate is formed by releasing co-enzyme A. This is an hydrolysis and so water is added to the thioester bond. In the process, energy contained in the bond is captured by GDP which is phosphorylated to GTP. This is equivalent to forming ATP because the energy in GTP is then used to convert ADP to ATP.

> **MAJOR POINT 47:** Hydrolysis of a thioester bond e.g. acetyl-CoA provides enough energy to convert ADP to ATP.

The next part of the cycle is to regenerate oxaloacetate. The pattern seen in these next reactions are exactly the same as are seen in fatty acid oxidation, which will be discussed in Chapter 11. Therefore, if you take note of the reactions here, then it is very easy to learn about fatty acid oxidation. In the first step, hydrogen (H_2) is removed from the 2 middle carbons of succinate.

What is the name of the enzyme which catalyses the removal of hydrogen from succinate to form fumarate?

Answer: *Succinate dehydrogenase is the logical answer, but it is not quite correct. This is a complex of enzymes much like was found for the decarboxylation of pyruvate and hence it is more correctly called succinate dehydrogenase complex.*

Removal of hydrogen, leads to the formation of a double bond between the 2 central carbons and fumarate is formed. Note that the hydrogen atoms attached to the carbon atoms of the double bond are on different sides of the molecule and hence they are trans rather than cis. Of course, this is an oxidation reaction because hydrogen is removed. The capturing agent of the hydrogen is not NAD^+ this time, but FAD. FAD is converted to $FADH_2$. This creates a problem in that $FADH_2$ is intimately associated with the enzyme complex and the enzyme complex cannot function until FAD is regenerated. This is done by transferring the energy (H_2) to another molecule called ubiquinone (Q). We will learn more about this molecule later.

CoA-SH   α-ketoglutarate NAD^+ α-ketoglutarate dehydrogenase complex   NADH CO_2 succinyl-CoA

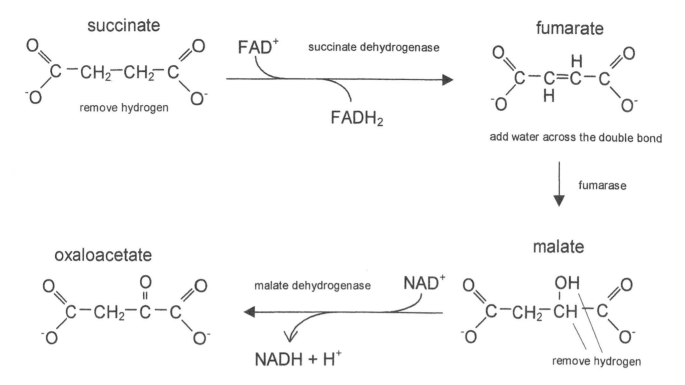

The next step adds water across the double bond to form malate. The enzyme that catalyses this reaction is fumarase. The final reaction is obvious knowing that oxaloacetate is the end-product. Hydrogen is removed from the α-carbon of malate to form the ketone group of oxaloacetate.

What is the name of the enzyme that catalyses the removal of hydrogen from malate to form oxaloacetate?

Answer: *Malate dehydrogenase.*

Q&A 5: Write the overall stoichiometry of the cycle. Remember that the reaction is essentially the processing of the acetyl group of acetyl-CoA to produce carbon dioxide and energy in the form of high energy compounds.

Tips for learning the cycle

Most importantly, the TCA cycle occurs in mitochondria <u>not</u> in the cytoplasm. Other focal points for study are:

- where CO_2 is produced
- where hydrogens are lost to NAD, FAD ie oxidation takes place and how many of each are produced
- where amino acids, glucose and fats may enter the cycle
- where the 1 GTP is formed.

- Start with oxaloacetate. Remember its structure from aspartate (**D**). This is doubly important because it also highlights that this is the point in the cycle where the carbon skeleton of aspartate (**D**) can enter the cycle.
- The ketone group of oxaloacetate is acetylated by a transfer of the acetate group from acetyl-CoA and it is the carbon of the methyl group (-CH₃) of acetate rather than the carboxyl group that is attached to the ketone group.
- At the same time, an hydroxyl group also forms at this carbon. The compound synthesised is citrate.
- At this point there are several ideas to associate with this process.
  - the origin of pyruvate: particularly alanine (**A**) and glucose, but also other amino acids
  - the origin of acetyl-CoA: from pyruvate, fats and the ketogenic amino acids. If it is from pyruvate, CO_2 and NADH are generated when pyruvate is converted to acetyl-CoA. This actual transition is not part of the TCAcycle. Do not forget that you obtain 2 molecules of pyruvate from 1 glucose
  - CoA cannot cross the mitochondrial membrane and so acetyl-CoA must be formed in the mitochondria.
- Focus on the idea that citrate is converted to α-ketoglutarate.
  - you should already know the structure of α-ketoglutarate from the structure of the amino acid glutamate (**E**).
  - it is 1 carbon less than citrate and hence it is a place where carbon dioxide is lost.
  - To form the ketone group, hydrogen must be removed and hence a high energy compound, NADH, is formed

citrate has the hydroxyl and carboxyl groups attached to the same carbon atom and must be rearranged so that the carboxyl and hydroxyl groups are attached to different carbons. Therefore, the intermediate step is the isomerisation of citrate where the hydroxyl group is transferred to an adjacent carbon.

- A second generation of CO_2 occurs in the next step as the 5 carbon compound α-ketoglutarate is converted to succinyl-CoA. Once again energy is obtained in this reaction in the form of NADH.
- Note that the reactions to and from α-ketoglutarate, CO_2 is lost and NADH is produced.
- Succinyl-CoA is the point of entry of the amino acids valine, isoleucine, threonine and methionine.
- Succinate is then cleaved from co-enzyme A by adding water. This releases some energy in the form of GTP.
- Next is the formation of oxaloacetate from succinate beginning with a dehydrogenation across the central 2 carbons. This forms fumarate which has a trans double bond. The energy is captured from this oxidation to form $FADH_2$.
- Fumarate is the entry point of the amino acids phenylalanine (**F**) and tyrosine (**Y**).
- Water is added across the double bond of fumarate to form malate and then hydrogen is removed to form the keto acid, oxaloacetate. The hydrogen is captured in the form of the high energy compound NADH.
- Once familiar with the overall pattern then learn the enzymes.

> **Q&A 6:** Using the study tips above, draw the whole cycle indicating points of entry for amino acids, fats and glucose. Also indicate where the high energy compounds are formed and carbon dioxide is released.

> **Q&A 7:** For each molecule of glucose which is processed through the TCA cycle, how many NADHs, $FADH_2$s and GTPs are formed?

Control of the citric acid cycle

To understand how the TCA cycle is controlled, we must consider its function.

What are the primary uses of the TCA cycle?

Answer: *(a) To shuffle carbon skeletons for the formation of new compounds, e.g. amino acids or fats. (b) To generate energy by oxidising the acetyl group of acetyl-CoA.*

It makes sense that the control of the TCA cycle depends upon its usage. Taking the first function, consider the situation where a cell is synthesising glutamic acid. The part of the cycle which is used is from oxaloacetate to α-ketoglutarate, at which point α-ketoglutarate is pulled out of the cycle to form glutamate by aminating it. In this case, oxaloacetate, the normal 'starting' point of the cycle, is consumed. Therefore, it is essential that more oxaloacetate is generated, otherwise the cycle would stop.

Compare this situation with the cycle being used to generate energy (NADH, $FADH_2$ and GTP), what would be consumed in this case?

Answer: *Acetyl-CoA would be consumed, and hence for the cycle to continue, acetyl-CoA would have to be regenerated. Less obvious in this case is that NAD^+ is also consumed to form NADH and hence NAD^+ would have to be regenerated.*

It is important to realise that the TCA cycle can only function as a generator of energy under aerobic conditions. The reason for this is that the cycle will run-down unless NADH is converted back to NAD^+. The process forming ATP from NADH and thus regenerating NAD^+ is called oxidative phosphorylation which is described in detail later, but here it is important to note that it uses oxygen, and ATP is one of the end-products.

> **MAJOR POINT 48:** The TCA cycle can only obtain energy from the acetyl group of acetyl-CoA under aerobic conditions.

Use of the TCA cycle by the liver

> **Nomenclature:** The term hepato means liver. I am sure you have all heard the term hepatitis which is inflammation of the liver. Hepatocytes are liver cells.

Which of the following do you think are the main functions of hepatocytes?
- *To consume lots of energy*
- *To store lots of energy in the form of fats*
- *To store lots of energy in the form of glucose*
- *To synthesise amino acids for its own use or to supply other organs or tissues with amino acids*

Answer: *Hepatocytes have many functions, but those relevant to the TCA cycle are that they produce or store glucose for other organs and synthesise amino acids. It may be a surprise that they do not need a great deal of energy to function and although they are critical in fat metabolism, they do not store energy in the form of fats. Fats stored in the liver is indicative of a diseased liver.*

> **MAJOR POINT 49:** The liver mainly uses the TCA cycle as a mechanism for converting the carbon skeletons from one source or another into new molecules.

To keep the TCA cycle running, liver cells can obtain the carbon skeletons from

- the breakdown of amino acids – these skeletons are pushed into the cycle at oxaloacetate, α-ketoglutarate, succinyl-CoA, fumarate and acetyl-CoA (see Chapter 8)
- fatty acids, which enter the cycle as acetyl-CoA (minimal usage in this regard)
- sugars, which enter the cycle as acetyl CoA from pyruvate (minimal usage in this regard)
- sugars which enter the cycle as oxaloacetate from pyruvate.

This last point is important because pyruvate can be channelled into the TCA cycle as acetyl-CoA or as oxaloacetate. That is, this point is a switch point which controls the main function of the cycle. If pyruvate is channelled to acetyl-CoA, then the cycle will generate mainly energy. If pyruvate is channelled into oxaloacetate, then its main function will be to produce carbon skeletons for amino acid, or fat synthesis.

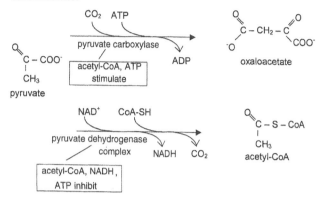

Clearly there is a priority. If the cell has no energy, it will be dead. Therefore, the switch to synthesis of oxaloacetate must occur when the cell has lots of energy.

Which compounds would be in high levels in a cell if it were in a high energy state?

Answer: *ATP, NADH, and acetyl-CoA.*

Such a cell would not need to produce more acetyl-CoA, and hence you might expect that these compounds would inhibit the activity of pyruvate dehydrogenase complex. This indeed is the case. In addition, the excess energy could be used for synthesising new molecules and hence it might be expected that these molecules would stimulate

pyruvate carboxylase so that more oxaloacetate is produced. This is almost the case: ATP is the energy source for driving the reaction and acetyl-CoA stimulates the activity of pyruvate carboxylase, but NADH has no direct role.

Therefore, high levels of acetyl-CoA inhibit the activity of pyruvate dehydrogenase, decreasing further synthesis of acetyl-CoA, and at the same time enhance the activity of pyruvate carboxylase, stimulating the synthesis of oxaloacetate.

Q&A 8: Under most conditions the liver is in a high energy state. Therefore, will it be converting pyruvate into oxaloacetate or acetyl-CoA so that it can enter the TCA cycle?

Consider the situation when the rest of the body is demanding glucose. This would occur when the person has not eaten for some time. The liver initially breaks down its glycogen stores to glucose and ships the glucose to the other organs. If that is not enough, it begins to convert carbon skel-

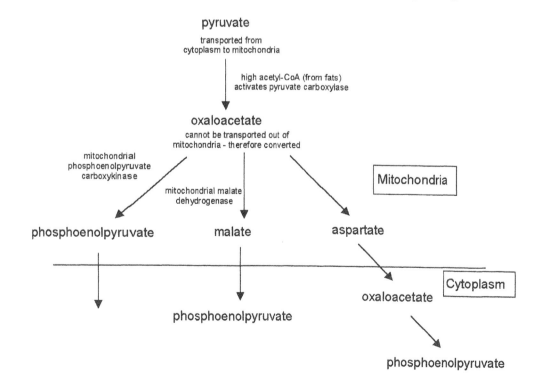

etons from amino acids into glucose. Therefore, amino acids are pushed into the TCA cycle e.g. glutamate at α-ketoglutarate, and taken out at oxaloacetate where they can be converted to glucose. Other signals cause the liver to begin to break down fatty acids to provide the acetyl-CoA for its own energy needs, and therefore it does not require glucose. This makes sense, since it is trying to make glucose for the rest of the body. Hence, under starvation conditions, the carbon skeletons of amino acids are fed into the TCA cycle at various points, which leads to an increase in oxaloacetate. Oxaloacetate is then converted to phosphoenolpyruvate which is in turn converted to glucose.

There is no transporter of oxaloacetate in the mitochondrial membrane and therefore oxaloacetate must be converted to another compound before it can be moved into the cytoplasm. The exact pathway for the conversion of oxaloacetate to phosphoenolpyruvate varies between species.

Several amino acids are channelled into the TCA cycle at succinyl-CoA. When this occurs it is convenient to reduce the flow through the first part of the cycle. This is achieved because α-ketoglutarate dehydrogenase, the enzyme which catalyses the conversion of α-ketoglutarate to succinyl-CoA, is strongly inhibited by its products, NADH and succinyl-CoA. The skeleton from the amino acids can then be used to synthesise glucose.

To summarise, in starvation the liver uses oxaloacetate obtained from the carbon skeletons of amino acids to produce phosphoenolpyruvate, which is converted to glucose. Oxaloacetate cannot be transported across the mitochondrial membrane and hence is converted to malate, aspartate or phosphoenolpyruvate firstly, which can cross the mitochondrial membrane. The exact pathway depends upon the species.

Use of the TCA cycle by the brain

In terms of metabolism, what is the role of the brain?
- *To use up lots of energy in the form of glucose*
- *To use up lots of energy in the form of fats*
- *To store lots of energy in the form of fats*
- *To store lots of energy in the form of glucose*
- *To supply other organs or tissues with amino acids*

Answer: Even if we are not using our brains for studying biochemistry, it still requires large amounts of energy. In fact, it has energy priority over the rest of the body, which is hardly surprising because if it does not obtain enough energy we die. This energy comes from glucose, although it can use ketones in starvation situations. Therefore, the brain uses glucose for energy, almost the opposite to the liver which is a provider of glucose. The brain uses a lot of fat but not for energy, it uses it for making membranes.

> **MAJOR POINT 50:** The main function of the TCA cycle in the brain is to produce energy from glucose.

The brain converts glucose to pyruvate which is pumped into the TCA cycle as acetyl-CoA. When starving, ketones are formed from many amino acids and fatty acids by the liver and these are released into the bloodstream, taken up by the brain and converted to acetyl-CoA.

Use of the TCA cycle by fat tissue

> **Nomenclature:** Another name for fat is lipid and hence fat cells are lipocytes; breakdown of fat is called lipolysis (lysis – disintegration); and the making of new fat is called lipogenesis.

What is the function of lipocytes?
- *To use up lots of energy*
- *To store lots of energy in the form of fats*
- *To store lots of energy in the form of glucose*
- *To supply other organs or tissues with amino acids*

Answer: I am sure that you picked the second point, but wondered about the others. Fat cells, like fat people, do not use up much energy. Compare this with muscle cells or muscular people that actually do lots of work. The relationship of lipocytes to glucose is not so obvious. The critical word here is store. *Fat cells use quite a lot of glucose, but not for energy production. They use it to provide the carbons for making fats, which they store.*

The liver, and to a small degree, the kidneys, are the only organs involved with amino acid production or degradation.

Even knowing the function of lipocytes, it is difficult to know which part of the TCA cycle the lipocytes would mainly use, unless you know about lipid synthesis (see Chapter 11). We will have a brief glimpse of fat synthesis here, if only to whet your appetite.

Essentially fats, fatty acids and cholesterol, are synthesised from acetyl-CoA in the cytoplasm, not in the mitochondria, of cells. The process is a little complicated and superficially illogical but bear with me and you too can make sense of the insanity. To make fatty acids, glucose is shunted into the mitochondria as pyruvate where it is converted firstly to acetyl-CoA (pyruvate dehydrogenase) and then to citrate (citrate synthase). The next bit is illogical: citrate is transported out of the mitochondria back to the cytoplasm where it is converted to acetyl-CoA and pyruvate. The acetyl-CoA in the cytoplasm is made into fatty acids and the pyruvate is transported back into the mitochondrion.

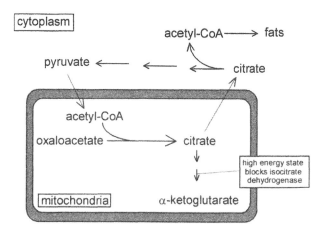

The reason this might happen is quite simple if you remember that the cell gives priority to energy production. If the energy state is low, the TCA cycle progresses to produce energy for the cell. If it is high, the cycle only progresses to citrate because isocitrate dehydrogenase is inhibited by NADH (activated by ADP). Citrate is then transported across to the cytoplasm where it is converted to acetyl-CoA and pyruvate. The acetyl-CoA is used for making fats. High levels of pyruvate, NADH, acetyl-CoA and ATP associated with this high energy state would lead to the pyruvate switch being activated and oxaloacetate being formed (pyruvate carboxylase activated) and hence a large amount of citrate. The end result is that the energy contained in glucose is converted to energy stored in fats.

> **MAJOR POINT 51:** The TCA cycle is used by lipocytes to convert mitochondrial acetyl-CoA to citrate which is then transported to the cytoplasm where it is converted to acetyl CoA and then fats.

Use of the TCA cycle by muscle

Fortunately we have other cells which can use fat. These are muscle cells. They preferentially use fat to produce energy and hence channel fats (fatty acids) into the TCA cycle in the form of acetyl-CoA. Fatty acids are transferred across the mitochondrial membrane from the cytoplasm and converted to acetyl-CoA. From there, they are oxidised via the TCA cycle. Note that this will only function if there is enough oxygen around to regenerate NAD from the NADH formed. Therefore, muscle can only use fats as an energy source if there is enough oxygen.

> **Q&A 9:** The general description for exercise which will use up fat stores is called a.................... exercise.

References and further reading

Bodner, G.M. (1986) The tricarboxylic acid (TCA) , citric acid or Krebs cycle. *J. Chem. Edu.* 63:673-677.

Devlin, T.M. (1992) *Textbook of Biochemistry with Clinical Correlations*, 3rd ed. New York: Wiley-Liss.

Greenstein, B. and Greenstein, A. (1996) *Medical Biochemistry at a Glance*. Oxford: Blackwell Science.

Mattevi, A., Obmolova, G., Schulze, E., Kalk, K.H., Westphal,. A.H., De Kok, A. and Hol, W.G.J. (1992) Atomic structure of the cubic core of pyruvate dehydrogenase. *Science* 255:1544-1550.

Patel, M.S. and Roche, T.E. (1990) Molecular biology and biochemistry of pyruvate dehydrogenase complexes. *FASEB J.* 4:3224-3232.

Wiegand, G. and Remmington, S.J. (1986) Citrate synthase: Structure, control amd mechanism. *Annu. Rev. Biophys. Biophys. Chem.* 15:97-117.

Q&A Answers

1 Pyruvate is derived from:
* the amino acid alanine (**A**) by a deaminination, and also glycine (**G**) and serine (**S**), which are converted to alanine (**A**)
* the breakdown of glucose (glycolysis).

2 To form acetyl-CoA the carboxyl group is removed from pyruvate and it is released as carbon dioxide.

3 To obtain the full energy available in glucose, it is converted to pyruvate and then to acetyl-CoA. The first enzyme in the pathway for converting pyruvate to acetyl-CoA is pyruvate decarboxylase which requires thiamine pyrophosphate as a cofactor. Thiamine pyrophosphate is a derivative of thiamine (vitamin B1). Lack of vitamin B1 therefore decreases energy available to the brain cells and decreases their function.

* The condition of lack of vitamin B1 is called beriberi. One of its characteristics is loss of neural function. Alcoholics quite often suffer from this condition because all of their calories are coming from alcohol and they become vitamin deficient.

4 Glutamate (**E**) is deaminated to α-ketoglutarate and can enter the TCA cycle at this point.

5 The overall equation for the citric acid cycle is
Acetyl-CoA + 3 NAD$^+$ + FAD + GDP + P$_i$ +2 H$_2$O $\rightarrow$ 2 CO$_2$ + 3 NADH + H$^+$ + FADH$_2$ + GTP + CoASH

6 See next page for this answer.

7 For each molecule of glucose 2 pyruvates are formed. These are converted to 2 acetyl-CoAs, each of which is broken down to 3 NADH, 1 FADH$_2$, and 1 GTP. Hence for 1 glucose molecule, 6NADH, 2FADH, and 2GTP are produced in the TCA cycle.

8   Under a high energy state, pyruvate dehydrogenase complex will be inhibited and pyruvate carboxylase will be activated and hence oxaloacetate will be synthesised.

9   The type of exercise needed for muscle to use up fat stores is aerobic exercise. Without oxygen it can only use glucose, which it converts to lactic acid.

6

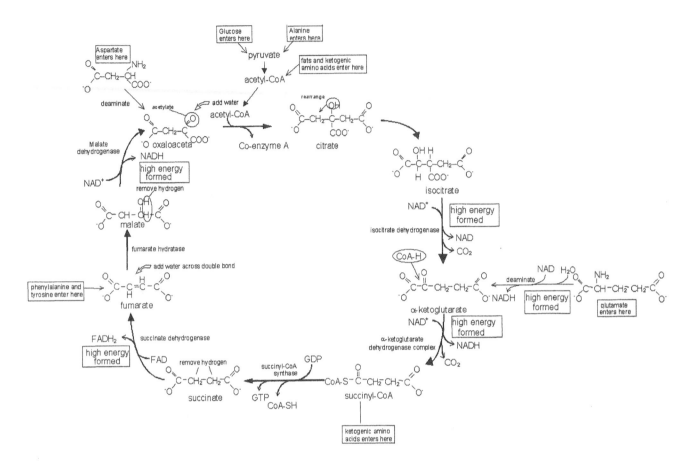

10 The Electron Transport Chain, Oxidative Phosphorylation and Photosynthesis

In this chapter you will learn:

- the location of the electron transport chain and oxidative phosphorylation
- that the electron transport chain and oxidative phosphorylation equates to oxidising NADH and FADH$_2$
- about compartments in mitochondria
- about mitochondrial enzyme complexes and their functions
- about electron transport along the mitochondrial complexes
- the functions of iron sulphur centres and ubiquinone
- the Q cycle
- how succinate dehydrogenase (complex II) directly links the electron transport chain with the TCA cycle
- the effects of uncoupling the electron transport chain
- the mechanisms of photosynthesis
- the structure of chloroplasts
- The Calvin cycle or dark reaction.

The electron transport chain and oxidative phosphorylation

The electron transport chain and oxidative phosphorylation are the mechanisms by which NAD$^+$ and FAD are regenerated so that they may be used again in the TCA cycle. Ultimately, the H atoms of NADH and FADH$_2$ are burnt (oxidised) to form water. The energy obtained from oxidising NADH and FADH$_2$ is captured in the high-energy phosphate bonds of ATP. Since the TCA cycle takes place in mitochondria, it makes sense to have oxidative phosphorylation occur in mitochondria as well. This is where the NADH and FADH$_2$ are formed and the regenerated NAD$^+$ and FAD are then immediately available for the TCA cycle again.

You will remember from earlier that mitochondria have 2 membranes, an outer and an inner membrane. The space within the inner membrane is called the mitochondrial matrix and the space between the 2 membranes is called the intermembranous space.

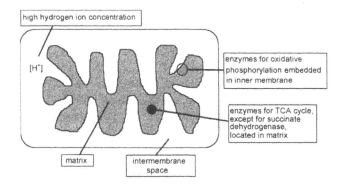

To oxidise NADH and FADH$_2$ to NAD$^+$ and FAD, electrons are transported along a series of enzymes (complexes) embedded in the inner mitochondrial membrane and eventually hydrogen is oxidised to form water. The energy contained in NADH or FADH$_2$ is gradually stripped away in a series of steps rather than one big step. The advantage of this is that there is less energy lost as heat than would occur if one big step were used to oxidise NA-+++++021\p0[DH or FADH$_2$. In this process, the energy is used to drive hydrogen ions from the matrix to the intermembrane space. This provides an electrical charge across the inner mitochondrial membrane which has the potential to do work. The hydrogen ions eventually flow back across the inner membrane and convert their potential energy to chemical energy by transforming ADP to ATP.

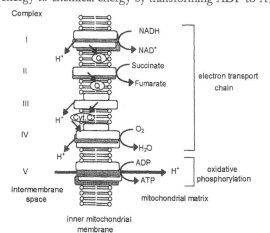

There are five enzyme complexes involved in these energy transfer processes. Unusually, they have been named systematically and hence are called mitochondrial complex I through to mitochondrial complex V.

What is an enzyme complex in oxidative phosphorylation?

Answer: *We do not know the details of a complex. What we do know is that each complex contains many polypeptide or protein subunits and several iron centres, which are the components that can be readily reduced or oxidised. These are all grouped together as a complex.*

Electron transport

Remember that a reduced compound contains energy and when it is oxidised, it releases energy. The overall reaction for the oxidation of NADH is $NADH + H^+ + \frac{1}{2}O_2 \rightarrow NAD^+ + H_2O$, but this takes place over a series of steps, enabling the body to 'capture' the energy for work rather than simply dissipate it as heat. In each of the steps, the starting compound is oxidised and the next compound is reduced. The energy transfer can be looked at chemically as transferring electrons, akin to energy transfer by electricity. The starting compound, NADH is oxidised by losing a pair of electrons, and eventually an oxygen atom is reduced by accepting a pair of electrons. The electrons are either transported from one molecule to the next as hydride ions (H^-), a proton with two electrons, or as free electrons.

> **MAJOR POINT 52:** Oxidation is loss of electrons. Reduction is gain of electrons.

Complex I

Complex I (NADH dehydrogenase; NADH/ubiquinone oxidoreductase) comprises 28 to 41 protein subunits depending upon the species, flavin mononucleotide (FMN) as a prosthetic group and about 7 iron-sulphur centres. Intimately linked to the complex is ubiquinone (Q). Hydride (H^-) from NADH, and H^+ from the matrix is transferred to FMN to form $FMNH_2$. This is an essential part of the process because NADH trans-

fers two electrons at one time to FMN, but FMN then transfers the electrons one at a time to the iron-sulphur centres.

NADH cannot do this directly because it must pass two electrons at once and the iron sulphur centres can only accept one electron. The electrons are eventually transferred to ubiquinone (Q) to form reduced ubiquinone (QH_2). QH_2 is at a slightly lower energy state than NADH and the energy obtained from the reaction is used to transport H^+ from the matrix to the intermitochondrial space. It is currently thought that $4H^+$ are transported for each NADH oxidised. Exactly how this transport occurs is unknown.

The iron sulphur centres and ubiquinone

Iron atoms in the iron-sulphur centres can bind or release electrons. In the Fe^{3+} form, they readily accept electrons to become Fe^{2+} and in the Fe^{2+} form, they readily give up electrons to become Fe^{3+}. The sulphur atoms stabilise the centre and bind to cysteine groups of the protein. Such centres either have 2 iron atoms bound to 2 sulphide atoms or 4 iron atoms bound to 4 sulphide atoms, and these are anchored to the protein by 4 cysteines in both cases.

Ubiquinone has the great advantage of having a long, flexible, lipid-soluble arm and hence can readily move through the inner membrane to transfer the electrons to the next enzyme in the sequence, complex III. Complex II will be dealt with later.

> **Q&A 1:** Given the structure of ubiquinone and dihydroubiquinone below, which part of its structure makes it lipid soluble?

ubiquinone   dihydroubiquinone

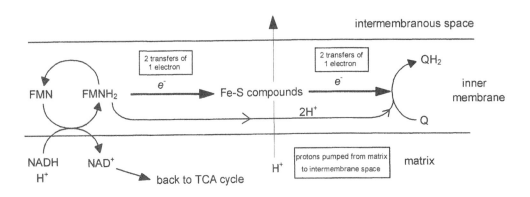

Complex III

Complex III, cytochrome bc_1 complex once again uses iron atoms to shuffle electrons within its structure. However, in addition to one iron-sulphur centre (2Fe-2S), other iron atoms are held within porphyrin groups. The iron together with the porphyrin is called a haeme group. In complex III, the proteins and their haeme groups are called cytochromes. The structures of the porphyrins vary slightly and this leads to different cytochromes. In complex III, there are two *b*-cytochromes, which differ slightly in their porphyrin ring structure, and one cytochrome c_1. These are very similar in structure to the prosthetic group of haemoglobin. The main difference is that in cytochromes, the iron group of the haeme is anchored on both sides, on one side by histidine (**H**) and on the other by methionine (**M**) for cytochrome *c* and by two histidines cytochrome *b*. In haemoglobin, it is only anchored on one side by histidine

Higher level

The Q cycle

The function of the Q cycle is to transfer two electrons, one at a time, to cytochrome *c*, and to transfer H^+ from the mitochondrial matrix across the inner mitochondrial membrane to the intermembranous space. Firstly, QH_2 transfers one electron to the iron sulphur centre and from there it is passed to cytochrome *c*. This means that ubiquinone is half oxidised and becomes ubisemiquinone ($Q^{\bullet -}$). At the same time the $2H^+$ are transferred to the intermembranous space. Importantly and logically, this reaction takes place at the intermembrane space side of the inner mitochondrial membrane. This creates a problem for $Q^{\bullet -}$ because it still has to lose an electron, but to do this it requires two protons from the mitochondrial matrix. Therefore, it somehow has to get to the other side of the membrane, but this is difficult because it is charged and hence not lipid soluble. To overcome this problem it transfers its electron to cytochrome *b*, and becomes the uncharged Q, which means that it can now travel to the inner side of the membrane. The oxidised cytochrome *b* transfers its electron to a second cytochrome *b*. Note that the protein containing the two *b*-cytochromes spans the membrane and this allows the electron to jump or tunnel from one side of the membrane to the other. It now can transfer it to Q which has moved from the outside of the membrane to the inside. The effect of this whole process is to transfer $Q^{\bullet -}$ from the outside of the inner mitochondrial membrane to the matrix side.

The problem now is that cytochrome *c* is only half reduced – it still requires another electron to become fully reduced. In the next step, the process is repeated with QH_2 donating an electron to the iron-sulphur centre and then onto the half oxidises cytochrome *c*. At this stage, cytochrome *c* is fully reduced. QH_2 is converted to $Q^{\bullet -}$, which transfers its electron to cytochrome *b*, which conveys this electron to the inner part of the inner mitochondrial membrane, and there it donates it to $Q^{\bullet -}$. At the same time $Q^{\bullet -}$ picks up $2H^+$ from the mitochondrial matrix to form QH_2.

The net effect is that cytochrome *c* is fully reduced, QH_2 becomes oxidised, and $4H^+$ are released into the intermembranous space.

Learning strategy
- Important is that the Fe-S centre can only accept electrons from QH_2 and not $Q^{\bullet -}$.
- However, it needs two electrons to fully reduce cytochrome *c* and therefore uses two molecules of QH_2.
- This process occurs on the outer part of the inner mitochondrial membrane. $Q^{\bullet -}$ can only be converted to QH_2 if it is located on the inner side of the inner mitochondrial membrane.
- Here there is a problem in that $Q^{\bullet -}$ is water and not lipid soluble (it is charged) and hence it cannot migrate from the outer part to the inner part of the membrane.
- Therefore, the electrons are jumped across the membrane via cytochrome *b* where they are added sequentially to the lipid soluble Q to form the water soluble $Q^{\bullet -}$ on the inner part of the membrane, and then to $Q^{\bullet -}$ to form the lipid soluble QH_2.

Q&A 2: Why can't $Q^{\bullet -}$ simply diffuse across to the other side of the membrane? After all, Q and QH_2 are very lipid soluble.

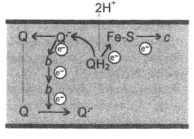

intermembrane space

first part of Q cycle

matrix

second part of Q cycle

intermembrane space

matrix

(H) enabling oxygen to interact with the 'free' side of Fe. Hence, haemoglobin can carry oxygen whereas the Fe of cytochromes can only carry electrons.

cytochrome c, e⁻ carrier   haemoglobin, O₂ carrier

In complex III, the electrons are eventually transferred from QH_2 to cytochrome *c* and $2H^+$ are released into the intermembranous space. This complicated process is called the Q cycle and comprises two distinct steps, details of which are fun to understand, and are outlined in the higher level section on the previous page.

Complex IV

Complex IV (cytochrome oxidase) contains two *a*-cytochromes *a* and *a₃* and two copper ions called Cu_A and Cu_B, which flip between Cu^{2+} and Cu^+ states as they transfer electrons. Eventually the two electrons of cytochrome *c* are transferred one at a time via the *a*-cytochromes and Cu to an oxygen atom. To reduce an oxygen molecule (O_2), four electrons are needed. In the process of reducing an oxygen atom, $2H^+$ are transferred to the intermembranous space and the oxygen atom consumes $2H^+$ from the matrix to form water. This adds to the potential difference across the inner mitochondrial membrane because not only are $2H^+$ transferred to the intermembranous space, but also another $2H^+$ are consumed from the matrix to form water.

Complex V

Complex V couples the re-entry of protons into the matrix with the formation of ATP. This complex uses the energy of the proton gradient to synthesise ATP from ADP. This protein complex comprises 2 major components, one which catalyses the synthesis of ATP (ATP synthetase) and the other forms a proton channel through the inner mitochondrial membrane. These 2 components are coupled. That is, you cannot have ATP synthesis unless the proton channel is operating. Interestingly, the condensation of phosphate to ADP to form ATP at the reactive centre of ATP synthetase occurs without any energy input. The problem is getting ATP synthetase to release ATP. Release oc-

curs due to a change in the structural conformation of ATP synthetase as protons pass through the proton channel and this change in shape of ATP synthetase causes the release of ATP.

Therefore, as electrons are passed down their oxidation gradient within the inner mitochondrial membrane, the energy produced is used to pump H^+ from the matrix into the intermembranous space. This provides a potential difference across the inner mitochondrial membrane which has the potential to do work. When H^+ is allowed to flow back across the membrane from to the matrix through complex V, work is performed by converting ADP to ATP.

Complex II

Complex II (succinate dehydrogenase) is succinate dehydrogenase of the TCA cycle. Unlike the other enzymes of the TCA cycle, which are located in the mitochondrial matrix, it is anchored in the inner mitochondrial membrane. Remember that to be embedded in a membrane, large portions of these enzymes must be lipophilic. Succinate dehydrogenase catalyses the removal of hydrogen from succinate and transfers it to FAD which becomes $FADH_2$. This in turn transfers its electrons to Q, which becomes QH_2. In this process, H^+ is not pumped from the matrix to the intermembranous space. QH_2 is then transferred to complex III for oxidation. Therefore, both complex I and II produce QH_2 which is channeled to complex III, but only complex I pumps protons across the inner mitochondrial membrane.

The fact that complex I pumps protons across the inner mitochondrial membrane and complex II does not means that when NADH is used as the substrate for oxidative phosphorylation, a bigger potential difference is created across the inner mitochondrial membrane than if $FADH_2$ were used. As a result, about 3 ATPs are synthesised from the oxidation of 1 NADH and only about 2 ATPs are synthesised when $FADH_2$ is used as a substrate.

> **Q&A 3:** If glucose is fully oxidised, how many ATPs can be produced? Remember that some ATP is produced when it is cleaved to pyruvate, and other high-energy compounds are produced as it enters the TCA cycle. NADH and FADH₂ are also formed in the TCA cycle and these pass into the oxidative phosphorylation pathway.

Learning strategy

- Oxidative phosphorylation takes place in the inner membrane of mitochondria.
- Electrons are transferred along the membrane through 4 protein complexes (complexes I-IV) and eventually to O_2 which becomes H_2O.
- As electrons flow along the membrane, protons (H^+) are pumped from the matrix to the intermembranous space of the mitochondria.
- The energy obtained by creating this charge gradient is used to phosphorylate ATP by allowing protons to flow back to the matrix through complex V.

- Except for the initial transfer of electrons from NADH to FMN, all electron transfers are as single electrons.
- Many of the electron transfers are achieved by flipping of iron from Fe^{3+} to Fe^{2+} and then back to Fe^{3+}.
- The iron is contained in Fe-S centres, or in porphyrins as cytochromes. These are static centres embedded in the protein units of complexes.
- Other electron transfers are accomplished by reducing ubiquinone (Q) to dihydroubiquinone (QH_2). These are lipid soluble compounds that move easily within the inner mitochondrial membrane.
- Both NADH (via complex I) and $FADH_2$ (via complex II) transfer their electrons to complex III, but only the transfer of electrons from NADH pumps protons from the matrix.
- Hence about 3 ATPs are obtained from NADH and only about 2 ATPs from $FADH_2$.

Mitochondrial facts

In some mitochondria, the electron transport chain is uncoupled from proton transport. In these cases, the energy generated is used to produce heat. In some plants, such as the *Arum* lily, this heat is used to attract insects by volatilising organic compounds. In brown adipose tissue, found in hibernating animals and in fat deposits in newborns, protons leak back across the inner mitochondrial membrane and thus generate heat. This is used to quickly restore the body temperature in hibernating animals and to maintain body heat in newborns.

Mitochondria contain their own DNA, mtDNA. This is packaged as closed circles of 16 569 base pairs and each mitochondrion contains 2 to 10 copies of the genome. This DNA encodes 13 of the 80 subunits of oxidative phosphorylation and has 22 transfer RNA genes and 2 ribosomal RNA genes. Mutations in mitochondrial DNA are much higher than with nuclear DNA because there are limited repair mechanisms and a high level of mutagenic oxygen free radicals. Mitochondria are inherited from the mother and hence mitochondrial DNA comes from the mother only.

Anecdote

Mitochondria are extraordinarily efficient in extracting energy. This is exemplified by the efficiency of heart tissue in mammals, and in particular the smallest mammal, a species of shrew. This animal is about the size of a cockroach and weighs in at a minuscule 2g. It has a heartbeat of 1500 beats/min and a total blood volume of 0.2mL. By contrast, the blue whale, which is the largest mammal (200 tonnes) has a heartbeat of 10 beats/min and an output of 5000L/min. The metabolic rate of the shrew is about 100 times that of the whale and the mitochondria are about 98% efficient for extracting energy. In the whale, as with most mammals, the mitochondria are 76% efficient in energy conversion.

Photosynthesis

The light cycle

No doubt, you already know that photosynthesis is the combination of carbon dioxide and water to form carbohydrates (sugars or starch). To occur, this process requires energy. It has many similarities to oxidative phosphorylation.

> **Q&A 4:** Where does the energy for photosynthesis come from and what is the other product besides carbohydrates?

> **Q&A 5:** Which organisms are capable of photosynthesis?

The organelles within plant cells which carry out photosynthesis, chloroplasts, are somewhat like mitochondria. They have an outer membrane and an inner membrane. Within the space enclosed by the inner membrane, the stroma, is another series of membranes called lamellae. The space enclosed by the lamellae is called the lumen. The enzymes for photosynthesis are embedded in the membranes of the lamellae, and during photosynthesis the enzymes pump H^+ into the lumen of the lamellae.

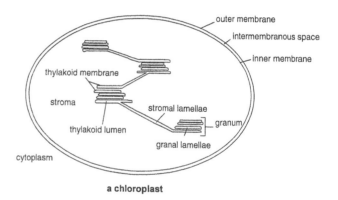

a chloroplast

Photosynthesis actually occurs in two distinct steps. The first is to oxidise water by removing two electrons which releases 2 protons and oxygen. The energy released in this process is used to reduce $NADP^+$ to NADPH and also to form an H^+ gradient across the lamellar membranes. This H^+ gradient is used for ATP synthesis, similar to oxidative phosphorylation. The second process is to use the energy contained in NADPH and ATP from the first process to form carbohydrates and in this process carbon dioxide is consumed.

As we all know, water is very stable and the idea of oxidising it or burning it is quite queer, but this is what happens. To accomplish this, electrons are enticed away from water. A protein containing a core of 4 Mn atoms encourages electrons away from 2 molecules of water. This process is not well understood, but the core of 4 Mn atoms can exist with valencies from +2 to +7 and so is a very good electron acceptor and donor. Although it is rela-

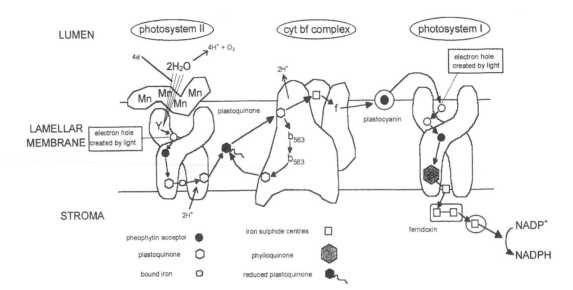

tively easy to encourage the electrons to come away from water, the process will stop unless they have a hole to fall into once they have been removed from water. Forming this electron hole is the nub of photosynthesis.

Q&A 6: What forms from two molecules of water when 4 electrons are removed?

The electron hole is dug in chlorophyll by light, i.e. light provides the energy to push electrons out of chlorophyll. This occurs in protein clusters called photosystem II and photosystem I. The whole process is illustrated in the diagram above and these protein clusters are anchored in the thylakoid membranes.

Chlorophylls are very similar in structure to the haeme groups of the cytochromes and haemoglobin. The main difference is that Mg, rather than Fe, is the central atom stabilised by the porphyrin ring. They also have a lipophilic phytol side chain. All green plants contain chlorophyll *a*. They also contain chlorophyll *b*, which differs slightly in the porphyrin ring structure and absorbs maximally at a slightly different wavelength of light.

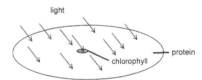

chlorophyll a

An important aspect of chlorophyll is that it is relatively easy to promote electrons of the central ring struc-

ture (containing the N atoms) to higher orbitals, but it is much more difficult to actually remove electrons from the ring. If chlorophyll is located in a protein core (photosystem I or II) then the local environment of the chlorophyll is sufficiently changed so that electrons can be removed from chlorophyll to be passed onto another electron acceptor. The problem with locating chlorophyll in a protein core is that it becomes inefficient in being excited by light since a large amount of the available area for the light to fall upon is taken up by protein.

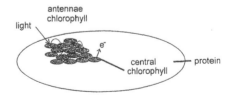

Plants have overcome this problem in a very clever manner. Many hundreds of chlorophylls are packed around the chlorophyll within the protein core (reaction centre). Although none of these can lose electrons, they can have their electrons pushed into higher orbitals by light energy. As these electrons return to ground state, the energy released is sufficient to excite the electrons in adjacent chlorophyll to higher orbitals. Therefore, the energy obtained from the initial photon jumps between chlorophylls (resonates) until by chance it jumps to the central chlorophyll. Here the energy is sufficient to push electrons out of the central chlorophyll of the reaction centre.

Therefore, the chlorophyll molecules not in the reaction centre act as an antenna, collecting the light and focusing the energy onto the central chlorophyll, which is oxidised by losing an electron. The electron hole, created in this manner, is filled by an electron obtained by the oxidation of water. The antenna system is even more clever because other pigments, such as carotenes, phycoerythrin and phycocyanin, also form part of this antenna system which means that the plant can make use of the whole visible spectrum rather than being limited to the blue and red ends by chlorophyll.

> **Q&A 7:** If green plants are green due to chlorophyll, what wavelengths (colours) of light must chlorophyll absorb?

Clearly, the electron driven from the reaction centre by light must go somewhere. Although it has theoretically just enough energy to convert NADP$^+$ to NADPH, it cannot do this in practice because energy transfer is never 100% efficient and the amount of energy lost in the transfer would be too much for the reaction to occur. Therefore, the electrons are captured by a series of oxidising agents (see below) to be eventually captured by plastoquinone, a molecule very similar in structure to quinone (recall oxidative phosphorylation above).

If plastoquinone is similar to quinone, would it be lipid soluble or water-soluble?

Answer: *Plastoquinone is lipid soluble and hence it can move freely within the thylakoid membrane.*

The electrons released from chlorophyll at the reaction centre are firstly captured by pheophytin and then passed to a plastoquinone which is bound to the proteins of photosystem II. This then passes its electrons to another bound form of plastoquinone, which is then released from the protein complex once reduced. It is reduced by capturing two electrons, one at a time, from the bound plastoquinone and binding 2H$^+$ from the stroma. It then becomes free from the protein complex and its lipid solubility allows it to move freely within the membrane to the cytochrome *bf* complex. Here, it transfers its electrons to cytochrome *b* one at a time in the manner previously described for the Q cycle. Reduced cytochrome *b* then passes its electrons to plastocyanin and 2 protons (2H$^+$) are dumped into the thylakoid lumen. Hence the two protons picked up by plastoquinone from the stroma are released into the lumen by cytochrome *bf*.

The electrons are passed onto yet another carrier, plastocyanin that once again needs somewhere to put its electrons. At this stage the whole system needs an energy boost because there is not enough energy to convert NADP$^+$ to

NADPH. This comes from another photosystem, photosystem I which is similar to photosystem II and has a bank of antenna chlorophylls that collect energy from light and focus it onto the chlorophyll at the reaction centre to dig an electron hole. One main difference is that it collects light from longer wavelengths than for photosystem II. This once again maximises use of energy from the whole visible spectrum. The electron hole is filled by electrons from phytocyanin. The electrons removed are passed along another chain of oxidants and reductants to be eventually transferred to NADP$^+$, which is reduced to NADPH. This second chain of oxidants and reductants consists of phylloquinone and then three iron sulphur clusters (compare with oxidative phosphorylation), then to ferredoxin and to NADP$^+$.

> **Nomenclature:** Photosystem I and photosystem II are named in the order they were discovered and not in the order of the electron transfer.

> **Q&A 8:** Could this part of photosynthesis, the oxidation of water and reduction of NADP$^+$, be carried out in the dark?

The energy levels in this system are schematically illustrated below.

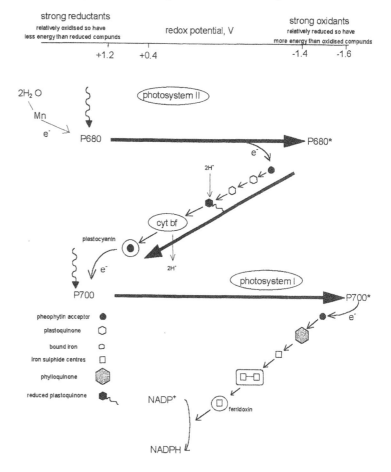

> **Q&A 9:** This reaction pathway is given a special name. Look at the arrows and think of a name that might be appropriate for the reaction.

The last step in this reaction sequence is to use the energy created by the H$^+$ gradient across the thylakoid membrane. Unlike mitochondria, where the H^+ concentration is lower inside (the matrix) the H^+ concentration is higher on the inside of the thylakoid membrane. Therefore, protons are allowed to pass from the inside to the outside of the thylakoid membrane through ATP synthetase and the energy is used to produce ATP. ·

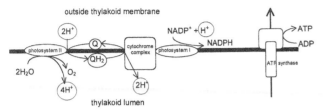

Both removal of H^+ from the outside of the thylakoid membrane and the deposition of H^+ into the thylakoid lumen creates the gradient which is used to generate ATP.

The result of the light reaction is that NADPH is formed and ATP is produced from an H^+ gradient across the thylakoid membrane. Both are energy sources and are used in the second part of photosynthesis to form hexoses (6 carbon sugars) by fixing carbon dioxide.

The dark cycle or the Calvin cycle

This part of photosynthesis occurs in the stroma of the chloroplasts. Remember that this is where the ATP is synthesised from the proton gradient created by the light reaction. The Calvin cycle is quite a complicated series of reactions that can be divided into two parts. In one part, 6 carbon dioxides are fixed from the atmosphere by attaching them to 6 ribulose 1,5-bisphosphates. These 6 carbons are then used for the synthesis of a hexose and in another series of reactions the 6 ribulose 1,5-bisphosphates are regenerated.

Draw the structure of ribulose 1,5-bisphosphate, remembering that it has phosphoesters at carbons 1 and 5 and that the 'ulose' ending means it is a ketone rather than an aldehyde and hence the keto group comes off C2.

Answer:

$$CH_2OPO_3^{2-}$$
$$|$$
$$C=O$$
$$|$$
$$H-C-OH$$
$$|$$
$$H-C-OH$$
$$|$$
$$CH_2OPO_3^{2-}$$

Carbon dioxide diffuses from the atmosphere and is attached, or fixed, to C2 of ribulose 1,5-bisphosphate. This is a carboxylation. Ribulose 1,5-bisphosphate is rearranged within the enzyme which catalyses this reaction before the carboxylation takes place.

What is the name of the enzyme that catalyses the carboxylation of ribulose 1,5-bisphosphate?

Answer: Ribulose 1,5-bisphosphate carboxylase. It also has another function, oxygenase activity, and hence an alternative name is ribulose 1,5-bisphosphate carboxylase/oxygenase, or rubisco.

The next step is to add water to the ketone group at C3 and then to cleave the molecule to form 2 molecules of 3-phosphoglycerate.

The two 3-phosphoglycerates are then phosphorylated and then reduced to glyceraldehyde 3-phosphate.

The first part of this reaction is a phosphorylation and therefore the enzyme that catalyses this reaction is a kinase, phosphoglycerate kinase. The energy comes from ATP. The enzyme that catalyses the second part of the reaction is named for the reverse reaction, glyceraldehyde 3-phosphate dehydrogenase. The energy for this reaction comes from NADPH.

Q&A 10: Where does the NADPH come from?

Q&A 11: How many molecules of NADPH are used for each carbon dioxide fixed?

It is important to note that the CO_2 is fixed one molecule at a time and hence 6 must be fixed to produce the equivalent of a glucose molecule. Therefore, to form one glucose, 12 ATPs and 12 NADPHs are required.

Glyceraldehyde-3-phosphate is then used as a precursor for the formation of sugars: hexoses as an energy source for the plant, and five carbon sugars to supply the ribulose for capturing the CO_2.

Higher Level

Regeneration of 1,5-ribulose bisphosphate is a very complicated pathway involving several ketolases and aldolases. It is inserted here in outline only.

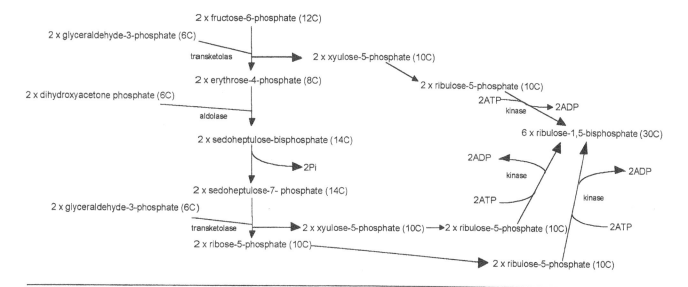

Six CO_2s must be fixed to form one hexose. How many glyceraldehyde-3-phosphates are produced if 6 CO_2s are fixed?

Answer: *There are 12 glyceraldehyde-3-phosphates produced for 6 CO_2s fixed.*

Only 2 of the 12 3-phosphoglycerates are used to produce a new hexose. The others pass through various pathways to regenerate six 1,5-ribulose bisphosphates.

Q&A 12: From the 6 CO_2 fixed, two of the glyceraldehyde-3-phosphates are used to make a hexose, and the rest are used to make new ribulose-1,5-bisphosphate. How many glyceraldehyde-3-phosphates are used to make ribulose-1,5-bisphosphate and how many ribulose-1,5-bisphosphates are formed? Hint: work out the total number of C in the remaining glyceraldehyde-3-phosphates and then calculate how many 5C compounds (ribulose-1,5-bisphosphates) could be made from this number of Cs.

Therefore the whole cycle can be written

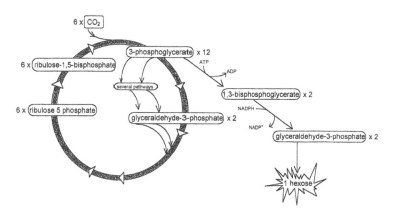

Q&A 13: Could this part of photosynthesis, the usage of NADPH and ATP to make a hexose, occur in the light? Could it occur in the dark?

Learning strategy

It is important to remember the 2 distinct parts of photosynthesis - the light reaction to generate NADPH and ATP, and the dark reaction or Calvin cycle to synthesise hexoses.

For the light reaction remember that

- it takes place on the thylakoid membranes of chloroplasts
- $2H_2O$ is cleaved into $4H^+$ and O_2 plus $4e^-$ which pass through a series of electron transfers to $NADP^+$ which is reduced to NADPH.
- an H^+ gradient is formed across the thylakoid membrane, which is used to synthesise ATP
- photons of light acting at different complexes are required to drive the process
- in both cases, the light digs an electron hole in chlorophyll at the reaction centre of the complexes, allowing the chlorophyll to accept electrons from other molecules
- the photons are directed by antennae chlorophylls towards the chlorophyll of the reaction centre, which loses electrons
- the electrons lost from chlorophyll at the reaction centre of complex II are replaced by those from water

- the electrons from complex II are passed to phylloquinone and then to cytochrome *bf* complex
- the electrons lost from the chlorophyll at the reaction centre of complex I are replaced by electrons from the cytochrome *bf* complex
- the electrons from complex I are passed onto iron-sulphur centres and eventually to $NADP^+$ to form NADPH

It is important to be able to compare the similarities and differences of photosynthesis and oxidative phosphorylation. It is also important to be aware that haeme groups and chlorophylls are similar with the main difference being that haemes have Fe at their reactive centres and chlorophylls have Mn.

For the dark reaction

- CO_2 is fixed to ribulose 1,5-bisphosphate, and then cleaved to two 3-phosphoglycerates.
- It is necessary to learn this pathway. In particular, carbon dioxide is attached to C2 and the ketone group is shifted to C3. Water is added to the ketone group (C3) and then it is cleaved into two 3-phosphoglycerates.
- For hexose production, the 3-phosphoglycerates (acids) are converted to aldehydes via a phosphorylation and then reduction. This is important because this is where some of the NADPH and ATP synthesised in photosynthesis is used.
- The glyceraldehyde-3-phosphate can be converted into fructose-1,6-bisphosphate as for gluconeogenesis (see Chapter 14).
- CO_2 is fixed to ribulose 1,5-bisphosphate one molecule at a time. It requires 6 ribulose 1,5-bisphosphates to synthesise 1 hexose.
- Each of these is converted to 3-phosphoglycerate (12 in total).
- The carbons of the 10 remaining 3-phosphoglycerates are shuffled between a variety of sugars to finally regenerate the original 6 ribulose-1,5-bisphosphates.

References and further reading

Babcock, G.T. and Wikström, M. (1992) Oxygen activation and the conservation of energy in cell respiration. *Nature* **356**:301-309.

Barber, J. and Andersson, B. (1994) Revealing the blueprint of photosynthesis. *Nature* **370**:31-34.

Calhoun, M.W., Thomas, J.W. and Gennis, R.B. (1994) The cytochrome oxidase superfamily of redox-driven proton pumps. *Trends Biochem. Sci.* **19**:325-328.

Devlin, T.M. (1992) *Textbook of Biochemistry with Clinical Correlations*, 3rd ed. New York: Wiley-Liss.

Govindjee and Coleman, W.J. (1990) How plants make oxygen. *Sci. Amer.* **262(2)**:50-58.

Greenstein, B. and Greenstein, A. (1996) *Medical Biochemistry at a Glance*. Oxford: Blackwell Science.

Hinkle, P.C., Resetar, M.A. and Harris D.L. (1991) Mechanistic stoichiometry of mitochondrial oxidative phosphorylation. *Biochemistry* **30**:3576-3582.

Slater, E.C. (1983) The Q cycle, a ubiquitous mechanism of electron transport. *Trends Biochem. Sci.* **8**:239-242.

Trumpower, B.L. and Gennis, R.B. (1994) Energy transduction by cytochrome complexes in mitochondrial and bacterial respiration: The enzymology of coupling electron transfer reactions to trans membrane proton translocation. *Annu. Rev. Biochem.* **63**:675-716.

Williams, R.J.P. (1995) Purpose of proton pathways. *Nature* **376**:643.

Youvan, D.C., and Marrs, B.L. (1987) Molecular mechanisms of photosynthesis. *Sci. Amer.* **256(6)**:42-48.

Q&A Answers

1   The long hydrocarbon chain makes ubiquinone lipid soluble. In humans the hydrocarbon tail comprises 10 repeats of an isoprenoid unit and hence it is called Q_{10}. In other species it has 6 or 8 units and therefore is called Q_6 or Q_8.

hydrocarbon tail

ubiquinone

2   Unlike Q and QH_2, $Q^{\bullet-}$ is water-soluble because it is charged. As such, it cannot readily cross from the outer part of the inner mitochondrial membrane to the inner part. Q and QH_2, being lipid soluble, can move freely within the lipid membrane.

3   Glucose is firstly split to form 2 pyruvates. This process yields a net of 2ATPs and 2NADHs. The 2NADHs are processed by oxidative phosphorylation to produce 3ATPs each, which is a total of 6ATPs. The two pyruvates are processed through the TCA cycle by firstly being converted to acetyl-CoA. The conversion of pyruvate to acetyl-CoA yields NADH. Hence conversion of 2 pyruvates to acetyl-CoA produces 2NADHs which is converted to 6ATPs. Each acetyl-CoA which passes through the TCA cycle produces 3NADHs (9ATPs after oxidative phosphorylation) and 1FADH$_2$ (2ATPs after oxidative phosphorylation) and 1ATP (or 1GTP). For two acetyl-CoAs this is 24ATPs. Therefore one molecule of glucose yields 38 ATPs when fully oxidised. Only 2ATPs are formed if there is no oxygen available because glucose is only converted as far as pyruvate.

4   The energy for photosynthesis comes from light and the product other than carbohydrates is oxygen.

5   Plants, algae and some bacteria are capable of photosynthesis.

6   When 4 electrons are removed from 2 molecules of water, one molecule of oxygen and 4 protons are formed.

$$2H_2O \rightarrow O_2 + 4H^+ + 4e^-$$

7   Green plants are green because they do not absorb green light, but instead reflect it. Therefore, chlorophyll must absorb the non-green wavelengths: blue and red.

8   No, of course not. It needs the energy from light to push electrons out of chlorophyll. This part of photosynthesis is thus called the light reaction.

9   The whole light reaction from water through photosystems II and I is called the Z reaction. The Z coming from the pattern formed by changes in energy levels associated with redox (reduction-oxidation) potentials.

10 The NADPH comes from the light reaction, as does the ATP.

11 Two molecules of NADPH are used for each CO_2 fixed.

12 10 glyceraldehyde-3-phosphates are used to form 6 ribulose-1,5-bisphosphates.

13 Yes, this part of photosynthesis can occur in the light, but it can also occur in the dark providing that there is enough NADPH and ATP available. Another name given for this part of photosynthesis is the dark reaction, but remember it can actually occur in the light as well as the dark.

11 Lipid Structure and Metabolism

In this chapter you will learn:
- the basic structures of glycerol, fatty acids, ester bonds, triacylglycerol, diacylglycerol
- what form fats take when they are to be stored
- the difference between saturated, monounsaturated and polyunsaturated fatty acids
- the nomenclature of fatty acids
- the numbering and naming of the carbons in fatty acids and shorthand notation of fatty acids
- the names of the essential fatty acids
- the relationship between structure and melting point of fats and oils
- how soaps are formed
- the structures of membrane lipids – phospholipids, cardiolipin, phosphatidylinositolbisphosphate, plasmalogens, platelet activating factor, sphingolipids
- the structure of micelles, vesicles, and membranes
- the different lipases and the reactions they catalyse
- metabolism of glycerol
- the carnitine shuttle
- breakdown of fatty acids (β-oxidation)
- ketone body formation from fatty acids
- transfer of acetyl-CoA across the mitochondrial membrane using the citrate shuffle
- synthesis of fatty acids
- understand the workings of the enzyme complex, fatty acid synthase
- the major differences and similarities between fatty acid synthesis and degradation
- structure of cholesterol, steroids and bile salts
- the standard nomenclature and numbering of the ring structure in cholesterol
- synthesis of cholesterol
- the mechanism of transport of lipids in the bloodstream – apolipoproteins and their nomenclature
- the formation of atherosclerosis

Overview of lipids

Remember from earlier that fats (lipids) and water do not mix. Therefore, lipids are:
- soluble in organic solvents, such as chloroform, ether and alcohols
- major elements of cell membranes
- the basis of some hormones (steroid hormones)
- the basis of bile acids
- second messengers
- involved with cell recognition

Structural features of lipids

Glycerides

The structural basis of one group of lipids is glycerol. It is an alcohol and one of the most important molecules to learn. We drew its structure earlier.

Draw the structure of glycerol and circle the hydroxyl groups.

Answer:

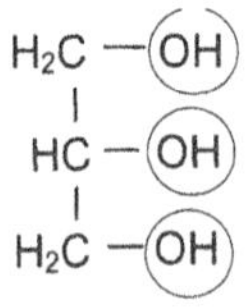

It is a 3C chain with an hydroxyl group at each carbon and the remaining bonds are saturated with hydrogens. The hy-

---

Condense a long chain carboxylic acid, e.g. 18C, 20C, 16C to each of the hydroxyl groups of glycerol. Do not forget that the carbon of the carboxyl group is counted as one of the carbons of the fatty acid. In fact, it is C1.

Answer:

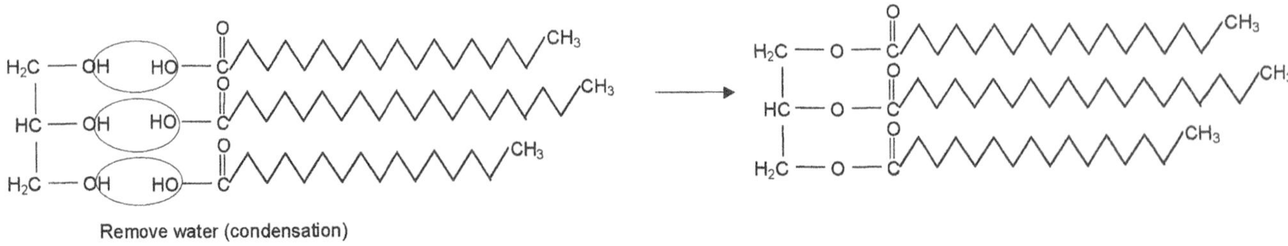

Remove water (condensation)

droxyl groups make it particularly important because it means that ester bonds could be formed at each of these.

> **Q&A 1:** What type of chemical would be linked to glycerol (an alcohol) to form an ester bond?

> **Q&A 2:** What substance is released when the ester bond is formed and what is the general name for such a reaction?

Remember that each point on the zigzag chain represents a CH_2. Note also that with standard structural notation, the fatty acid at C2 should be to the left of the molecule and not to the right. The way it is drawn here and elsewhere in this chapter with the fatty acid at C2 projecting to the right is for convenience, but strictly speaking it is incorrect.

A long chain of carbons is called an acyl group and above there are 3 acyl groups attached to glycerol. What would be a good name for this structure?

Answer: *A triacylglycerol.*

Triacylglycerols have no charge and hence are lipid soluble. In fact, these are fats. We use this type of fat as an energy source and store it in our fat deposits. Triacylglycerols are a whole family of fats, not just one. Different triacylglycerols occur by changing the fatty acids attached (esterified) to the glycerol backbone. It is therefore important to be aware of the common fatty acids and to know the common nomenclature associated with fatty acids.

Fatty acids

Understanding the nomenclature associated with fatty acids makes this part of biochemistry quite easy. Without this understanding, it is all but impossible. It is really worthwhile spending extra time on this section.

Fatty acids are long carbon chain with a carboxyl group (-COOH) at one end. Some have no double bonds between the carbons in their structure and others have double bonds (one or more but not usually more than 4). If there are no double bonds, then the molecule has as many hydrogen atoms attached to it as is possible. Therefore, it is saturated with hydrogens and is called a saturated fatty acid. Saturated fatty acids have the general structure of $CH_3(CH_2)_n COOH$. They are named after the parent hydrocarbon with *oic* replacing the final *e*. e.g. C_{18} parent hydrocarbon is octadecane and the saturated fatty acid is octadecanoic acid. If they have one or more double bonds in their structure, then theoretically they could accommodate more hydrogens if the double bonds were converted to single bonds. Therefore, they are not saturated with hydrogen atoms.

What would be an appropriate general name for a fatty acid with one or more double bonds?

Answer: *Such fatty acids are called unsaturated. If they have only 1 double bond in their structure they are called monounsaturated fatty acids; if more than 1 double bond, then they are called polyunsaturated fatty acids. I am sure that you are familiar with this term from food products such as margarines.*

> **MAJOR POINT 53:** A fatty acid with no double bonds is called saturated; with double bonds mono- or polyunsaturated.

If a carbon chain has a double bond in its structure it is an alk*ene* and hence has the term *ene* meaning double bond in its name. Octadecanoic acid with 1 double bond becomes octadeca*ene*oic acid (*octa-deca-een-oic*), and with 2 double bonds octadeca*diene*oic acid (*octa-deca-die-een-oic*). The diene means 2 double bonds.

> **Q&A 3:** Given that a 20C saturated fatty acid is called eicosanoic acid, and that tetra means 4. What would be the name of a 20C fatty acid with 4 double bonds?

The naming of the carbon atoms in fatty acids is also important. They are sequentially numbered from the carboxyl end with the carbon of the carboxyl group being C1. They are also named with Greek letters but here C2 is the αC and C3 the βC. The carboxyl group, which contains C1, is referred to as the carboxyl group of the αC. The last letter of the Greek alphabet is omega (ω) and therefore the last carbon in a fatty acid (the carbon of the methyl group) is called the ωC (omega carbon) irrespective of the length of the fatty acid.

> **Q&A 4:** In the following fatty acid indicate C1, βC and ωC.

Often a shorthand notation is used to describe fatty acids. Firstly, the number of carbons is indicated followed by the number of double bonds. If designated 18:0 = no double bonds 18:2 = 2 double bonds. We also need to know the position of the double bonds and whether the hydrogen atoms are in the cis or trans position. The position of a double bond is represented by the symbol Δ with a superscript number, e.g. cis-Δ^9 means there is that the 9th bond in the fatty acid is a double bond.

> **Q&A 5:** Given that the first bond in a fatty acid is between C1 and C2, between which carbons would the double bond occur in a fatty acid designated 18:1, cis-Δ^9?

Some final points to note are that
- fatty acids are ionised at physiological pH and therefore they are commonly referred to as their carboxylate form, e.g. palmitate, hexadecanoate.

Carbon skeleton	Structure	Systemic name	Common name (derivation)
12:0	$CH_3(CH_2)_{10}COOH$	Dodecanoic acid	Lauric acid (laurel plants)
14:0	$CH_3(CH_2)_{12}COOH$	Tetradecanoic acid	Myristic acid (nutmeg genus)
16:0	$CH_3(CH_2)_{14}COOH$	Hexadecanoic acid	Palmitic acid (palm tree)
18:0	$CH_3(CH_2)_{16}COOH$	Octadecanoic acid	Stearic acid (hard fat)
20:0	$CH_3(CH_2)_{18}COOH$	Eicosanoic acid	Arachidic acid (arachius is legume genus)
22	$CH_3(CH_2)_{20}COOH$	Docosanoate	Behenic acid (*Moring pterygosperma* - Ben tree
24:0	$CH_3(CH_2)_{22}COOH$	Tetracosanoic acid	Lignoceric (lignum is wood)
16:1(Δ^9)	$CH_3(CH_2)_5CH{=}CH(CH_2)_7COOH$	*cis*-Δ^9-hexadecenoate	Palmitoleic
18:1(Δ^9)	$CH_3(CH_2)_7CH{=}CH(CH_2)_7COOH$	*cis*-Δ^9-octadecenoate	Oleic (oleum is oil)
18:2($\Delta^{9,12}$)	$CH_3(CH_2)_4CH{=}CHCH_2CH{=}CH$ $(CH_2)_7COOH$	*cis,cis*-$\Delta^{9,12}$-octadecadienoate	Linoleic (linon is flax)
18:3($\Delta^{9,12,15}$)	$CH_3CH_2CH{=}CHCH_2CH{=}CHCH_2$ $CH{=}CH(CH_2)_7COOH$	all *cis*-$\Delta^{9,12,15}$-octadecatrienoate	Linolenic
20:4($\Delta^{5,8,11,14}$)	$CH_3(CH_2)_4CH{=}CHCH_2CH{=}$ $CHCH_2CH{=}CHCH_2CH{=}CH(CH_2)_3COOH$	all *cis*-$\Delta^{5,8,11,14}$-eicosatetraenoate	Arachidonic

- normally, fatty acids found in plants and animals have carbon chains of even numbers with C16 and C18 the most common.
- the double bonds in polyunsaturated fatty acids are separated by at least 1 methylene group.
- the double bonds are almost always in the *cis* form.

Commonly found fatty acids are listed in the table above, and most have common names as well as scientific names.

You are not expected to learn all of the fatty acids, however some are important and worthwhile committing to memory. These are: myristic (14), palmitic (16) stearic (18), arachidonic (20:4) linoleic (*lin-oh-lay-ic*) (18:2), and linolenic (*lin-oh-lean-ic*) (18:3) are the most important. Myristic, palmitic and stearic are the most common saturated fatty acids encountered. Arachidonic acid is an important precursor for several bioactive compounds such as the prostaglandins, and linoleic and linolenic acids are not synthesised in mammals, and therefore are essential dietary requirements. A mnemonic for this group is 'MPs (members of parliament) are laid leanly': M=myristic, P=palmitic, S=stearic, are=arachidonic, laid = linoleic, leanly=linolenic.

If these chemicals are liquid at room temperature, then they are called oils and if they are solid, then they are called fats. Fats have high melting points and oils have low melting points. The melting point of oils and fats is determined by the length of the fatty acids and the number of double bonds. The longer the fatty acid the higher the melting point (fats are made from long chain fatty acids, 10C or more). Increasing the number of double bonds decreases the melting point (oils have fatty acids with many double bonds). Molecules are not free to rotate around double bonds and *cis* double bonds cause an inflexible bend in an otherwise flexible straight molecule. This is best appreciated by building a model. If the molecules are all linear then they pack easily together and form a rigid structure (a fat). If they have double bonds they are kinked and cannot pack tightly and hence tend to remain as a liquid (oils) because they keep moving to try to pack in better.

> **Q&A 6:** What would you predict for the structure of vegetable oils, and butter in terms of number of double bonds in their fatty acid structure?

Soaps

Soaps are salts of fatty acids. To make soap, one boils fat with an alkali such as KOH or NaOH. The triacylglycerides are hydrolysed by the boiling and the fatty acids form insoluble salts - the K^+ or Na^+. You have probably heard of hard water. In hard water, soaps will not lather. It has divalent cations present such as Ca^{2+} and Mg^{2+}, which cause them to be converted into their insoluble calcium or magnesium salts.

add heat and water (hydrolysis)   fatty acids released   salt of fatty acids formed soap

Draw a diacylglycerol and then esterify phosphoric acid to C3.

Answer:

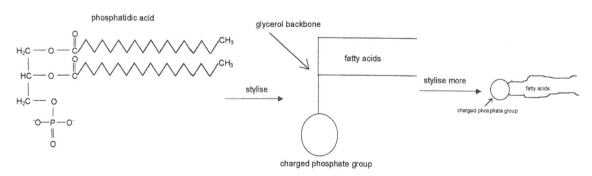

Membrane Lipids

Phospholipids

The most common components of cell membranes are variants of triacylglycerols, the phosphoglycerides or glycerophospholipids. Instead of a fatty acid at C3 they have a charged group esterified to this position.

The product formed is diacylglycerol phosphate, commonly referred to as phosphatidic acid. The fatty acids esterified at C1 and C2 of glycerol can vary in chain length and saturation and hence there is not 1 phosphatidic acid, but a family of phosphatidic acids. Usually the acid in position 1 is saturated and position 2 is unsaturated.

This molecule is very exciting because one part of it is charged (water-soluble) and the other part is uncharged (lipid soluble). Identify the charged and uncharged parts of the molecule.

Answer:

phosphatidic acid

uncharged part (lipid soluble)

charged part (water soluble)

| **Nomenclature:** A molecule, which has a charged and an uncharged portion, is called an amphipathic molecule.

Imagine taking a teaspoon of phosphatidic acids and mixing them in water. What would happen? Hint, think about the fact that lipid soluble parts will tend to go away from water and the water-soluble parts will tend to go towards water.

Answer: *To answer this question, it is firstly convenient to draw a simplified cartoon of phosphatidic acid.*

Using this cartoon of a phosphatidic acid we can show there are several possibilities, all based on the fact that the lipid soluble parts will tend to coalesce together.

miscelle

vesicle

membrane

A micelle is formed from a single layer of phosphatidic acid with the lipid soluble tails pointing towards each other. A vesicle forms from a bilayer of lipids. It encloses water and hence the phosphate groups of one layer face the centre and those of the other layer face the water surrounding the vesicle. Large sheets of 2 layers of phospholipids can also form. These are membranes. Most high power cartoons of cell membranes schematically show the membranes as phospholipid bilayers. You should now be able to interpret these diagrams properly. Therefore, membranes can form spontaneously from amphipathic lipids (polar and non-polar portions). Diacylphospholipids are more likely to form membranes than micelles, but lysophospholipids (one fatty acid attached to the phosphoglycerol backbone) and free fatty acids form micelles rather than bilayers because of their small hydrophobic tails.

The phosphate group on phosphatidic acid makes it doubly exciting because it can form more than 1 ester bond: all that is needed is an hydroxyl group. This leads to several important variations on phosphatidic acid that you should be aware of. If you let your imagination run wild, well at least assess molecules we already know to have hydroxyl groups, these can be esterified to the phosphate group of phosphatidic acid. This varies the charge, size and polarity of the phosphate group. One example is glycerol. (Note R1 and R2 represent chains of carbons with hydrogens attached. They may or may not have double bonds in their structure).

phosphotidylglycerol

Two other examples are ethanolamine and choline. It is relatively easy to learn their structures. To begin with we have to know the structure of ethanol, not just how it tastes. Ethanolamine, as the name suggests, is ethanol with an amine group.

ethanol   ethanolamine

Choline is ethanolamine, but containing 3 methyl groups, rather than 3 hydrogen atoms attached to the nitrogen.

ethanolamine   choline

Esterified to phosphatidic acid they form phosphatidylethanolamine and phosphatidylcholine respectively.

phosphatidylethanolamine   phosphatidylcholine

Clearly, the charge has changed on the polar group. Both have a negative charge on the phosphate and a positive charge on the amine group. This means that there is no net charge, but it is still polar (water soluble), because it carries charge. The choline carries 3 methyl groups and therefore is less polar than the ethanolamine group.

> **Q&A 7:** Esterify serine (S) to phosphatidic acid, name the molecule and specify the charge and the net charge.

> **Q&A 8:** For something a little more difficult, esterify phosphatidylglycerol to phosphatidic acid. Identify the charge on this molecule. These molecules are given a common name: cardiolipins.

One of the most important variations is phosphatidylinositolbisphosphate. You need to learn this. It is important because the inositol part of the molecule is a second messenger and will be discussed in detail in Chapter 15.

Inositol is a sugar alcohol (like glycerol) which means that it is like a sugar but has only hydroxyl groups and no ketone or aldehyde group. It has 6 carbons in a ring and each carbon has an hydroxyl group attached.

myoinositol

The particular form shown, myoinositol, is the most common form of inositol encountered in biochemistry. Others have the hydroxyl groups in different positions relative to the ring (above or below). The hydroxyl groups at 2 and 4 are down relative to the ring for myoinositol. Inositol 4,5-bisphosphate has 2 phosphate groups esterified to 2 hydroxyl groups. Phosphatidylinositol 4,5-bisphosphate has a third hydroxyl group of inositol esterified to the phosphate of phosphatidic acid.

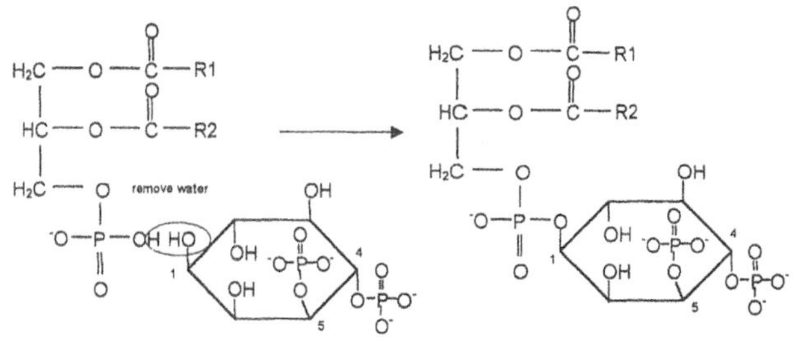

myoinositolbisphosphate → phosphatidylinositolbisphosphate

The most common form of inositolbisphosphate is shown with phosphates at C4 and C5. Phosphatidylinositol 4,5-bisphosphate has a massive -5 charge associated with it.

> **Nomenclature:** Note that this is *bis*phosphate rather than *di*phosphate. The *bis* indicates that the phosphates are linked to 2 different carbons on the ring where *di* indicates that there are 2 phosphates connected to each other (pyrophosphate) and these are esterified to 1 carbon, e.g. as in adenosine diphosphate.

> **Q&A 9:** Sometimes phosphoglycerides lack a carboxylic acid substitution at C2. These are called lysophospholipids. Draw a lysophosphatidylethanolamine.

Higher level

A variation on phosphoglycerides are the plasmalogens. Instead of an ester bond at C1 they have an α, β-unsaturated ether linkage in the *cis* configuration (a vinyl ether linkage).

a vinyl ether linkage

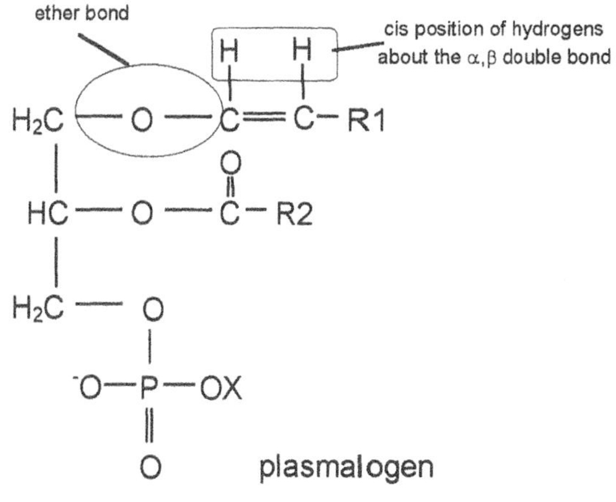

plasmalogen

The substitution on the phosphate (-X) is ethanolamine, choline or serine.

Plasmalogens comprise 23% of the glycerophospholipids in the human central nervous system, but a specialised physi-

ological function for them is still to be determined. However, the function of a closely related substance, platelet activating factor (PAF), is known. This molecule has an ether bond at C1, acetate attached to C2 and choline esterified to the phosphate group at C3.

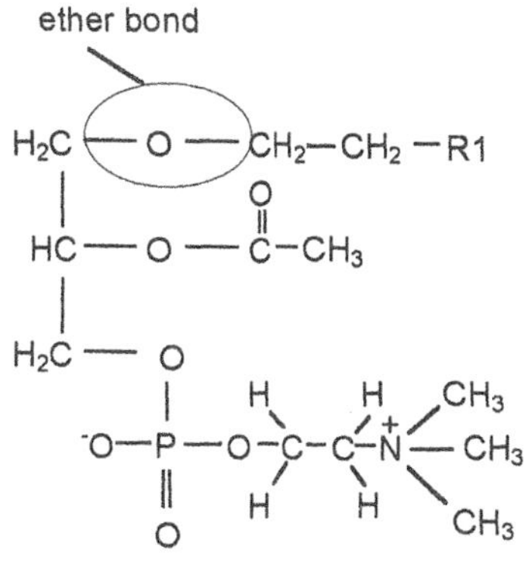

platelet activating factor

PAF causes platelets to aggregate, which is necessary to cause proper blood clotting. It causes smooth muscle to contract which helps stem the blood flow by constricting arterioles. Studies associated with *in vitro* fertilisation indicate that newly fertilised eggs that secrete high amounts of PAF are more likely to implant into the uterine wall and become embryos than those which secrete small amounts.

Sphingolipids

Another major group of lipids in cell membranes is the sphingolipids. The structural basis for these lipids is sphingosine (*trans*-4-sphingonine). This is an 18C chain with hydroxyl groups at C1 and C3, an amine group at C2 and the fourth bond is a *trans* double bond.

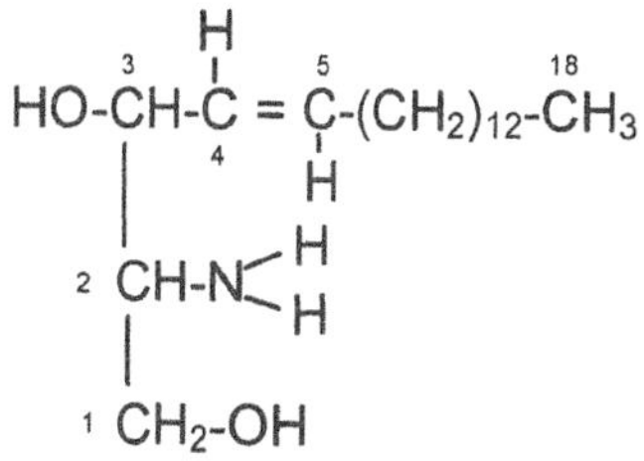

To remember the structure of this molecule relate it back to glycerol. It is glycerol with an amine group at C2 rather than an hydroxyl group. At C3, in addition to the hydroxyl group it has a 15C chain attached to it with a *trans* double bond between the first 2 carbons of the chain, otherwise it is saturated. This makes it very similar to a monoacylglyceride. The amine group at C2 provides the opportunity to form an amide bond with an acid. The hydroxyl group at C1 can also form ester or ether bonds. The fats formed by such bonds are all sphingolipids. They have the general structure of

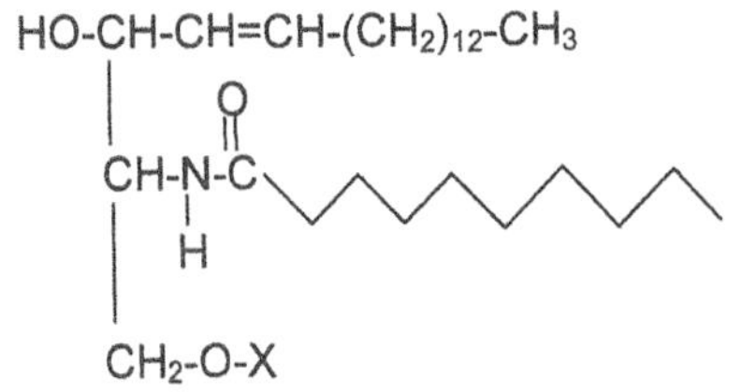
HO-CH-CH=CH-(CH2)12-CH3
CH-N-C
H
CH2-O-X

MAJOR POINT 54: Major lipids of cell membranes are the phospholipids, phosphoglycerides and sphingomyelin, glycosphingolipids and cholesterol.

An important aspect of phospholipids in cell membranes is that they can be acted upon by phospholipases. These catalyse the hydrolysis of phospholipids at particular sites. Phospholipase A1 catalyses the hydrolysis of phospholipids at C1, phospholipase A2 at C2, phospholipase B at C1 and C2, phospholipase C catalyses the hydrolysis at the glycerol side of the phosphate group and phospholipase D on the phosphate side of the phosphate group. A specific example is given on the next page: the hydrolysis of phosphatidylinositolbisphosphate catalysed by phospholipase C results in the formation of diacylglycerol and inositoltrisphosphate. Both are important intracellular signaling chemicals (Chapter 15).

Cholesterol is another type of fat that is a component of cell membranes and will be dealt with later.

Fats as an energy source

Fats/oils are a tremendous source of energy and indeed, 40% of our calories are provided by fatty acids that come from phospholipids and triacylglycerols. Fats have the distinct advantage over sugars in that they are less oxidised and do not require water for storage. The latter means that less weight is required to store the equivalent amount of energy as fats than as sugar. This is strange because we

Higher level

The sphingolipids are given special names depending upon the substitution for X.

A ceramide is formed when X is an H

A sphingomyelin is formed if X is phosphocholine.

A cerebroside is formed if X is a sugar.

A ganglioside is formed if X is a chain of sugars which include *N*-acetylneuraminic acid.

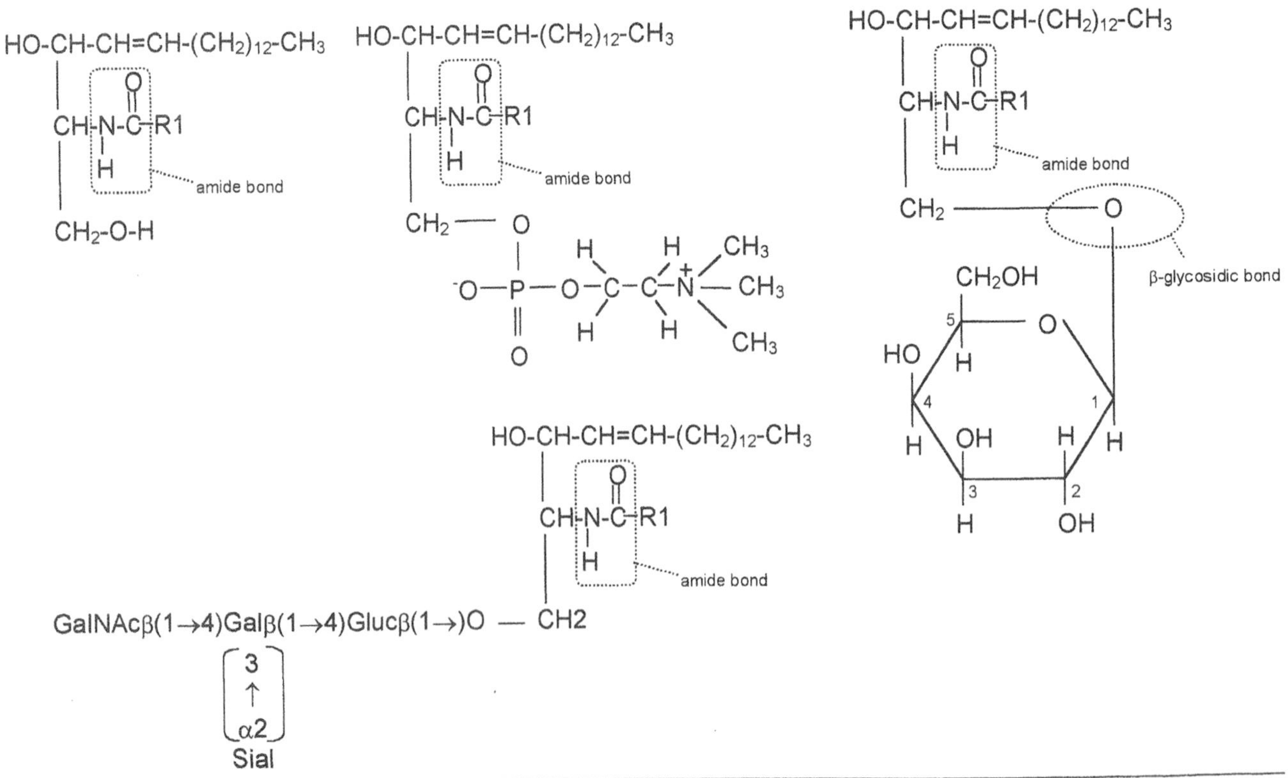
HO-CH-CH=CH-(CH2)12-CH3
CH-N-C-R1
amide bond
CH2-O-H
CH2-O
β-glycosidic bond
CH2OH
GalNAcβ(1→4)Galβ(1→4)Glucβ(1→)O — CH2
α2
Sial

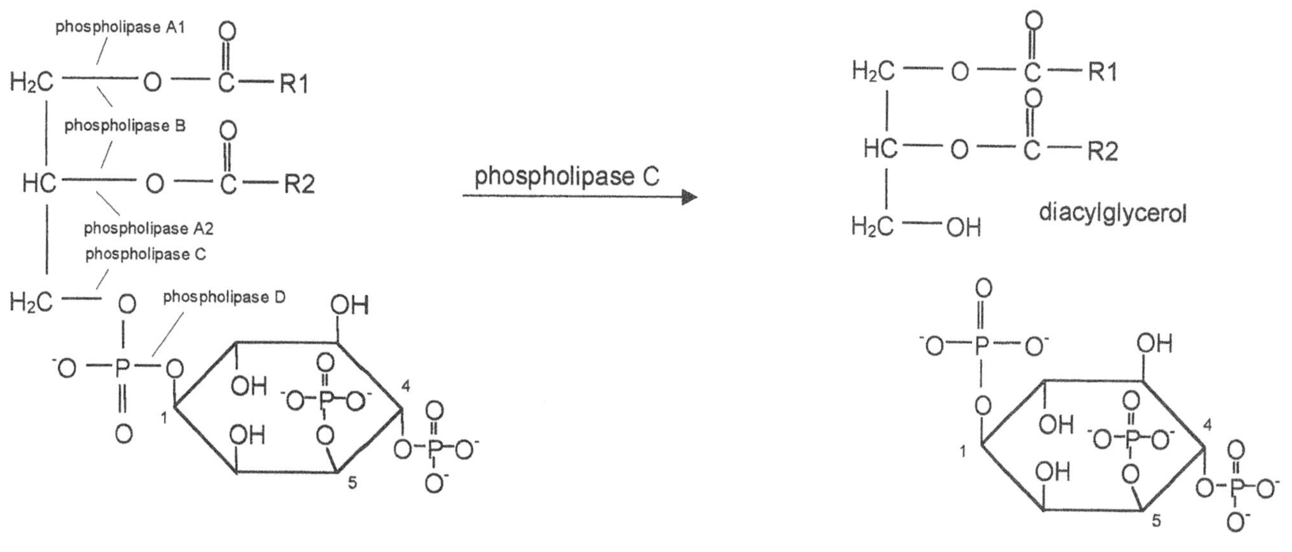

phosphatidylinositolbisphosphate

phospholipase C →

diacylglycerol

inositoltrisphosphate

always associate fats with overweight. However, by not being in water, they have the disadvantage of not being readily accessible. This is reflected by the fact that we prefer to use sugars and proteins as energy sources before we use fats.

The first thing to realise about using fats as an energy source is that fat tissue stores the fat and other tissues (muscle mainly) use it for energy. This creates an interesting problem: how does the fat get from the fat stores to the other tissues? Clearly it is by the blood, but how? The blood is mainly water and fats and water do not mix. If fats were simply released into the bloodstream they would all float and we would end up with a pool of fat on the tops of our brain - a fat head. Therefore, how are the fats transported to the tissues that need them? I will not answer this question at this stage, but allow you to invent your own solution to this problem.

When fats are needed, lipases are activated in lipocytes and these catalyse the separation (hydrolysis) of the fats into glycerol and fatty acids. These are taken by the bloodstream to the liver (glycerol and some fatty acids) and muscle (fatty acids). Energy is extracted from glycerol in a different way from fatty acids.

Glycerol metabolism

Glycerol is mainly metabolised by the liver where it is converted to dihydroxyacetone phosphate which enters the glycolytic or gluconeogenesis pathway.

Compare the structures of glycerol and dihydroxyacetone phosphate, and determine the differences in their structures.

Answer:

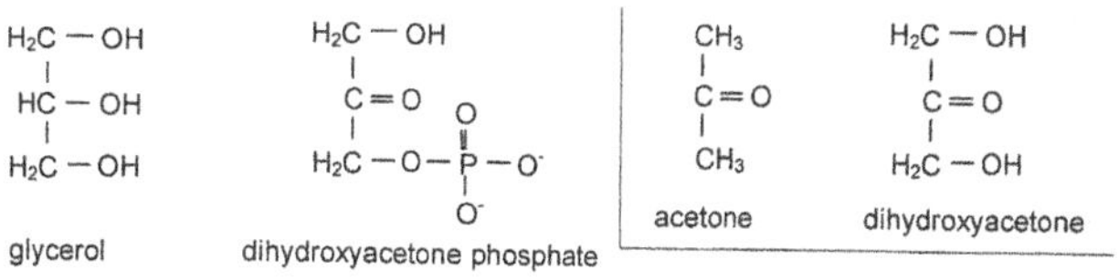

glycerol · dihydroxyacetone phosphate · acetone · dihydroxyacetone

Clearly, dihydroxyacetone phosphate differs from glycerol in that it is phosphorylated and it has lost hydrogen (dehydrogenated).

Therefore it becomes obvious that the metabolic pathway for the synthesis of dihydroxyacetone phosphate from glycerol involves a phosphorylation and a dehydrogenation. ATP is the donor of the phosphate and NAD^+ is the acceptor of the hydrogen.

> **Q&A 10:** The phosphorylation reaction occurs firstly to form glycerol-3-phosphate, and this is followed by the dehydrogenation. Remembering that enzymes that catalyse the phosphorylation of molecules are called kinases, draw the reaction and name the 2 enzymes involved.

Fatty acid degradation (oxidation)

Muscle takes up fatty acids into its cytoplasm from the bloodstream. Firstly, they are activated by formation of a thioester linkage between the fatty acid and CoA-SH. The product formed is an acyl-coenzyme A, and therefore the name of the enzyme which catalyses this reaction is acyl-CoA synthase. This reaction requires a great deal of energy and in the process ATP is converted to AMP.

Draw the esterification of a fatty acid to CoA. Remember what is removed when an ester bond is formed.

Answer

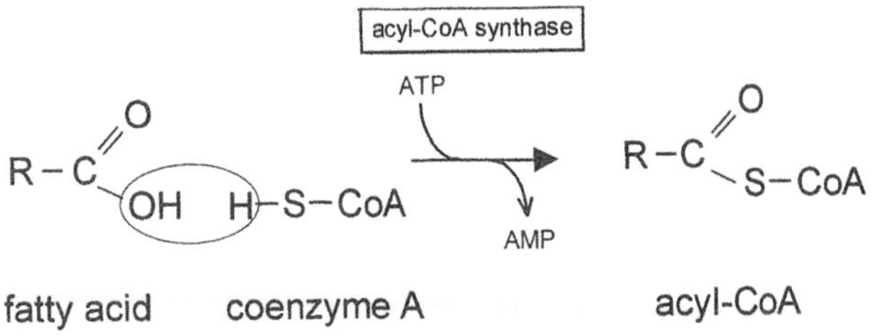

fatty acid   coenzyme A   acyl-CoA

The carnitine shuttle

Further breakdown of the acyl chain occurs in the mitochondrial matrix. This creates a problem because acyl-CoA, like acetyl-CoA, cannot cross the mitochondrial membrane. To overcome this problem, the acyl group is transferred to a carrier, carnitine, located in the mitochondrial membrane. The first part of this process occurs in the outer mitochondrial membrane and the acyl group is transferred to carnitine. The structure of carnitine is illustrated here for interest only.

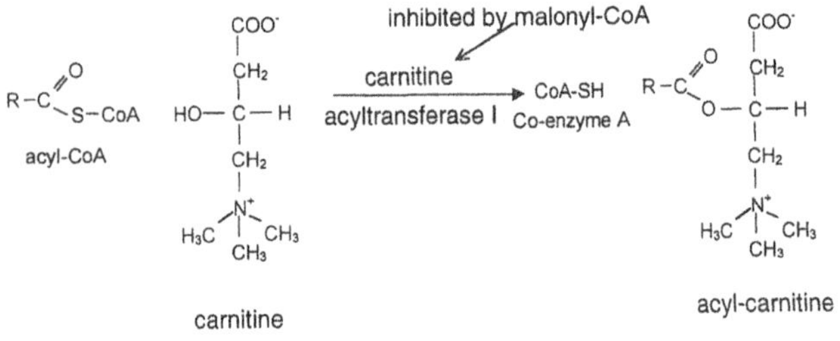

The enzyme which catalyses this reaction, carnitine acyltransferase I, is inhibited by malonyl-CoA. This is important because malonyl-CoA is a substrate for fatty acid synthesis. It does not make sense to have fatty acid synthesis occurring at the same time as fatty acid degradation and hence a substrate for fatty acid synthesis inhibits fatty acid degradation (in this case by preventing the acyl group being transferred to the mitochondrial matrix).

Acylcarnitine is then translocated across the inner mitochondrial membrane in exchange for carnitine. This is catalysed by acylcarnitine translocase. Finally, carnitine is exchanged for CoA in the inner mitochondrial matrix. This is catalysed by carnitine acyltransferase II

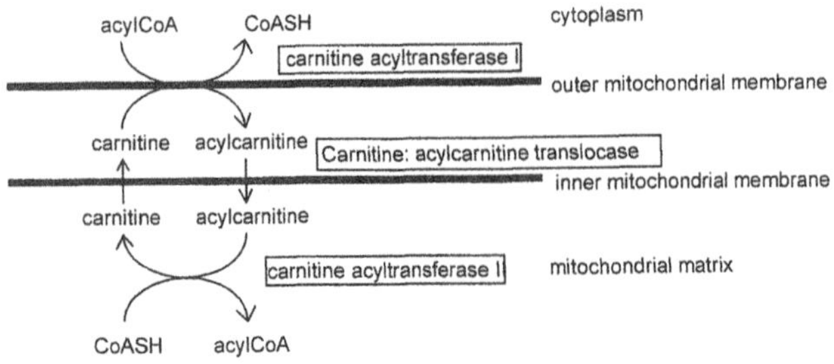

β-oxidation

Once in the mitochondrial matrix, the fatty acids (in the form of acyl-CoA) are oxidised at the β-carbon (β-oxidation) and then degraded into acetic acid and acyl-CoA with 2 less carbons. The acetic acid formed is not free but bound to CoA (acetyl CoA). This process is repeated until all of the carbons of the acyl chain have been degraded.

> **MAJOR POINT 55:** Oxidation of fatty acids occurs in the mitochondrial matrix.

The first reaction in the breakdown of fatty acids is an oxidation which occurs across the α and β carbons by a dehydrogenation, and a double bond is formed between these 2 carbons.

> **Q&A 11:** Recall the nomenclature of fatty acids and apply this to the fatty acid attached to CoA. In the diagrams below indicate the α and β carbons and bonds 1 and 2 of the fatty acid or the acyl group of acyl-CoA.

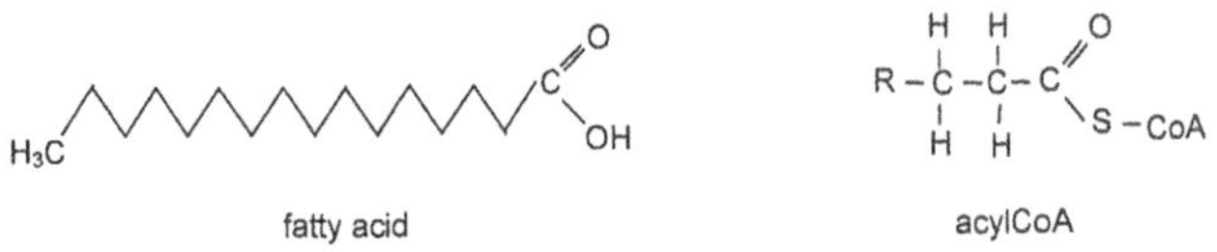

fatty acid   acylCoA

Oxidise an acyl-CoA by removing hydrogen across the α and β carbons to form a trans double bond between them.

Answer:

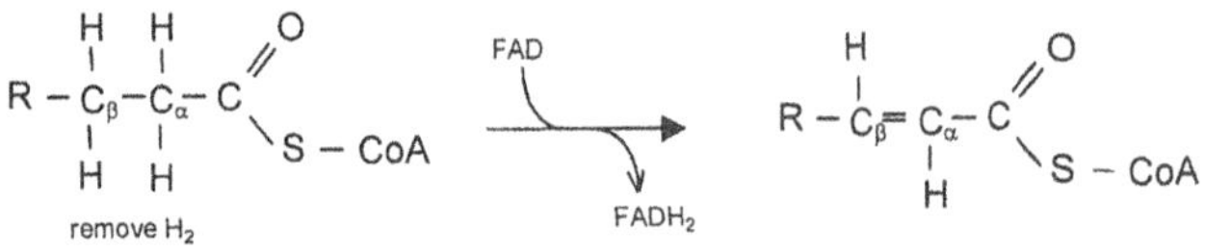

The energy obtained from this oxidation is captured by FAD. The $FADH_2$ formed is passed along the oxidative phosphorylation pathway to produce 2ATPs.

What would be appropriate names for the enzyme and the product? Remember for the product that it has a double bond (an ene) and in naming the molecule it is necessary to state which bond is the double bond and whether or not it is a trans or cis double bond.

Answer: *The name of the enzyme which catalyses this reaction is acyl-CoA dehydrogenase. In fact there are 3 different acyl-CoA dehydrogenases which show different specificities: short, medium and long chain fatty acids. The product is a trans-Δ2-enoyl-CoA.*

Water is then added across the double bond.

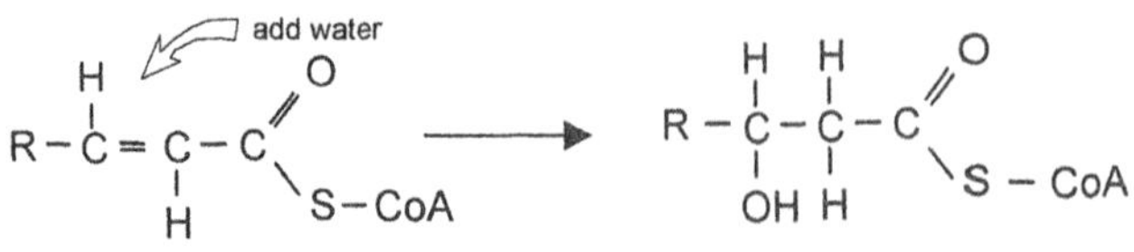

What is the name of the enzyme that catalyses the hydration of enoyl-CoA and what is the name of the product? Remember the product is now an acyl-CoA with an hydroxyl group, but you must state where that hydroxyl group is located.

Answer: *The name of the enzyme that catalyses this reaction is enoyl-CoA hydrase. There are 2 enoyl hydrases: one preferring short chain fatty acids; and the second long chain. The name of the product is L-3-hydroxy-acyl-CoA. Note that the hydroxyl group is drawn below the βC. This is important because in the synthesis of fatty acids a similar molecule is formed as an intermediate but it is in the D conformation.*

The next step is oxidation of the βC by removing hydrogen (H_2) from it to form a ketone at the βC.

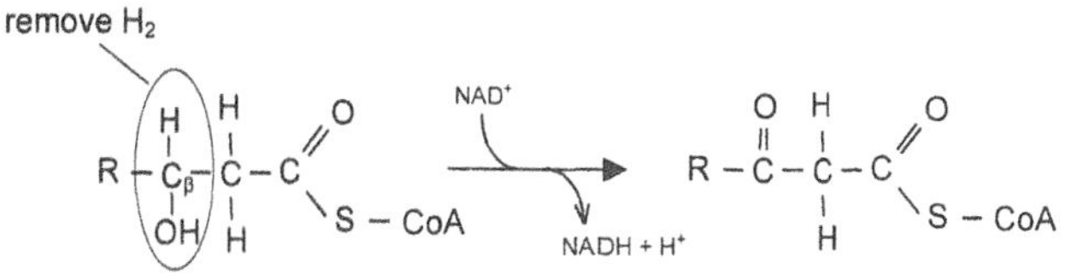

remove H_2

The energy obtained from this reaction is transferred to NAD^+ and then passed onto the oxidative phosphorylation pathway to form 3ATPs.

What is the name of the enzyme that catalyses this reaction and the name of the product?

Answer: *The enzyme is 3-hydroxy-acyl-CoA-dehydrogenase and the product is β-keto-acyl-CoA.*

β-keto-acyl-CoA is next cleaved into acetyl-CoA and a new acyl-CoA with 2 carbons less than previously by adding coenzyme A to bond 2.

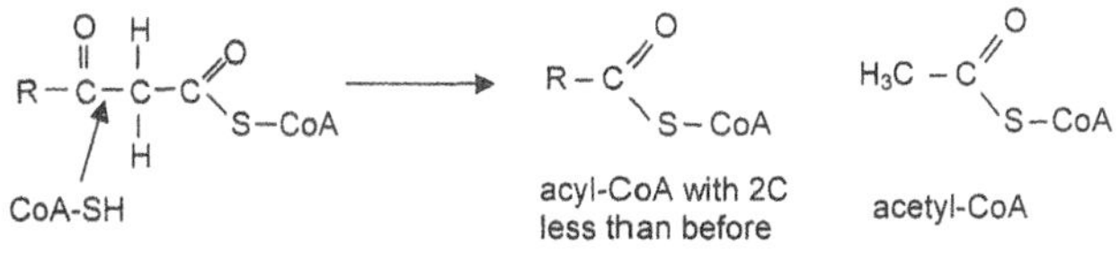

acyl-CoA with 2C less than before   acetyl-CoA

The enzyme that catalyses this reaction is thiolase. The acyl-CoA formed can proceed through the reaction again. This continues until butyryl-CoA is formed and this is converted to acetoacetyl-CoA, which is cleaved into 2 acetyl-CoAs.

MAJOR POINT 56: Each cycle of β-oxidation produces acetyl-CoA, acyl-CoA, $FADH_2$, and NADH.

The acetyl-CoA formed is combined with oxaloacetate to form citrate and then is oxidised through the TCA cycle. This

of course yields a further $3NADH^+$, $1FADH_2$ and 1ATP which becomes 12 ATPs once the $NADH^+$s and $FADH_2$s are subjected to oxidative phosphorylation. If the tissue is lacking in oxygen, oxidative phosphorylation does not occur and hence β-oxidation cannot occur.

MAJOR POINT 57: β-oxidation only occurs in the presence of oxygen, which means aerobic exercise is required to break down fats.

Note that the oxidation, hydration, oxidation seen in β-oxidation is identical to that seen in the transformation of succinate to oxaloacetate in the TCA cycle. This means that it only has to be learnt once and applied twice (see below).

Higher level

Unsaturated fatty acids are also oxidised in mitochondria. This occurs in the same way as saturated fatty acids until a double bond is approached. Remember that fats are usually in the *cis* form. This has to be converted to a *trans* form. If the double bond occurs at bond 3 rather than bond 2 of the acyl chain then an additional step must occur. The double bond must be 'shifted' from bond 3 to bond 2. This isomerisation is catalysed by enoyl-CoA isomerase. β-oxidation can then resume. Since 1 round of dehydrogenation is missed when a double bond is encountered, the conversion of FADH to $FADH_2$ does not occur and hence 2 fewer ATPs are made.

Ketone body formation

The liver has relatively low energy requirements compared with muscle and therefore it would make sense if some of the energy formed from β-oxidation could be transferred to other tissues such as muscle and brain. Ideally this would be in the form of glucose, but this is not the case. Once carbons are in the form of acetyl-CoA they cannot be converted to glucose and hence fatty acids cannot be converted to glucose.

MAJOR POINT 58: Glucose cannot be synthesised from fatty acids.

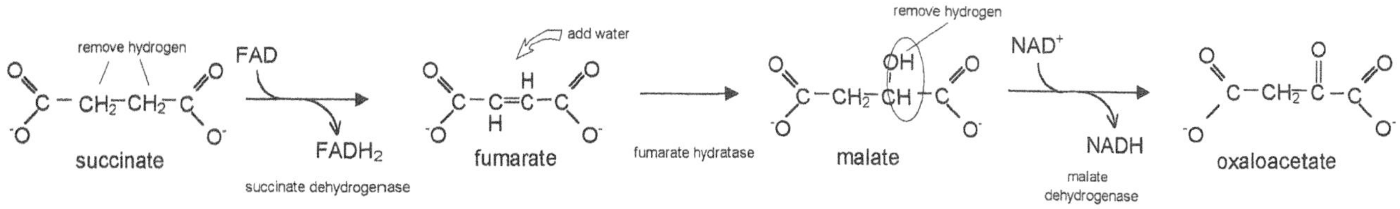

This means that fatty acids must be converted into another substance or energy form, other than glucose, in the liver so that this energy can be used by muscle and brain. To accomplish this, the acetyl-CoA formed from β-oxidation is converted to ketone bodies. These are acetoacetate, β-hydroxybutyrate and acetone and this process takes place in the mitochondrial matrix.

The first enzyme in the reaction, thiolase, would normally catalyse the cleavage of acetoacetyl-CoA into 2 acetyl-CoAs. However, when the levels of acetyl-CoA are high, the direction of the reaction is reversed and acetoacetyl-CoA is formed releasing CoA-SH.

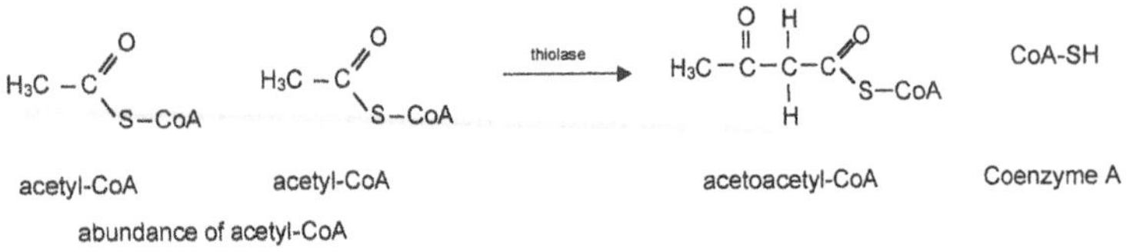

acetyl-CoA   acetyl-CoA   acetoacetyl-CoA   Coenzyme A

abundance of acetyl-CoA

A third molecule of acetyl-CoA is joined to the acetoacetyl-CoA to form 3-hydroxy-3-methylglutaryl-CoA (HMG-CoA). This reaction is catalysed by HMG-synthase. Water is also needed in this reaction to cleave CoA from acetyl-CoA.

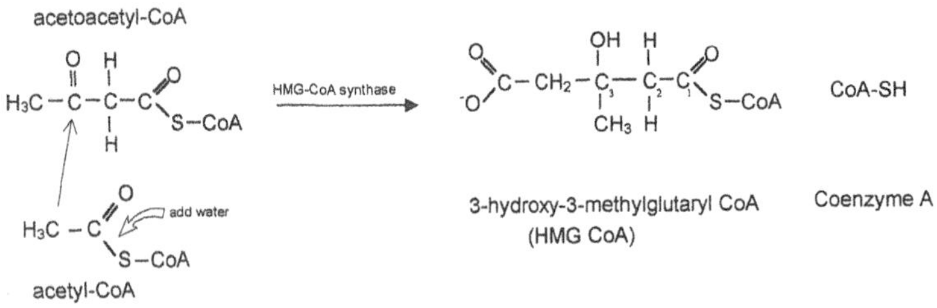

acetoacetyl-CoA

3-hydroxy-3-methylglutaryl CoA   Coenzyme A
(HMG CoA)

acetyl-CoA

Study tip

To help understand this reaction, it is important to realise that the C of the methyl group rather than the carboxyl group of acetyl-CoA is linked to the βC (or C3) of acetoacetyl-CoA. In the formation of acetoacetyl-CoA it was the carbon of the carboxyl group that was added. Their reactions parallel other reactions that occur in different pathways.

- The reactions described above occur in the mitochondrial matrix. The same reactions in the cytosol lead to the production of cholesterol (see below).

- The formation of citrate parallels the formation of HMG CoA. The carbon of the methyl group of acetyl-CoA is attached to the ketone group of oxaloacetate.

> **Nomenclature:** The term 'gluta' indicates 5 carbons, e.g. glutamic and glutaric acids, α-ketoglutarate. Hence above, we have glutaryl-CoA with an hydroxyl and a methyl group substituted at C3, and therefore its name 3-hydroxy-3-methylglutaryl-CoA). It has an alternative name, β-hydroxy-β-methylglutaryl CoA, because C3 is also the βC.

The next reaction is the cleavage of the carbon chain to form acetoacetate and release acetyl-CoA. This is catalysed by HMG-CoA lyase. Acetoacetate can be processed further by reducing it to β-hydroxybutyrate. The enzyme that catalyses this reaction, β-hydroxybutyrate dehydrogenase, is located on the inner mitochondrial membrane. Acetoacetate can also be decarboxylated to form acetone but this is normally of minor importance.

Acetoacetate, β-hydroxybutyrate and acetone are collectively called ketone bodies. These are mostly synthesised in the liver because only in the liver is mitochondrial HMG-CoA synthase in large quantities. In diabetes melitis, there is a lack of glucose for use in the cells so fatty acids are used for energy. This leads to the liver making ketones and hence the blood has abnormally high ketone levels (ketosis).

Main points of fatty acid degradation

β oxidation is the process of converting acyl-CoA to several acetyl-CoAs. Firstly there is an oxidation (by removal of hydrogen) of the acyl

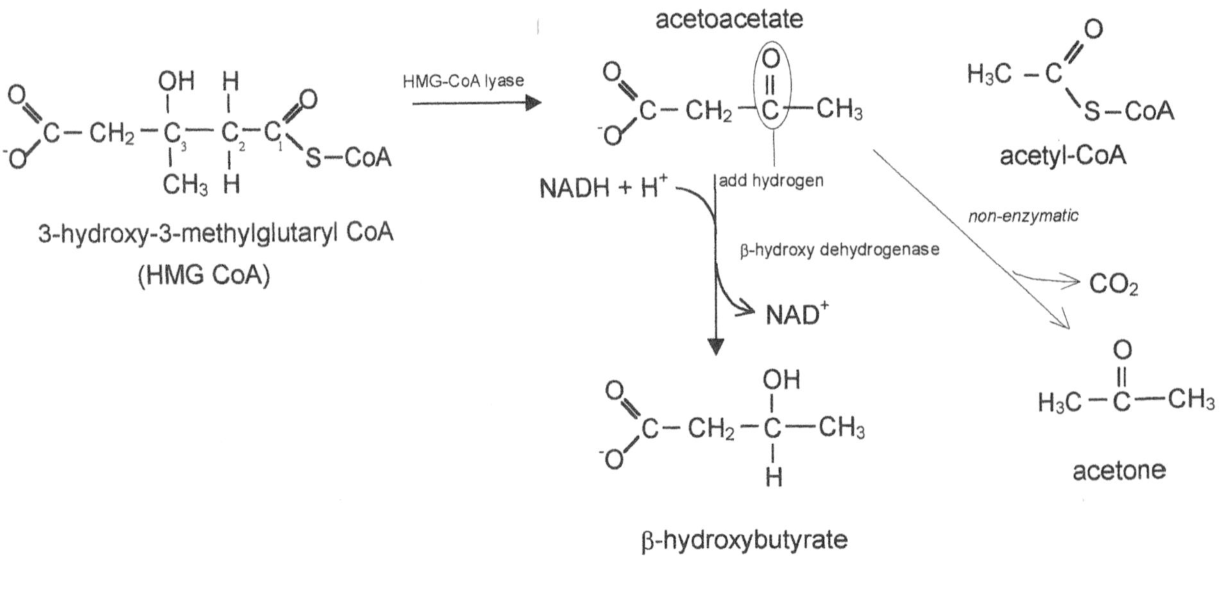

3-hydroxy-3-methylglutaryl CoA
(HMG CoA)

acetoacetate

acetyl-CoA

CO_2

acetone

β-hydroxybutyrate

group, followed by an hydration (add water) and a second oxidation (removal of hydrogen) at the β carbon. There are other enzymes to break the double bonds in polyunsaturated fatty acids. The final step is cleavage of acetyl-CoA – a thiolysis.

The acetyl-CoA and the immediate products of the oxidation the reduced co-enzymes $FADH_2$ and NADH, supply energy through the TCA cycle and the respiratory chain. If excessive fatty acid degradation takes place then one has the formation of ketone bodies in order to generate CoA.

> **MAJOR POINT 59:** Breakdown of fats is different from breakdown of fatty acids because breakdown of fats includes the processing of the glycerol backbone as well as the oxidation of fatty acids.

Fatty acid synthesis

Fatty acids arise from diet and *de novo* synthesis. In fact, the body prefers to use fatty acids from the diet rather than make them because to make them uses an enormous amount of energy. The effect of using fatty acids from the food we eat, rather than synthesising them, is that we begin to smell like the animals we eat. The fatty acids in fish (fish oils) smell like fish, those in cows smell like cows.

In this section, we will only be dealing with synthesis of fatty acids. In general, they are synthesised from carbohydrates and only to a limited degree from amino acids. Lipocytes (fat cells) are the major cells in the body which carry out this function and therefore they are specialised for synthesis and storage of triacylglycerols.

Fatty acids (with few exceptions) have multiples of 2C atoms and hence it is hardly surprising to learn that they are synthesised from successive addition of 2C units. The elongation of the fatty acid chain stops upon formation of palmitate

(16C). Further elongation and insertion of double bonds are carried out by other enzyme systems.

In terms of nutritional state, when do you think we would synthesise fats?

Answer: *The signal for fat synthesis is a high nutritional state and therefore high levels of proteins and glucose indicate fatty acid synthesis should start.*

However, glucose may be needed for energy, and proteins may be needed for building tissue, so other signals are also required. These signals develop if the TCA cycle is loaded: high levels of pyruvate and acetyl-CoA in mitochondria.

Fatty acid synthesis occurs in the cytoplasm of cells, but when the body is in energy excess, the high levels of acetyl-CoA are in the mitochondrial matrix. Therefore, the first problem is to transport the acetyl-CoA from the mitochondria into the cytoplasm. This is difficult because the mitochondrial membrane is not permeable to CoA.

> **MAJOR POINT 60:** Coenzyme A, and hence acetyl-CoA, cannot cross the mitochondrial membrane. Therefore there are 2 independent pools of Coenzyme A: a cytoplasmic; and a mitochondrial pool.

The citrate shuttle

What is really required is that the acetyl group of acetyl-CoA is transferred out of the mitochondria rather than acetyl-CoA. To accomplish this the acetyl group is of mitochondrial acetyl-CoA is firstly transferred to oxaloacetate to form citrate and citrate is transported across the mitochondrial membrane. In the cytosol it is then converted back to pyruvate and acetyl-CoA. This is called the citrate transport system.

Despite its frightening appearance, it is really worthwhile committing this pathway to memory. Note that the part occurring in the mitochondrial matrix is the first part of the TCA cycle and therefore it links these 2 pathways together. The key to learning the citrate transport system is to know the structure of 2 amino acids – aspartate (**D**) and alanine (**A**).

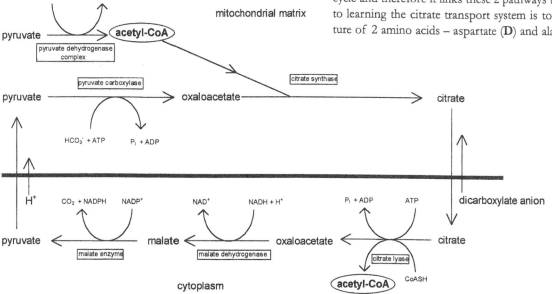

*Draw the structures of alanine (**A**) and aspartate (**D**) and then deaminate them and add a ketone.*

Answer:

$$H_3C-\underset{\underset{COOH}{|}}{\overset{\overset{NH_2}{|}}{CH}} \qquad \underset{O^-}{\overset{O}{\parallel}}C-CH_2-\underset{\underset{COO^-}{|}}{\overset{\overset{NH_3^+}{|}}{CH}}$$

alanine   aspartate

$$H_3C-\overset{\overset{O}{\parallel}}{\underset{\underset{COOH}{}}{C}} \qquad \underset{O^-}{\overset{O}{\parallel}}C-CH_2-\overset{\overset{O}{\parallel}}{\underset{\underset{COOH}{}}{C}}$$

What is the name of the keto acids formed?

Answer: *These are pyruvate and oxaloacetate. If there is an excess of proteins, then there would be an excess of the amino acids alanine (**A**) and aspartate (**D**), which can be converted to oxaloacetate and pyruvate and then to fats via acetyl-CoA.*

If there is also an excess of glucose, then excess pyruvate will be formed from glycolysis. Its transfer into the mitochondria is catalysed by pyruvate translocase. It would make sense in dietary excess of glucose if half of the pyruvate in the mitochondria could be converted to oxaloacetate instead of acetyl-CoA. This would enable citrate to form and promote fatty acid synthesis. Indeed, this is what happens.

To understand the conversion of pyruvate to oxaloacetate compare their structures and circle the difference.

Answer:

$$H_3C-\overset{\overset{O}{\parallel}}{\underset{\underset{COOH}{}}{C}} \qquad \underset{O^-}{\overset{O}{\parallel}}C-CH_2-\overset{\overset{O}{\parallel}}{\underset{\underset{COOH}{}}{C}}$$

pyruvate   oxaloacetate

Oxaloacetate has an additional CO_2. Therefore to convert pyruvate to oxaloacetate, pyruvate is carboxylated. This is catalysed by pyruvate carboxylase and the reaction requires en-

ergy, which comes from the conversion of ATP to ADP. Remember that this is not a problem because there is excess ATP available when the body needs to store energy.

Other pyruvate molecules are converted to acetyl-CoA in a reaction catalysed by pyruvate dehydrogenase complex. This is exactly the same as for the TCA cycle. Carbon dioxide and energy are released and the energy is captured by NAD^+ which becomes NADH.

Acetyl-CoA is combined with oxaloacetate to form citrate and CoA is regenerated. This is also exactly the same as for the TCA cycle. Remember that it is C of the methyl group of acetyl-CoA that attaches to the C of the ketone group of oxaloacetate to form citrate. The oxygen of the ketone group acquires a hydrogen in the process and therefore is transformed into an hydroxyl group.

The citrate formed has a choice between progressing further in the TCA cycle or diffusing across the mitochondrial membrane in exchange for a dicarboxylate anion (this keeps the charge difference across the mitochondrial membrane the same). Remember from the TCA cycle that isocitrate dehydrogenase, which catalyses the conversion of isocitrate to α-ketoglutarate, is inhibited by NADH and activated by ADP. Therefore in a high energy state when NADH would be high and ADP low, the pathway for the progression of citrate into the TCA cycle is blocked and hence it builds up in the mitochondrial matrix. This excess of citrate is then passed into the cytoplasm.

In the cytoplasm, the reverse of the previous reaction occurs. Coenzyme A is supplied and citrate is cleaved into acetyl-CoA and oxaloacetate. This process is catalysed by citrate lyase and requires energy which is supplied by the conversion of ATP to ADP. At this stage we have effectively transferred acetyl-CoA from the mitochondrial matrix to the cytoplasm. Oxaloacetate is converted to pyru-

What molecule is formed when oxaloacetate is reduced at the carbon with the ketone? Recall the TCA cycle in reverse?

Answer:

vate to be cycled back into the mitochondrial matrix. Oxaloacetate is firstly reduced by adding hydrogen to the carbon with the ketone.

The reduction of oxaloacetate to malate is catalysed by malate dehydrogenase and uses NADH as the proton donor. Malate is then converted to pyruvate by removing hydrogen and CO_2. The enzyme that catalyses this reaction has the simple name malic enzyme. The hydrogen is captured by $NADP^+$. Pyruvate is then translocated across the mitochondrial membrane to be cycled again.

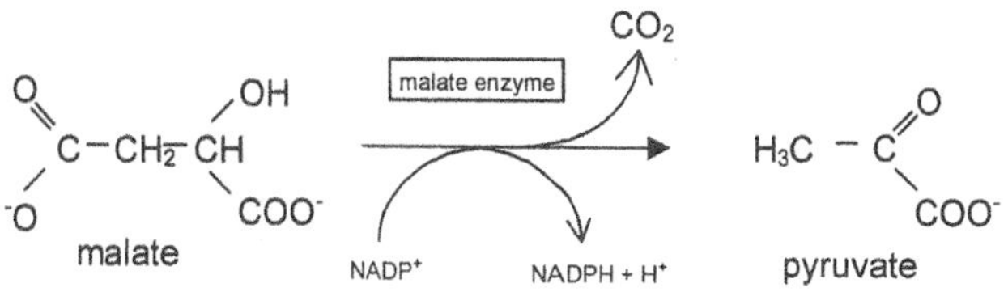

Conversion of acetyl-CoA into fatty acids

Once acetyl-CoA has been transferred to the cytoplasm, the first step in fatty acid synthesis is the conversion of acetyl-CoA to malonyl-CoA by the addition of carbon dioxide to acetyl-CoA. This reaction is catalysed by acetyl-CoA carboxylase. This enzyme has similar properties to pyruvate carboxylase, which is important in the gluconeogenesis pathway. It needs biotin as a cofactor and it is a control point in fatty acid biosynthesis.

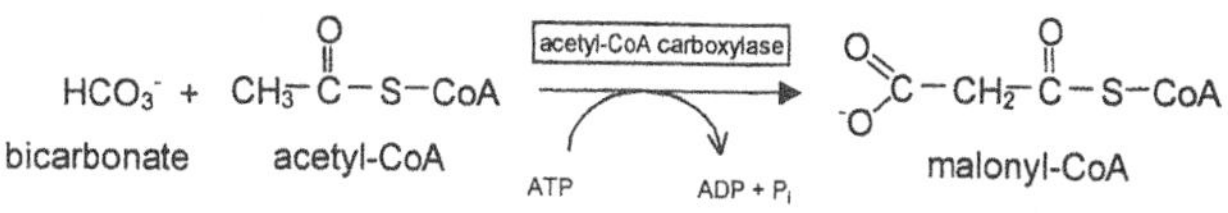

Higher level

In animal tissues acetyl-CoA carboxylase exists as a dimer of 2 functional units, but in the dimeric form has low activity. Incubation with citrate leads to activation of the enzyme, and the dimer aggregates to a polymer. The enzyme is deactivated and depolymerised by the product malonyl-CoA and by the end-product of fatty acid synthesis, palmitoyl CoA.

In mammals, the next steps take place on a protein that has multiple enzymatic sites in which subsequent reactions take place. Such proteins are called multi-enzyme complexes. This particular multienzyme complex is called fatty acid synthase and has a variety of domains, each of which carries out the function of an enzyme. In bacteria, individual enzymes carry out the function of the different domains of fatty acid synthase. Functional fatty acid synthase comprises 2 identical protein subunits.

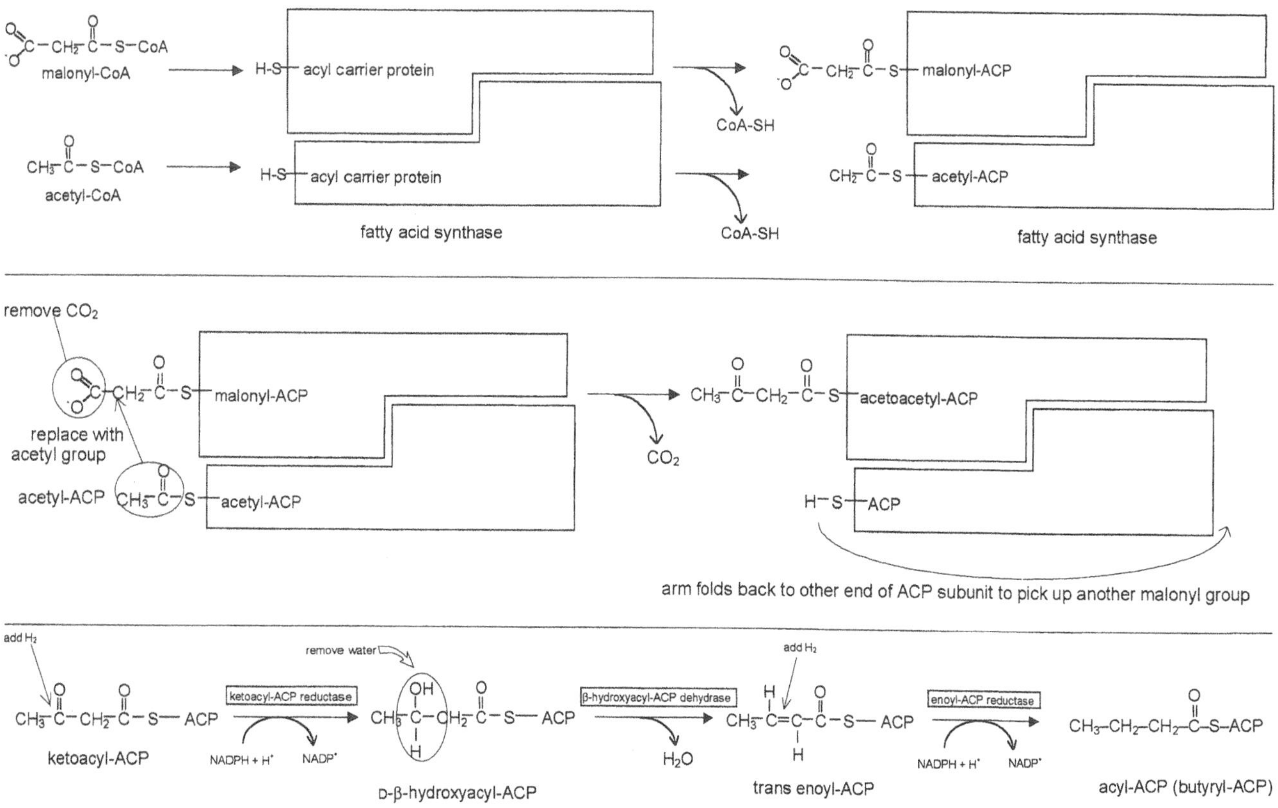

One domain of each subunit carries an acyl group (the fatty acid chain) and hence this domain is called acyl carrier protein. In the very first round of fatty acid synthesis, one of the acyl carrier proteins is loaded with an acetyl group and the other with a malonyl group via thioester bonds. The attachment site is a long arm comprising phosphopantetheine group phosphorylated to a serine. This is identical to the phosphopantetheine group of coenzyme A. These are illustrated below for interest only.

phosphopantetheine group of acyl carrier protein

$$HS-CH_2-CH_2-\overset{O}{\underset{H}{\overset{||}{N}}}-C-CH_2-CH_2-\overset{O}{\underset{H}{\overset{||}{N}}}-C-\overset{CH_3}{\underset{OH}{\overset{|}{CH}}}-\overset{CH_3}{\underset{CH_3}{\overset{|}{C}}}-CH_2-O-\overset{O}{\underset{O^-}{\overset{||}{P}}}-O-serine$$

$$HS-CH_2-CH_2-\overset{O}{\underset{H}{\overset{||}{N}}}-C-CH_2-CH_2-\overset{O}{\underset{H}{\overset{||}{N}}}-C-\overset{CH_3}{\underset{OH}{\overset{|}{CH}}}-\overset{CH_3}{\underset{CH_3}{\overset{|}{C}}}-CH_2-O-\overset{O}{\underset{O^-}{\overset{||}{P}}}-O-\text{attached to the 5' phosphate of 3',5'-adenosine bisphosphate}$$

phosphopantetheine group of coenzyme A

The enzyme that catalyses the transfer of the malonyl group is malonyl-CoA-ACP transacylase and the enzyme that catalyses the transfer of the acetyl group is acetyl-CoA-ACP transacylase (if the fatty acids to be synthesised have an odd number of carbons then proprionyl-CoA is attached to ACP instead of acetyl-CoA).

In the next step, CO_2 is removed from the malonyl group of malonyl-ACP and replaced by the acetyl group of acetyl-ACP. The decarboxylation provides the energy for this reaction. The enzyme that catalyses this reaction is one of the domains of the fatty acid synthase called ketoacyl-ACP synthase which makes sense since it catalyses the synthesis of a ketoacyl-ACP.

The next steps in fatty acid synthesis resemble the reverse of fatty acid degradation: a reduction at the βC, a dehydration to form a *trans* double bond between the αC and the βC and a second reduction across the double bond to saturate it. This leads to the formation of the 4C butyryl chain.

When looking at this sequence of reactions, think about the structures and the enzymes and how the names describe the structures and the enzyme activities that catalyse the reactions. The ketoacyl-ACP is in fact acetoacetyl-ACP. It is then reduced (ketoacyl-ACP reductase). To describe the hydroxyacyl ACP formed, we need to indicate which C is hydroxylated and you will remember from fatty acid nomenclature that it is C3 or the βC. Hence it becomes β-hydroxyacyl-ACP. Note that the D-epimer is formed, D-β-hydroxyacyl-ACP, which differs from fatty acid degradation in which the L-epimer forms. Water is then removed from the βC (β-hydroxy-ACP dehydrase) to form a double bond between the αC and βC. The double bond is *trans* and the term 'ene' indicates a double bond. Hence the product is a *trans* enoyl-ACP (*trans* butenoyl-ACP). This is reduced across the double bond (*trans* enoyl-ACP reductase) to form an acyl-ACP (butyryl-ACP). An important point to note is that the supplier of hydrogen in the 2 reduction steps is NADPH and not NADH (see bottom of previous page).

> **MAJOR POINT 61:** In general terms NADP is associated with synthetic pathways and NAD is associated with degradative pathways.

All of these enzymatic steps are carried out in different domains of fatty acid synthase with the flexible arm of the phosphopantetheine group moving the acyl group from one active site to the next. This has the advantage of limiting dif-

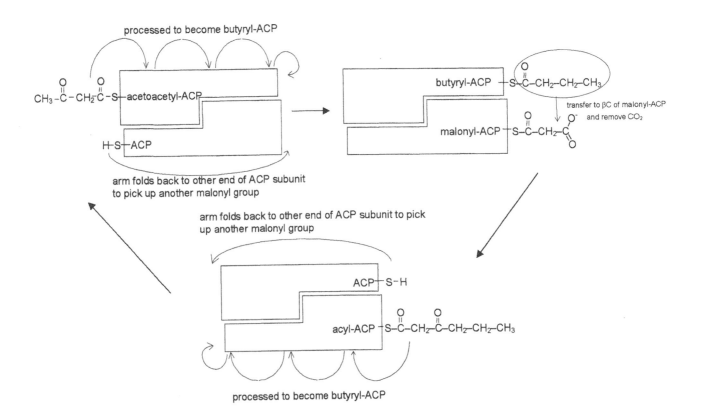

fusion of substrates and the intermediates are protected from competing reactions. When it gets to the last part of the sequence, the new acyl group (here butyrate) is passed to the ACP of the second part of the dimer. This ACP of the second part of the dimer has now been reloaded with malonate. CO_2 is released and the cycle starts again but the acyl group is now 2C longer.

Note that pairs of carbons from the malonyl group are added at the carboxyl (C1) end and <u>not</u> the methyl (ωC) end of the growing chain. The cycling continues until the acyl group is 16C long (palmitoyl-ACP) and this is not a substrate for the condensing enzyme, ketoacyl-ACP synthase. It is, however, a substrate for thiolase, which catalyses the hydrolysis (add water) of palmitoyl-ACP to release palmitate and regenerate HS-ACP.

The stoichiometry for the synthesis of palmitate is

$$8\text{acetyl-CoA} +7\text{ATP}+14\text{NADPH} \rightarrow \text{palmitate} +14\text{NADP}$$
$$+8\text{CoA} + 6H_2O + 7\text{ADP} + 7P_i + 7H^+$$

Higher level

Mono and polyunsaturated fatty acids

Biosynthesis of monounsaturated fatty acids can occur either by an anaerobic or aerobic pathway. The aerobic pathway is in the liver and the anaerobic pathway is used by bacteria. In the endoplasmic reticulum of liver, cells have desaturases, which are enzymes specific for catalysing the desaturation of particular bonds. They use NAD as the hydrogen acceptor.

Biosynthesis of polyunsaturated fatty acids occurs mainly in eurkaryotes. Mammals do not have the enzymes for desaturation between C9 and the end of the acyl chain. Plants can desaturate at bonds 12 and 15 and thus mammals have a dietary requirement for linoleic acid (18:2,9,12) and linolenic acid (18:3,9,12,15). These are essential components of phospholipids and eiconosoids. In animal tissues, there are enzymes that catalyse the desaturation at the 5 position if the molecule is desaturated at the 8 position and the 6 position if the molecule is desaturated at the 9 position.

Q&A 12: Compare the similarities and differences between fatty acid synthesis and fatty acid degradation. This is a favourite exam question.

Cholesterol

Despite popular belief, cholesterol is not a poison, but a very necessary part of our cell membranes and the basis of several hormones such as the androgens and oestrogens. Cholesterol is only a problem if it is in excess and in

this respect we do not need cholesterol in our diets because we can synthesise it.

Structure

Cholesterol differs considerably in structure from the fatty acids. It is a complex ring structure. By learning its structure and the nomenclature associated with the ring, understanding steroid hormone biochemistry is very much easier. The basis of its structure is a perhydrocyclopentanophenanthrene (*perhydro-cyclo-pentaino-fenanthreen*) ring. What a mouthful. This is a saturated (perhydro) phenanthrene ring with a cyclopentane attached.

perhydrocyclopentanophenanthrene

phenanthrene

The 4 rings that make up perhydrocyclopentanophenanthrene are systematically named alphabetically from left to right. Cholesterol is a modification of this ring with an hydroxyl (OH) group at C3, methyl groups at C10 and C13, a Δ^5 double bond and an 8C branched alkyl group attached to C17. This makes a total of 27C.

To learn the structure attach 2 benzene rings side to side then add a third benzene ring to the top right hand side. Finally attach a pentane house to the side of the third benzene ring. Each of the 'valleys' at the top of the molecule has a methyl group attached to it and the top of the house has a TV antenna in the shape of a YVY coming from it. An hydroxyl group is attached to the bottom left of the molecule. There is a double bond at the bottom left of the B ring Δ^5. The numbering of the carbons takes a little practice, but it is surprising how quickly it is learnt if you practise it a couple of times.

attach third benzene ring   attach house   attach methyl groups to valleys

attach TV antenna to house YVY

attach OH to bottom left and double bond to bottom left of ring B   number ring

The real beauty of cholesterol cannot be appreciated by a flat stick model like this one. Best of all is to make the structure yourself with a modeling kit. The point is that the rings are not flat but have bends in them. In the diagram above, the methyl groups at C10 and C13, and the hydroxyl group at C3 are drawn with solid lines. This means that they lie above the plane of the ring. Like sugars, above the plane of the ring is a β conformation, e.g. there is a β-OH attached at C3. Note that the H at C5 is drawn with a broken line indicating that it sits below the plane of the ring. This is important as it means that for cholesterol, rings A and B are joined in the *trans* configuration. In a closely related molecule, coprostanol (a bile acid found in faeces) they are in the *cis* configuration, which gives a very different spatially organised molecule.

Cholesterol is an essential part of cell membranes in eukaryotes. The ring structures are lipid soluble and the hydroxyl group at C3 is hydrophilic. The rings make it a much more rigid structure than the fatty acids and it is believed to alter the stiffness of the membrane and its permeability. Prokaryotes and plants do not contain cholesterol, but do have similar molecules to cholesterol in their membranes. In plants, these are stigmasterol and β-sitosterol which differ only in the alkyl group attached to C17.

stigmasterol   β-sitosterol

Biosynthesis of cholesterol

Complex ring structures are difficult to synthesise. Cholesterol is actually synthesised from a simple molecule, acetyl-CoA. An outline of cholesterol synthesis is

- 2 acetyl-CoAs are converted to β-hydroxy-β-methyl-glutaryl-CoA (HMG-CoA)
- another acetyl-CoA is attached to HMG-CoA to form mevalonate (6C)
- mevalonate is converted into the branched 5C isoprenoid unit, isopentyl pyrophosphate by phosphorylation and decarboxylation
- isopentenyl pyrophosphates are joined to form the 15C compound farnesyl pyrophosphate

- 2 farnesyl pyrophosphates are joined to form the 30C squalene
- squalene is cyclised to lanosterol
- 3 methyl groups are removed and various other rearrangements are made to form cholesterol (27C)

The key thoughts are:

acetate →mevalonate →isopentenyl pyrophosphate→squalene→cholesterol

| C_2 | C_6 | C_5 | C_{30} | C_{27} |

You need to know the basics of cholesterol synthesis as given above, but the details are not too difficult although they border on higher level.

The first part of the reaction sequence is to form β-hydroxy-β-methylglutaryl-CoA by combining 3 acetyl-CoAs. These reactions take place in the cytoplasm (not in the mitochondria).

Firstly, 2 acetyl-CoAs are joined with the acetyl group attached to the methyl group.

What would be the name of the product given that a 4C acid is butyric acid and there is a ketone at the βC? What would be the name of the enzyme?

Answer: *The name of the product is β-ketobutyryl-CoA (also acetoacetyl-CoA) and the enzyme which catalyses this reaction is β-ketobutyryl-CoA synthase.*

A third acetyl-CoA is added to the βC of β-ketobutyryl-CoA to form HMG-CoA and the enzyme that catalyses this reaction is logically HMG-CoA synthase.

CoA-SH is then removed from HMG-CoA to form mevalonate. Compare the structures of HMG-CoA and mevalonate and note the difference to determine what has been added in the reaction.

β-hydroxy-β-methylglutaryl-CoA   mevalonate

Clearly, 4H are required to complete this reaction. These come from 2NADPH + 2H$^+$. Therefore, it is a reduction of HMG-CoA.

> **Q&A 13:** What is the name of the enzyme which catalyses the conversion of HMG-CoA to mevalonate?

This reaction is critical in controlling the synthesis of cholesterol. The cell has to be in a high energy state to supply NADPH and the activity of HMG-CoA reductase is controlled indirectly by the hormone glucagon, which increases its activity and the hormone insulin, which decreases its activity. Glucagon activates HMG-CoA reductase kinase kinase. Remember a kinase adds a phosphate

mevalonate   5-phosphomevalonate

5-pyrophosphomevalonate   3-phospho-5-pyrophosphomevalonate

group (large negative charge) to a protein. Once activated HMG-CoA reductase kinase kinase phosphorylates HMG-CoA reductase kinase. HMG-CoA reductase kinase is activated by this phosphorylation and in turn phosphorylates HMG-CoA reductase. This is activated and catalyses the conversion of HMG-CoA to mevalonate. Insulin inhibits these activations and phosphorylations. Cholesterol inhibits the transcription of HMG-CoA reductase and increases its breakdown.

Note that another fate for HMG-CoA is to be cleaved to acetyl CoA and acetoacetate to form ketone bodies. This is by a different enzyme which is located in the mitochondria of liver cells.

Mevalonate is then phosphorylated thrice. The phosphate group in each case comes from ATP and hence 3 ATPs are converted to ADPs in this series of reactions.

Remembering that an enzyme which catalyses a phosphorylation is called a kinase, what would be the names of the enzymes involved?

Answer: *Mevalonate kinase catalyses the phosphorylation of mevalonate to form 5-phosphomevalonate, and phosphomevalonate kinase catalyses the phosphorylation of 5-phosphomevalonate to form 5-pyrophosphomevalonate. Pyrophosphomevalonate kinase catalyses the phosphorylation of 5-pyrophosphomevalonate to form 3-phospho-5-pyrophosphomevalonate.*

The carboxyl group of 3-phospho-5-pyrophospho-mevalonate is then removed. Logically the enzyme that catalyses this reaction is 3-phospho-5-pyrophosphomevalonate decarboxylase and the product is isopentenyl pyrophosphate.

3-phospho-5-pyrophosphomevalonate   isopentenyl pyrophosphate

> Nomenclature: Isopentenyl pyrophosphate describes its structure. It has 5 carbons which are in a branched carbon chain (*isopent*). There is a double bond (*ene*) and there is a pyrophosphate group (2 phosphates) attached to the chain.

An isomer of isopentenyl pyrophosphate is readily formed by shifting the double bond. This forms dimethylallyl pyrophosphate.

isopentenyl pyrophosphate   dimethylallyl pyrophosphate

These 2 isomers are then fused to form the 10C geranyl pyrophosphate and to this another isopentenyl pyrophosphate is fused to produce farnesyl pyrophosphate.

dimethylallyl pyrophosphate   isopentenyl pyrophosphate   geranyl pyrophosphate

geranyl pyrophosphate   isopentenyl pyrophosphate   farnesyl pyrophosphate 15C

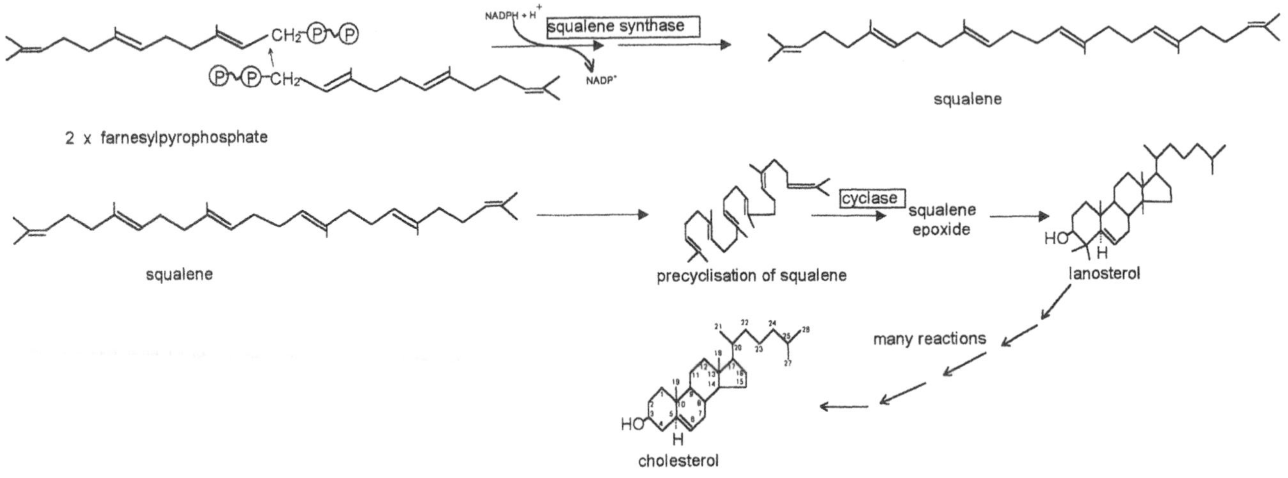

2 x farnesylpyrophosphate

squalene

squalene

precyclisation of squalene

squalene epoxide

lanosterol

many reactions

cholesterol

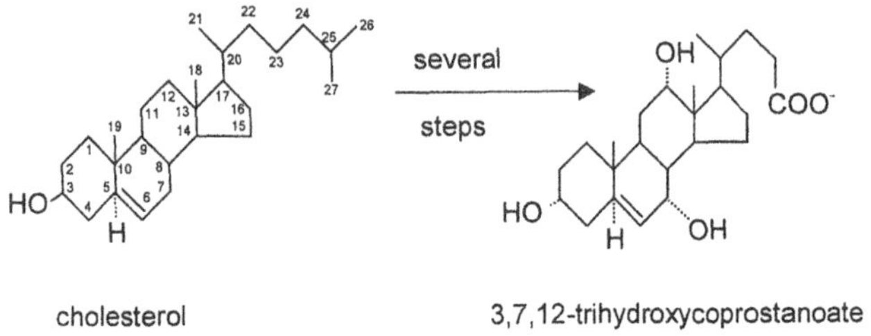

A 30C chain, squalene, is then formed by fusing 2 farnesyl pyrophosphates head to head. This is then cyclised to form an intermediate squalene epoxide and then lanosterol. NADPH and oxygen are needed to produce squalene epoxide. Lanosterol is then modified through a series of reactions (about 20) to form cholesterol.

Bile salts

These are polar derivatives of cholesterol and are detergents as they contain polar and non-polar regions. They are synthesised in the liver and stored in the gall bladder. They solubilise dietary lipids so that they can be broken down (hydrolysed) and absorbed. Bile salts are also the major breakdown products of cholesterol.

Through a series of steps, cholesterol is made more polar by the addition of 2 more hydroxyl groups to the ring and the alkyl chain becomes carboxylated. Note that the additional hydroxyls and the existing one all become below the plain of the ring, i.e. α-conformation. The structure formed is trihydroxycoprostanate. The amino acid glycine is then added to the carboxyl group via an amide bond to form glycocholate which is the major bile salt.

> **Q&A 14:** Add glycine to the carboxyl group of trihydroxycoprostanate by forming an amide bond. This forms glycocholate which is the major bile salt.

cholesterol

3,7,12-trihydroxycoprostanoate

3,7,12-trihydroxyprostanoate

glycine

form an amide bond

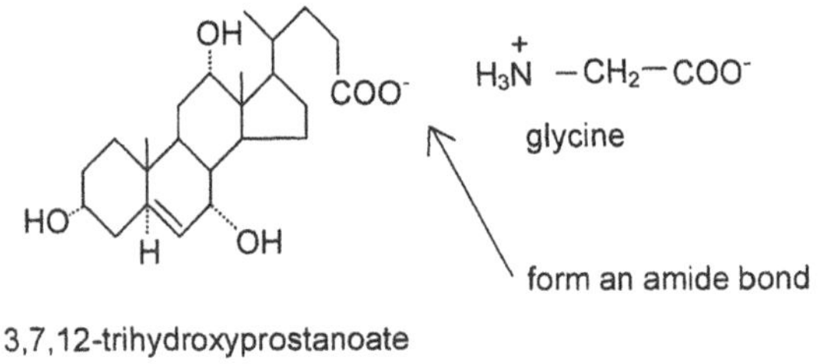

Steroid synthesis

Steroids are formed in a variety of tissues (adrenal cortex and gonads) from cholesterol esters which are hydrolysed and transported into the mitochondrial matrix. Here there is modification of the acyl group to form a ketone and the hydroxyl group at C3 is dehydrogenated to form a ketone as well.

Further modification of progesterone leads to the formation of androgens (19C) and oestrogens, which have 18C.

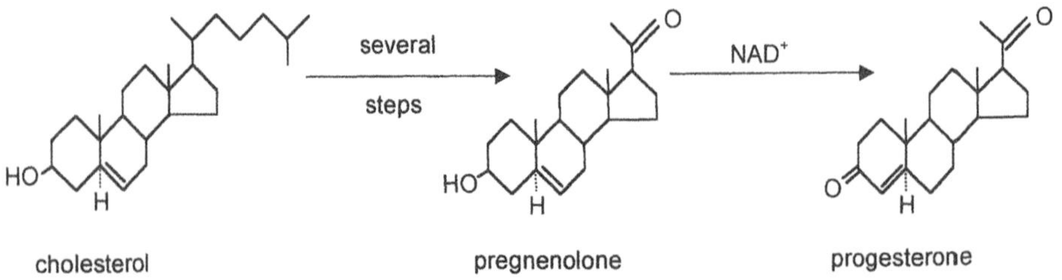

cholesterol

several steps

pregnenolone

NAD^+

progesterone

progesterone

several steps

androstenedione

aromatase

estrone

reductase

reductase

dihydrotestosterone

testosterone

aromatase

estradiol

Lipoproteins and lipid transport

The problem with transporting lipids around in the bloodstream is that the blood is mainly water and lipids and water do not mix. To overcome this problem, the lipids are embedded in proteins. The proteins are constructed so that they have regions of amino acids that are hydrophobic and regions that are hydrophilic. In addition to specialised proteins, albumin also carries lipids in the bloodstream. The liver and intestinal epithelial cells synthesise these proteins. When mixed with lipids, the whole structure is called a lipoprotein. They have a lipid core made of triglycerides and cholesterol entwined with the hydrophobic parts of the protein, and the surface comprises the hydrophilic part of the protein and some phospholipids.

The protein parts of lipoproteins are called apolipoproteins and each of these is given an alphabetic name. These are abbreviated to apo-A, apo-B, apo-C, apo-D, apo-E with some subgroupings e.g. apo-B48 which is a fragment of the apo-B100 molecule but about half the size.

In human plasma, there are 6 main classes of lipoproteins which differ in terms of size, density, the relative proportions of triglycerides and cholesteryl esters in the core, and in the nature of the apoproteins on the surface. Each lipoprotein has a defined tissue of origin and role as are outlined in the table on the next page.

Lipids obtained from our diet are transported from the intestine to other tissues after being incorporated into a lipoprotein. The epithelial cells of the intestines form lipoproteins known as chylomicrons. These are the largest of all lipoproteins (80–500 nm diameter) and are secreted into the lymph and from there are transported to the bloodstream. When they reach capillaries of adipose tissue and muscle, they are digested by an enzyme, lipoprotein lipase, that is bound to the surface of the endothelial cells. It hydrolyses the triglycerides in the core of the chylomicrons and the liberated fatty acids cross the endothelium and enter the underlying adipocytes or muscle cells. They are then either esterified again to form triglycerides for storage or oxidised to provide energy. After the triglycerides have been removed from the chylomicron in this fashion, the lipoprotein re-enters the circulation but its size has been reduced. It is now a chylomicron remnant (diameter 30–50 nm).

The liver cells have a receptor that recognises parts of the protein structures of apo-E and apo-B48, which then enables the liver cells to take up the chylomicron remnants (receptor-mediated endocytosis). Within the hepatocyte, the remnant is digested in lysosomes and the cholesteryl esters are cleaved to generate free cholesterol. The free cholesterol is used to form membranes, or is stored as cholesteryl esters, or is excreted in bile as cholesterol or bile salts.

Some of the cholesterol and lipids, including those synthesised by the liver, are used to form very low density lipoproteins (VLDL, diameter 30–80 nm), which are secreted into the blood. The major stimulus is a high calorie intake, especially carbohydrates, which induce the liver to assemble triglycerides for export and storage in adipose tissue. The triglycerides of VLDL are cleaved within capillaries by the same lipoprotein lipase that digests chylomicrons. Digestion produces a VLDL remnant analogous to the chylomicron remnant that is designated as intermediate density lipoprotein IDL (diameter 25–35 nm).

Lipoprotein Class	Major Core Lipid	Major Apoproteins	Origin of Apoproteins	Transport Function	Mechanism of Lipid Delivery
chylomicrons	Dietary triglycerides	A-1, A-2, A-4, B-48	Small intestine	Dietary triglyceride	Hydrolysis by lipoprotein lipase
chylomicron remnants	Dietary cholesterol esters	B-48, E	Chylomicrons	Dietary cholesterol	Receptor-mediated endocytosis in liver
VLDL (very low density lipoprotein)	Endogenous cholesteryl esters	B-100, C, E	Liver and small intestine	Endogenous triglyceride	Hydrolysis by lipoprotein lipase
IDL (intermediate density lipoprotein)	Endogenous cholesteryl esters	B-100, E	VLDL	Endogenous cholesterol	Receptor mediated endocytosis in liver (50%) or conversion to LDL (50%)
LDL (low density lipoprotein)	Endogenous cholesteryl esters	B-100	IDL	Endogenous cholesterol	Receptor mediated endocytosis in liver or extrahepatic tissue
HDL (high density lipoprotein)	Endogenous cholesteryl esters	A-1, A-2	Liver and small intestine	Facilitates removal of cholesterol from extrahepatic tissues	Cholesteryl ester transfer to IDL and LDL

In the plasma, the IDL particles have 2 metabolic fates. Some are cleared rapidly by the liver by receptor mediated endocytosis. The receptor that binds to IDL also binds to LDL because it binds either apo-B100 or apo-E. About half of the IDL particles are not cleared by the liver and their triglycerides are removed in the plasma until they contain mainly cholesterol. At this stage they are known as low density lipoproteins (LDL, 18–28nm diameter). LDL circulates for a relatively long time with a half life of 1.5 days. The particles are degraded by binding to LDL receptors in the liver and certain extrahepatic tissues including the endothelial cells of arteries. LDL comprises the major source of circulating cholesterol in human plasma. If the liver requires more cholesterol, it synthesises more LDL receptors. The converse is also true.

As cells die and cell membranes undergo turnover, free cholesterol is continually released into the plasma. This cholesterol is immediately adsorbed onto high density lipoprotein, (HDL, diameter 5–12nm) and in this location it is esterified with a long chain fatty acid by an enzyme in the plasma (lecithin: cholesterol acyl transferase (LCAT). The newly formed cholesteryl esters are rapidly transferred from HDL to VLDL or IDL particles by cholesteryl ester transfer protein in plasma. The IDL particles are ultimately taken up by the liver or converted to LDL. The formation of HDL is enhanced by the formation of and secretion of apo-E by peripheral tissue.

Cholesterol and atherosclerosis

Atherosclerosis is a condition where fatty deposits occur in the walls of arteries. The walls of arteries have a very strong muscular coat, and therefore as the deposits get larger they push towards the lumen, rather than expand outwards. This eventually clogs the artery. When this occurs in the vessels of the heart (coronary arteries), the cardiac muscle becomes starved of blood and hence oxygen. This can lead to death of the cardiac muscle cells which is known as a heart attack. In other large arteries such as the aorta or the carotid arteries, parts of the fatty deposits protruding into the arterial lumen can be knocked off by the fast moving blood. They then travel in the bloodstream (an embolus) and become lodged in a smaller artery downstream. When this smaller artery is in the brain, the brain tissue fed by this artery dies. This is commonly called a stroke.

It has been found that a major component of the fatty deposits is cholesterol. Therefore, one of the causes of this condition is not cholesterol per se but too much cholesterol in the blood. Most often this occurs because the dietary intake of cholesterol is too high. The reality is that we do not require any cholesterol in out diets because we can make it. Other people have a genetic defect whereby they do not make enough LDL receptor. Hence apo-B100 cannot bind to the endothelial cells and therefore cholesterol cannot be removed from the blood and processed by the tissue. The excess cholesterol is then pushed between the endothelial cells into the arterial wall. The 'pushing' will be greatest where the pressure is greatest, which is in the arteries closest to the heart, coronary arteries, aorta and carotid arteries. The 'pushing' will also be greater in people with high blood pressure and therefore high blood pressure predisposes to atherosclerosis. Smoking tends to increase blood pressure, as does anxiety, and both of these are also predisposing factors for atherosclerosis. Treatment is directed at these causes: reduce cholesterol in the diet, reduce uptake of cholesterol from the gastrointestinal tract, stop smoking, reduce stress and reduce blood pressure. You will remember that HDL takes cholesterol away from the tissues back to the liver for excretion and therefore a further strategy is to increase HDL levels.

References and further reading

Special review edition of membrane chemistry. (1992) *Science* **258**:917-969.

Bieber, L.L., (1988) Carnitine. *Annu. Rev. Biochem.* **88**:261-283.

Chao, W. and Olson, M.S. (1993) Platelet activating factor:Receptors and signal transduction. *Biochem. J.* **292**:617-629.

Fordhuchinson, A.W., Gresser, M. and Young, R.N. (1994) 5-Liopoxygenase, *Annu. Rev. Biochem.* **63**:383-417.

Greenstein, B. and Greenstein, A. (1996) *Medical Biochemistry at a Glance.* Oxford: Blackwell Science.

Gurr, M.I. and Harwood, J.L. (1991) *Lipid Biochemistry: An Introduction,* 4th ed.. London: Chapman and Hall.

Higgins, C.F. (1994) Flip-flop: the translocation of lipids. *Cell* **79**:393-395.

Jacobson, K., Sheets, ED and Simson, R. (1995) Revisiting the fluid mosaic model of membranes. *Science* **268**:1441-1442.

Lardy, H. and Shrango, E. (1990) Biochemical aspects of obesity. *Annu. Rev. Biochem.* **59**:689-710.

Nilsson-Ehle, P., Garfinkel, A.S. and Schotz, M.C. (1980) Lipolytic enzymes and plasma lipoprotein metabolism. *Annu. Rev. Biochem.* **49**:667-693.

Snyder, F. (1995) Platelet-activating factor and its analogs: Metabolic pathways and related intracellular processes. *Biochem. Biophys. Acta* **1254**:231-249.

van Echten, G and Sandhoff, K. (1993) Ganglioside metabolism: enzymology, topology and regulation. *J. Biol. Chem.* **268**:5341-5344.

Q&A Answers

1   To form an ester bond an acid is added to an alcohol.

2   Water is released when an ester bond is formed. It is a condensation reaction.

3   A 20C fatty acid with 4 double bonds is called eicosatetraenoic acid.

4   C1, βC and the ωC are indicated below.

5   The double bond would occur between carbons 9 and 10.

6   Vegetable oils are oils and have many double bonds, hence polyunsaturated appears on the labels of most vegetable oils. Butter, on the other hand, is a fat and hence would be expected to have saturated fatty acids, no double bonds.

7   Phosphatidylserine is formed by the esterification of serine to phosphatidic acid. Note the net charge on the polar part of the molecule is -1 but the actual charge is 2- and 1+.

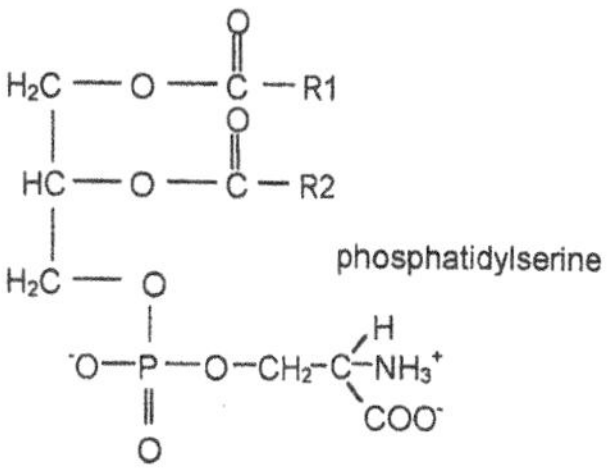

phosphatidylserine

8   Cardiolipins are formed by linking 2 phosphatidic acids with a glycerol bridge. Both non-polar parts project into the membrane and the polar part faces the cytoplasm of the cell. The charge on this polar group is 2-.

cardiolipin

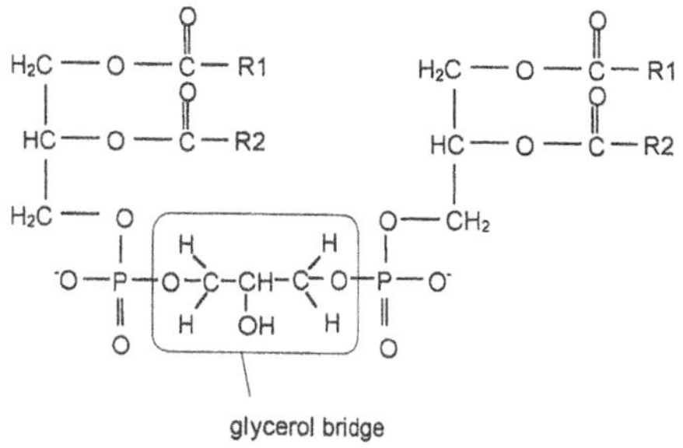

glycerol bridge

9   Lysophospholipids lack an acid group attached to C2, leaving it with the hydroxyl group.

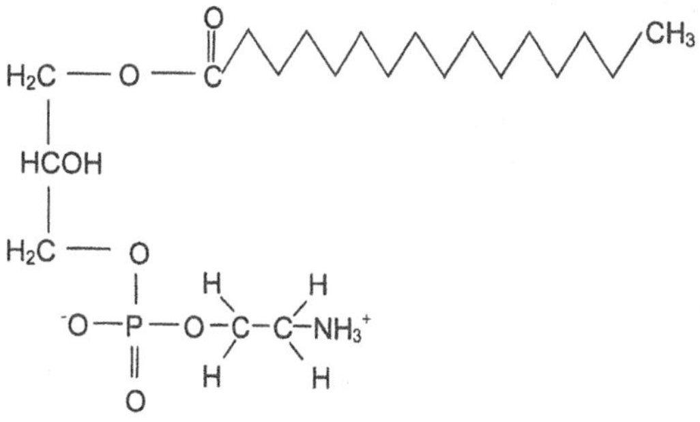

lysophosphatidylethanolamine

10 The conversion of glycerol to dihydroxyacetone phosphate is a 2-step reaction with a phosphorylation catalysed by glycerol kinase, followed by a dehydrogenation catalysed by glycerol-3-phosphate dehydrogenase. This takes place in the cytoplasm of cells (particularly the liver). It yields energy since 1NADH is equivalent to 3ATPs (see diagram overpage).

glycerol   glycerol-3-phosphate   dihydroxyacetone phosphate

11 The comparative naming of carbons in a fatty acid and acyl-CoA is given below.

fatty acid   acylCoA

12 This table lists the main differences between synthesis and degradation of fatty acids.

Synthesis	Degradation
Occurs in cytosol	Occurs in mitochondrial matrix
Acyl carrier = acyl carrier protein	Acyl carrier = coenzyme A
Reductant is NADP	Reductant is NAD, FAD
The formation of the hydroxyl group at the βC forms a D-epimer	The formation of the hydroxyl group at the βC forms an L-epimer
2C units obtained from malonyl-CoA	2C units form acetyl-CoA
Synthesised in 2C units	Degraded in 2C units
2C units added to the αC	2C units taken from the αC end

13 The name of the enzyme which catalyses the conversion of HMG-CoA to mevalonate is HMG-CoA reductase.

14 Glycine is added to the carboxyl group of trihydroxycoprostanate by forming an amide bond. This forms glycocholate, which is the major bile acid.

3,7,12-trihydroxyprostanoate   glycoholate

12 Nucleotides and their Derivatives

In this chapter you will learn:

- the structure of nucleotides and nucleosides
- the numbering of the ring structures, particularly how the prime designation comes about, and the phosphate groups
- the nomenclature of nucleotides and nucleosides
- the names and structures of the purines and pyrimidines
- the synthesis of purines
- the synthesis of pyrimidines
- synthesis of deoxyribonucleotides
- salvage of nucleotides – purine salvage and pyrimidine salvage
- catabolism of bases – degradation to uric acid

Nucleotides: synthesis, salvage and functions

Nucleotides are extremely important. They are not only the basic unit of DNA and RNA, but also the basis of high-energy compounds. The best known is ATP, but GTP is used as an energy source in protein synthesis and a few other reactions. UTP is used for activating glucose and galactose when these molecules are to be attached to other molecules such as in the synthesis of polysaccharides and proteoglycans. CTP is an energy source in lipid metabolism. Other high-energy compounds which act as co-enzymes in many reactions, NADH, NADPH and $FADH_2$ have nucleotides as part of their structure. Nucleotides also have an important role as intracellular second messengers (cAMP and cGMP) and as such activate or inhibit protein activity.

Structure of nucleotides

To understand nucleotides, you firstly need to become familiar with the nomenclature, which of course, is associated with their structure.

> **MAJOR POINT 62:** Nucleotides comprise a nitrogenous base, a sugar (ribose or deoxyribose) and one or more phosphate groups.

To learn how to draw their structure, focus on ribose, the sugar group.

Draw β-D-ribose in its furanose form. Remember that it is a 5C sugar.

Answer:

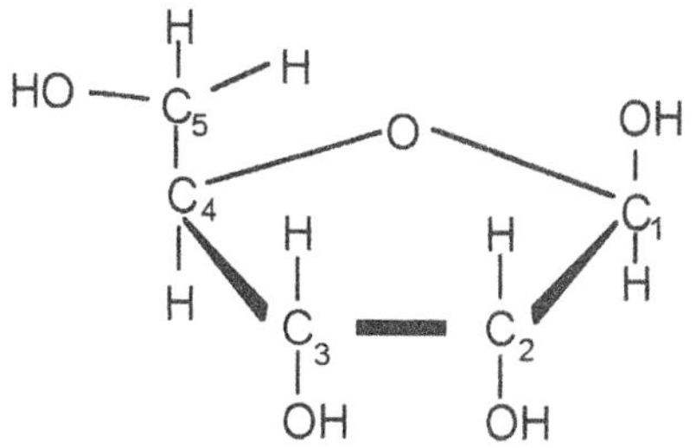

Attach a phosphate to C5. Remember, to do this, phosphoric acid is esterified to the ribose by removing water.

Answer:

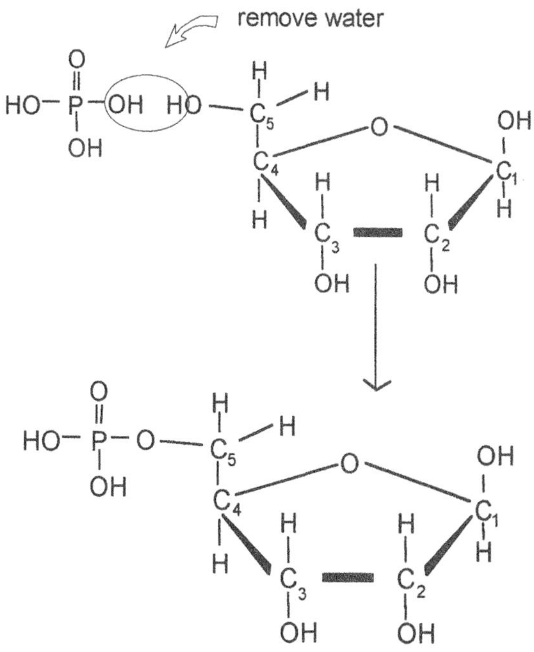

Add a base (e.g. adenine) to the ribose at C1. When bases are added, the C1 of ribose is attached to an amine group of the base. Guess what is removed to form the N-glycoside bond with the base?

Answer:

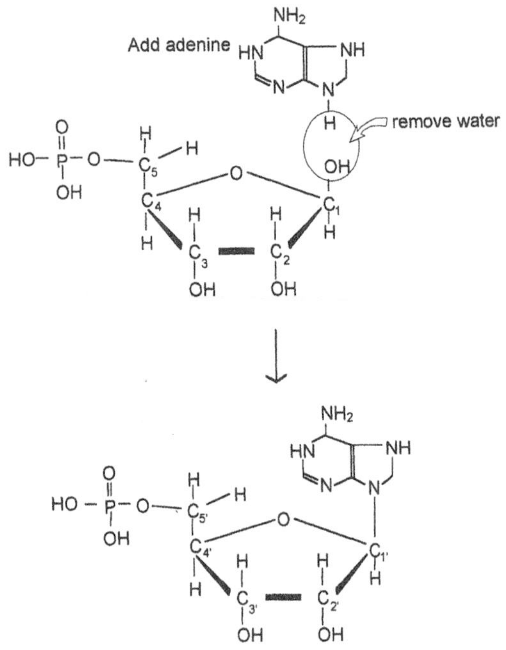

Add adenine

remove water

↓

adenosine monophosphate (AMP)

Note that the carbons of the ribose take on a prime designation e.g. C1'. This is to distinguish them from the atoms of the ring of the base which are numbered without the prime designation.

A base attached to ribose or deoxyribose alone is called a nucleoside. 'S' for simple as it does not have phosphates attached to it.

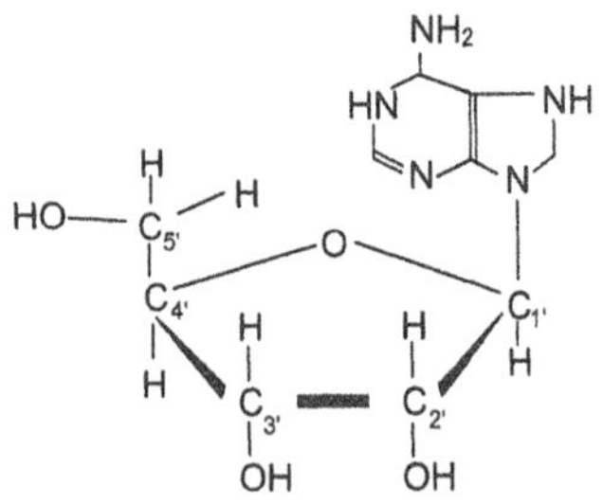

A nucleoside

A nucleoside with phosphates attached to the sugar is called a nucleotide.

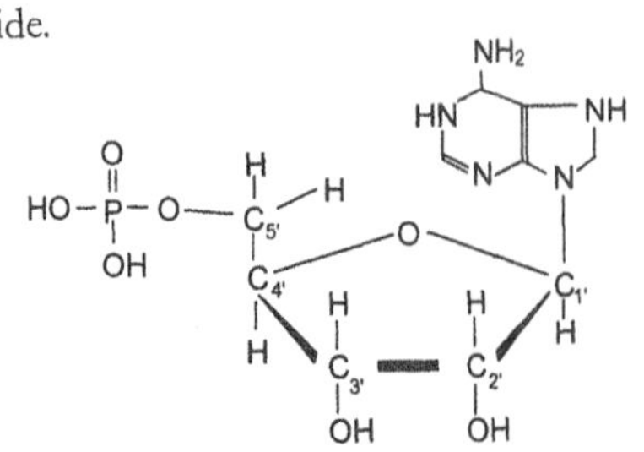

adenosine monophosphate (AMP)
a nucleotide

Therefore, we speak of nucleoside monophosphates, nucleoside diphosphates, and nucleoside triphosphates, which are all nucleotides.

The phosphates are given Greek letters α, β and γ, with the α phosphate directly attached to the ribose.

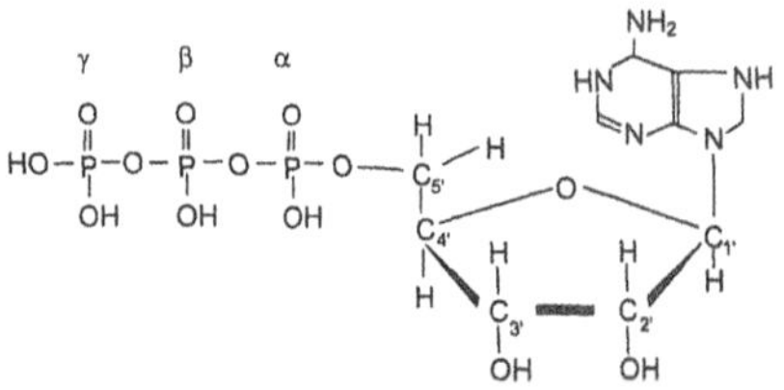

adenosine triphosphate (ATP)

Naming the nitrogenous bases

Adenine, guanine, uracil, thymine and guanine are the common nitrogenous bases that make up DNA or RNA.

- Adenine and guanine are purines.
- Cytosine, uracil and thymine are pyrimidines.

Mnemonic: Stones are CUT to build pyramids – (just remember that it is pyrimidines and not pyramidines)

Nomenclature:
- The names of purine nucleosides end in -osine: guanosine and adenosine.
- The names of pyrimidine nucleosides end in -idine: cytidine, uridine and thymidine.
- The common names of the monophosphate nucleotides end in -ylate: adenylate (AMP), guanylate (GMP), uridylate (UMP) thymidylate (TMP) and cytidylate (CMP).

Synthesis of the bases

I find it quite fascinating that most of the food we eat (meat, vegetables, fruit) contain bases and yet, despite this abundance, we obtain very little of our bases from the diet. We synthesise our own or conserve (salvage) bases that have already been synthesised but no longer necessary such as in dying cells or from red blood cells which eject their nuclei upon maturation. The amount of DNA being turned over is just amazing. Consider red blood cells (erythrocytes) alone.

How many erythrocytes do you think that we replace each day: 350 000; 3 500 000; 350 000 000; or 35 000 000 000 000.

Answer: *Unbelievably, 35 million million red blood cells are replaced each day (give or take a few million). Hence the need for bases is enormous.*

Purine synthesis

Q&A 1: Name 2 purine bases.

The purines bases are not just synthesised as such, but are assembled on ribose.

MAJOR POINT 63: Purines are built directly onto ribose-5-phosphate.

Firstly, ribose must be activated. It is activated at the carbon where the base will be attached. This is C1. It is activated by placing it in a high-energy state, and often a high-energy state is achieved by attaching phosphate groups to molecules. Therefore, the expectation is that phosphate will be attached (esterified) to C1 of ribose.

Almost right, but not quite! Firstly, it is not ribose that is used, but ribose-5-phosphate instead, and secondly, the energy in a single phosphate group attachment is not enough to enable the next step, which is the transfer of an amine group to the C1 of ribose-5-phosphate. Therefore, not 1, but 2 phosphate groups must be transferred to C1 of ribose-5-phosphate. The 2 phosphates are linked together and hence it is a pyrophosphate (PP_i). The pyrophosphate comes from ATP. This reaction takes place in nearly all cells of the body.

Draw α-D-ribose-5-phosphate and then transfer a pyrophosphate to C1 of ribose. Remember that in the process ATP is converted to AMP.

Answer:

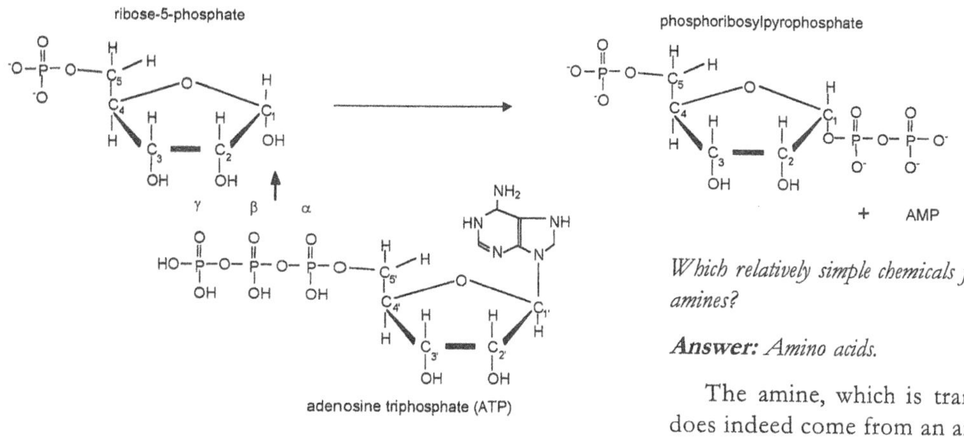

The product formed is phosphoribosylpyrophosphate, which is an appropriate name for a molecule that has a phosphate attached to a ribose attached to a pyrophosphate.

Draw the transfer of an amine from glutamic acid to C1 of phosphoribosylpyrophosphate.

Answer:

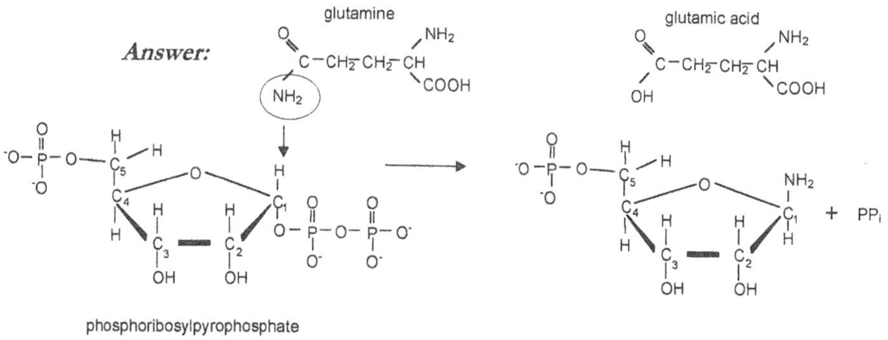

What would be a suitable name for the enzyme that catalyses the transfer of a pyrophosphate to ribose-5-phosphate? Hint: an enzyme, which transfers a phosphate from ATP, is called a kinase, but an enzyme that catalyses the transfer of a pyrophosphate is called a pyrophosphokinase.

Answer: *Ribose-5-phosphate pyrophosphokinase. Easy, wasn't it?*

The rest of the formation of purines occurs mainly in the cytoplasm of liver cells where the enzymes for these reactions occur as a macro-molecular aggregate. Recall that we are trying to build a purine ring onto ribose-5-phosphate.

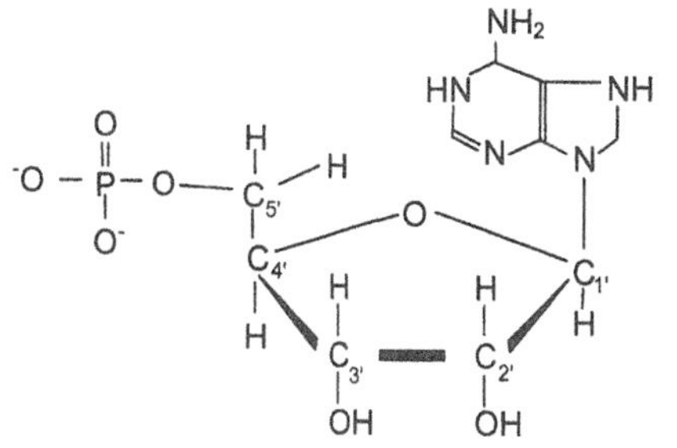

To begin, an amide bond must be formed at C1 of the ribose. For this to occur a source for the amine is needed.

Which relatively simple chemicals found in abundance in the body have amines?

Answer: *Amino acids.*

The amine, which is transferred to the C1 of ribose, does indeed come from an amino acid. Not just any amino acid, but 1 of the 2 amino acids which can lose an amine and be converted into another amino acid by doing so. The amino acid is glutamine (**Q**). When it loses its amine it becomes glutamic acid (**E**). This is almost the identical reaction which occurs in the synthesis of glucosamine and galactosamine (Chapter 6).

The obvious questions now arise, the name of the product and the name of the enzyme that catalyses the reaction. Let us be very posh and name the product precisely giving consideration to the ribose in terms of whether it is α or β and whether it is L or D form. Therefore, it is 5-phospho-β-D-ribosylamine Note that in this reaction, the α form of ribose is transformed into the β form of ribose (mutarotation has occurred). The enzyme has catalysed the transfer of an amide from glutamine (**Q**) to phosphoribosylpyrophosphate. The expectation would be that the name of the enzyme would describe this and it does. The enzyme is glutamine phosphoribosylpyrophosphate amide transferase. As an aside, this is one of several methods for synthesising glutamic acid (**E**) from glutamine (**Q**).

> **MAJOR POINT 64:** Transfer of an amide from glutamine to phosphoribosylpyrophosphate is the commitment and rate-limiting step in purine synthesis.

Next, 2 carbons and a nitrogen are added in 1 step. Once again the nitrogen comes from an amino acid, glycine (**G**). The glycine (**G**) is attached to the amine group of 5-phosphoribosylamine.

How could an amino acid be attached to an amine group? Hint – it involves removing water.

Answer: *An amide bond is formed between the α-carboxyl group of the glycine (**G**) and the amine of 5-phospho-β-D-ribosylamine.*

Notice the name of the product 5'-phosphoribosylglycinamide. A complex but descriptive name. It now receives the prime designation to distinguish the atoms of the furan ring from those of the growing chain. The enzyme involved is given a trivial name of a synthase. It should be noted that energy is required for this reaction to occur and this comes from ATP, which is converted to ADP.

The next step is really interesting from a dietary and physiological point of view, but difficult to grasp chemically. A formyl

(-CHO) group is attached to the amine group derived from glycine.

Nomenclature: Formyl comes from formaldehyde.

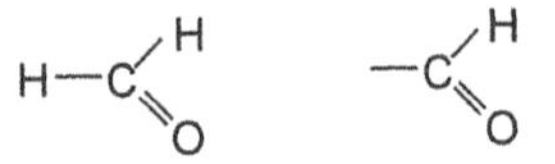

formaldehyde   formyl

The name formaldehyde comes from the animals that contains a large amount of this chemical - ants. The Latin for ants is *formica*. One can talk of people formicating, which means that they are behaving like ants such as when they leave a football stadium. This differs from people fornicating which means that they are behaving like rabbits.

This group is simply transferred to the growing chain. The carrier is N^{10}-formyltetrahydrofolate, which becomes tetrahydrofolate after the transfer.

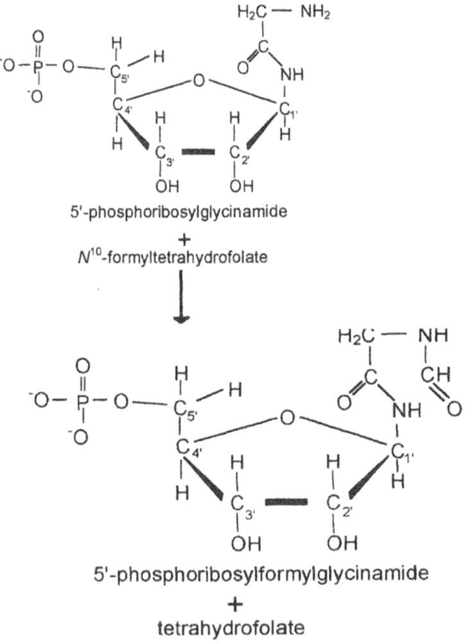

5'-phosphoribosylglycinamide
+
N^{10}-formyltetrahydrofolate

5'-phosphoribosylformylglycinamide
+
tetrahydrofolate

Since the formyl group has been transferred to 5'-phosphoribosylglycinamide the name of the enzyme that

---

Draw glycine and 5-phospho-β-D-ribosylamine and then form an amide bond between them.

Answer:

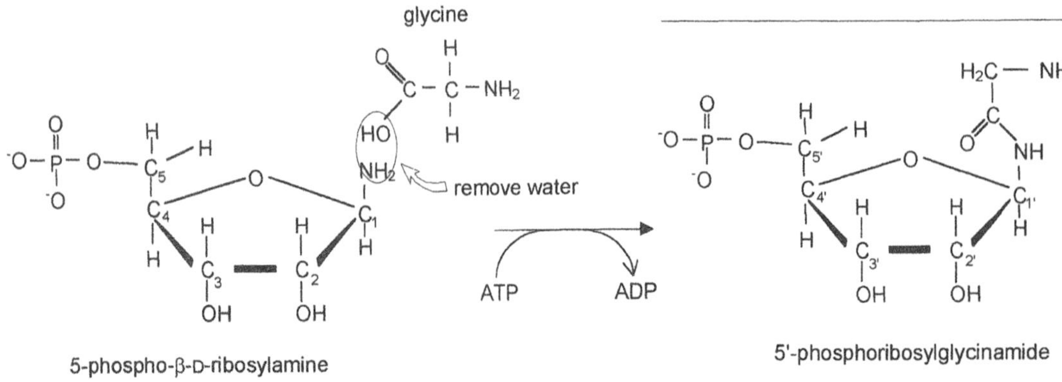

glycine

remove water

ATP   ADP

5-phospho-β-D-ribosylamine

5'-phosphoribosylglycinamide

catalyses the formation of 5'-phosphoribosylformylglycinamide is phosphoribosylglycinamide transformylase.

That is the biochemistry, but what is the exciting dietary and physiological importance of this reaction? Folic acid is one of the vitamins essential for our well-being. We must obtain folic acid from our diet so that we can make tetrahydrofolate and from this make purines. Without purines we cannot make new DNA and without new DNA we cannot make new cells. When you consider that we need to make 35 000 000 000 000 new red blood cells per day, someone deficient in folic acid would not be able to make this many and would become anaemic. Furthermore, people who have large unusual masses of growing cells, such as people suffering from cancer or pregnant women, often require additional folic acid because the rapidly growing cells are using the supplies to synthesise purines. Alcoholics obtain the majority of their calories from alcohol and therefore begin to become vitamin deficient. It is not uncommon for alcoholics to be anaemic because of the lack of folic acid.

> **MAJOR POINT 65:** Tetrahydrofolate, essential for purine (e.g. adenine, guanine) synthesis, is obtained from the vitamin folic acid.

> **Nomenclature:** Folic acid is obtained from green leafy vegetables and its name reflects this - foliage, derived from the Latin, *folium*, meaning leaf.

> **Q&A 2:** Name another stage in our lives when folic acid supplements may be important. Hint: think about situations when there will be a great demand on DNA synthesis, or rapid cell growth.

The next step in purine synthesis resembles the transfer of an amine from glutamine (**Q**) to the keto group of α-ketoglutarate to form glutamic acid (**E**) (2 glutamic acids actually). Do you remember this reaction? In the case of purine synthesis, the amine transfer is from glutamine (**Q**) to the keto group of 5'-phoshoribosyl-*N*-formylglycinamide. Let us compare these 2 reactions side by side.

Draw the conversion of α-ketoglutarate to glutamate.

Answer:

α-ketoglutaric acid / glutamine / glutamic acid / glutamic acid

Draw the transfer of an amine group to the keto group of 5'-phoshoribosyl-N-formylglycinamide.

Answer:

5'-phosphoribosyl-*N*-formylglycinamide 5'-phosphoribosyl-*N*-formylglycinamidine

These 2 reactions differ in that the amine is transferred as an 'imide' rather than an amide.

> **Nomenclature:** An imide has a double bond to nitrogen, unlike an amide which has a single bond to nitrogen.

The transfer of the amine from 5'-phosphoribosyl-*N*-formylglycinamide to 5'-phosphoribosyl-*N*-formylglycinamidine is catalysed by an enzyme that is simply called 5'-phosphoribosylformylglycinamidine synthase.

Note that energy is required for this reaction to occur and this energy is obtained by the conversion of ATP to ADP.

This has set up the molecule for ring closure between the aldehyde group and the nitrogen attached to C1 of the ribose. In the process, water (H_2O) is lost from the whole ring structure. This is just one of those things which is a little difficult to learn and takes practice.

5'-phosphoribosyl-*N*-formylglycinamidine 5'-phosphoribosyl-5-aminoimidazole

The enzyme that converts 5'-phosphoribosyl-*N*-formylglycinamidine to 5'-ribosyl-5-aminoimidazole is once again given the trivial name 5'-ribosylaminoimidazole synthase. Like the previous reaction, it requires energy which it obtains from ATP which is converted to ADP.

To help learn the names of these complicated molecules, try to picture each part of the molecule as you are saying the name. It helps you to remember the structure and how it is formed.

> **Nomenclature:** The 5 in the name refers to the atom (C) of the imidazole ring to which the amine is attached (5-aminoimidazole). Atom 1 of this ring is attached to the ribose, and of course is N.

The next step is the addition of carbon dioxide (as a carboxyl group) to the ring. This reaction is remarkable because it does not require energy, nor does it require a carrier of carbon dioxide such as biotin. The enzyme that catalyses this reaction is a carboxylase, 5'-phosphoribosyl-5-aminoimidazole carboxylase.

5'-phosphoribosyl-5-aminoimidazole   5'-phosphoribosyl-5-aminoimidazole-4-carboxylate

The next process is to attach an amine to the carboxyl group (-COO) and hence form an amide bond. This is a 2-step process involving the attachment of aspartate (**D**) to the carboxyl group via an amide bond and then cleavage of the amine group to release fumarate.

5'-phosphoribosyl-5-aminoimidazole-4-carboxylate   5'-phosphoribosyl-4-(N-succinocarboxamide)-5-aminoimidazole   5'-phosphoribosylaminoimidazole

The addition of aspartate to the carboxyl group forms 5'-phosphoribosyl-4-(N-succinocarboxamide)-5-aminoimidazole and the enzyme that catalyses this reaction is a synthase, 5'-phosphoribosyl-4-(N-succinocarboxamide)-5-aminoimidazole synthase. This step requires energy and hence ATP is converted to ADP in the process.

5'-phosphoribosylaminoimidazole   5'-phosphoribosyl-4-carboxyamide-5-formamidoimidazole   inosine-5'-monophosphate (IMP)

Nomenclature: The term succinate in the name comes from the fact that aspartate with amine group removed becomes succinate. The succinocarboxyamide comes from the fact that succinate is bound to a carboxyl group via an amide bond.

$$^-OOC-CH_2-CH_2-COO^-$$
succinate

The cleavage of the carbon skeleton of aspartate from 5'-phosphoribosyl-4-(*N*-succinocarboxamide)-5-aminoimidazole to form 5'-ribosyl-4-carboxyamide-5-aminoimidazole is catalysed by adenylosuccinate lyase.

The transfer of an amine group from aspartate and the resulting formation of fumarate is not unique to purine synthesis and occurs in other processes in the cell.

> **Q&A 3:** What is another common biochemical pathway in which aspartate is deaminated to form fumarate?

The final atom of the purine ring, a carbon, is brought into the ring as a formyl group. The advantage of this is that it potentiates ring closure just as it did for the formation of the imidazole ring (see bottom of previous page).

Which cofactor will carry the formyl group? Hint: it is the same cofactor that carried the formyl group for the synthesis of the imidazole ring.

Answer: N^{10}-*formyltetrahydrofolate*

This is followed by ring closure which produces the purine nucleotide, inosine-5'-monophosphatexe inosine-5'-monophosphate synthesis (IMP).

The name of the purine formed in this series of reactions is inosine. To form adenosine monophosphate and guanosine monophosphate, the purine ring has to be altered. This will be discussed later.

> ★ **Logic process:** I am sure at this stage you are thinking how on Earth will I remember this pathway, so let us tackle this problem.

To learn this pathway you must know

- the structure of ribose, ribose-5-phosphate, α and β anomeric forms of ribose
- the structure of glutamine and its relationship to glutamate
- the structure of aspartate
- the structure of an aldehyde group
- the formation of an amide bond

Do not even attempt to learn this pathway without this information under control. Revise the chapter on sugars and amino acids if necessary

- You must also know how to draw a purine ring. This information is given later

Focus then on the way the purine is formed and do not worry initially about the names of the structures which are formed, nor the enzymes involved.

Now to conquer the pathway

- Draw a purine ring (on a scrap of paper). This is your target
- Draw ribose-5-phosphate from memory – comit this structure to memory as it helps eliminate mistakes by numbering the carbons. It is important to remember that this ribose is in the α form.
- Remember that the purine will be formed on C1 of ribose and hence this must be activated by adding a pyrophosphate (PP$_i$) to it. (5-phosphoribosylpyrophosphate)
- The pyrophosphate comes from ATP which is converted to AMP
- Look at the purine structure and see that the first step has to be to add an N at the bottom of the 5-sided ring to C1 of ribose
- Think that N groups commonly come from amino acids and one of the amino acids that donates N is glutamine (**Q**)
- We are now ready to build the purine so it is upwards (β-anomer) from here
- The carboxyl group of glycine (**G**) then forms an amide bond with the amine attached to ribose
- Then remember diet, folic acid, formyl group and add a formyl group to the end of the growing chain
- Although it is tempting to close the ring at this stage, don't! The ring must be further activated before it can be closed
- An amine must be transferred to replace the ketone group (C=O)
- Once again glutamine (**E**) is the carrier of the amine group, but in fact an imine (=N) is formed
- Close the 5-sided ring
- Then we work on the top of the ring. Carbon dioxide is added
- Now we add an amine to the carboxyl group formed. This amine comes from another amino acid that readily transfers amine groups – aspartic acid (**D**)
- But this occurs in 2 steps. Firstly an amide bond forms, then the carbon skeleton of aspartate (**D**) is cleaved to give fumarate
- Then we prepare for ring closure again by adding a formyl group as above using formyltetrahydrofolate
- close the ring and we have inosine-5'-monophosphate.

Once you are familiar with these steps begin to incorporate the names, which in fact describe the structures. Then learn the enzymes

Drawing a purine

We now digress to learn how to draw a purine, in this case uric acid. You must remember it is a 5-sided ring and a 6-sided ring joined together.

Starting with the right hand side N at the bottom of the 5-sided ring, draw towards the left - nitrogen, carbon, nitrogen, carbon, nitrogen, carbon.

Starting again with the right hand N at the bottom and draw towards the right – nitrogen, carbon nitrogen, carbon. Complete the ring structures to form the basis of a purine. Add an =O to each of the CH_2 groups to form uric acid.

Start with this nitrogen   add carbon   nitrogen   carbon   nitrogen   carbon

Start again with with this nitrogen   add carbon   nitrogen   carbon   Complete the bonds

This is essentially a purine ring   Uric acid   complete by adding H to each N and =O to each C

To number the atoms of a purine ring is a memory task and starts with the left most nitrogen of the 6-sided part of the ring, goes anti-clockwise and then begins with the right most nitrogen of the 5-sided ring and goes clockwise.

Q&A 4: What is the origin of each of the numbered atoms of the purine?

Q&A 5: How many ATPs are used in forming inosine monophosphate from ribose-5-phosphate? Remember that the conversion of ATP to AMP counts as 2 ATPs being consumed.

The most ironical aspect of this whole pathway, for which I have no answer, is where did the ATP come from? ATP is needed for the synthesis of purines and purines are needed for the formation of ATP. It is a bit like the chicken and the egg question.

Having formed inosine-5'-monophosphate, the question arises as to how the more common purine nucleotides, adenosine-5'-monophosphate and guanosine-5'-monophosphate, are formed from inosine-5'-monophosphate.

Firstly compare the structures of inosine monophosphate with adenosine monophosphate and determine the difference.

inosine -5'-monophosphate (IMP)   adenosine-5'-monophosphate (AMP)

Ribose-5'-phosphate   Ribose-5'-phosphate

Answer: *An amine group has been added to the ring to replace the keto group (=O)*

From experience, there are 2 places where amine groups can come from, aspartate (**D**) or glutamine (**Q**). In this case, the amine is transferred to inosine from aspartate in a 2-step reaction.

What is extremely important is that the energy for this reaction comes from GTP. The enzymes involved in this pathway are adenylsuccinate synthase and adenylsuccinate lyase (see bottom of page).

Repeat the above process to determine the conversion of inosine monophosphate (IMP) into guanosine monophosphate (GMP).

Firstly draw inosine monophosphate and compare its structure with guanosine monophosphate.

inosine -5'-monophosphate (IMP)   guanosine monophosphate (GMP)

Ribose-5'-phosphate   Ribose-5'-phosphate

Answer: *The difference is the addition of an amine group to C2 of the purine ring.*

From experience, an amine cannot be simply attached to the ring. Normally it needs a ketone (=O) and this case is no different. Hence it is a 2-step reaction with first reaction being to attach a keto group to C2.

Draw the transfer of an amine from aspartate to inosine in a 2-step reaction. Hint: look at the transfer of the amine from aspartate above when 5'-phosphoribosyl-4-(N-succinocarboxamide)-5-aminoimidazole was formed.

Answer:

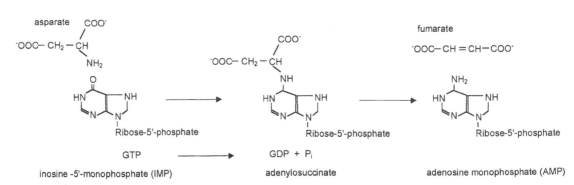

inosine -5'-monophosphate (IMP)   xanthosine monophosphate (XMP)   guanosine monophosphate (GMP)

The addition of the keto group actually generates energy because the oxygen comes from water and NAD^+ is reduced to NADH. The next step is to add an amine to this molecule (xanthosine monophosphate) and in this case the amine is added by a transfer of the amine from glutamine (**Q**). Importantly, ATP is required to generate the energy for the conversion of xanthosine monophosphate to guanosine monophosphate.

The enzymes involved in these reactions are inosine monophosphate dehydrogenase and guanosine monophosphate synthase. Adenosine monophosphate and guanine monophosphate are readily converted to the di- and triphosphates.

Mnemonic: The amine donor for <u>a</u>denosine monophosphate synthesis is <u>a</u>spartate: The amine donor for <u>g</u>uanosine monophosphate synthesis is <u>g</u>lutamine.

In cells, the levels of adenine and guanine in DNA are about the same. This balance is maintained in a relatively simple way. If the cell has an excess of adenine (ATP excess) then GMP is preferentially synthesised and if it has an excess of guanine (GTP excess) then AMP is preferentially synthesised. Hence these 2 purines stay in balance. Feedback inhibition also controls the proportions because GMP inhibits the conversion of IMP to XMP and AMP inhibits the conversion of IMP to adenylosuccinate.

> **MAJOR POINT 66:** Adenosine triphosphate (ATP) is the energy source for guanosine monophosphate (GTP) synthesis and guanosine triphosphate is the energy source for adenosine monophosphate synthesis.

Pyrimidine synthesis

As for purine synthesis, pyrimidine synthesis occurs in the cytoplasm of cells, and in man, the enzyme activities are part of a multifunctional protein. In multifunctional proteins, the product from one reaction is not actually released into the cytoplasm but remains attached to the enzyme complex. This prevents wastage of products due to degradation and diffusion.

Tissues most capable of making pyrimidines are the spleen, thymus, gastrointestinal tract and testes. Fortunately, pyrimidine synthesis is simpler than purine synthesis. A major difference is that the pyrimidine ring is formed firstly and then attached to the 5-phosphoribosylpyrophosphate. The formation of a pyrimaidine ring begins with the synthesis of carbamoyl phosphate from the amide group of glutamine (**Q**) and bicarbonate. Aspartate (**D**) is then joined to carbamoyl phosphate to form the rest of the atoms of the pyrimidine ring.

> **MAJOR POINT 67:** A pyrimidine nucleotide is synthesised by firstly forming the pyrimidine ring and then attaching it to ribosyl phosphate.

Synthesis of carbamoyl phosphate

Another very important pathway in which carbamoyl phosphate is used is the synthesis of urea, but this pathway uses a different enzyme from the enzyme used for pyrimidine synthesis. The enzyme used for urea synthesis is located in mitochondria, where for pyrimidine synthesis, <u>it is in the cytoplasm.</u> Carbamoyl phosphate synthase II (CPS II) is used in purine synthesis and it prefers glutamine to free ammonia. The formation of carbamoyl phosphate requires 2 ATPs, one to provide the phosphate group and the other to provide the energy for the formation of the amide bond.

Formation of orotic acid

Carbamoyl phosphate condenses with aspartate (**D**) in the presence of aspartate transcarbamoylase to yield *N*-carbamoylaspartate, which is then converted to dihydroorotate. Note the numbering of the pyrimidine ring.

carbamoyl phosphate   *N*-carbamoylaspartate   dihydroororate

The next step is the oxidation of the ring. Like many oxidations this does not involve addition of oxygen, but instead loss of hydrogen. This hydrogen is transferred to NAD^+, and hence produces energy. Interestingly, the enzyme is not in the cytoplasm like the other enzymes, but is located on the outer face of the inner mitochondrial membrane.

What would be a suitable name for an enzyme that catalyses the removal of hydrogen from dihydroorotate?

Answer: *dihydroorotate dehydrogenase.*

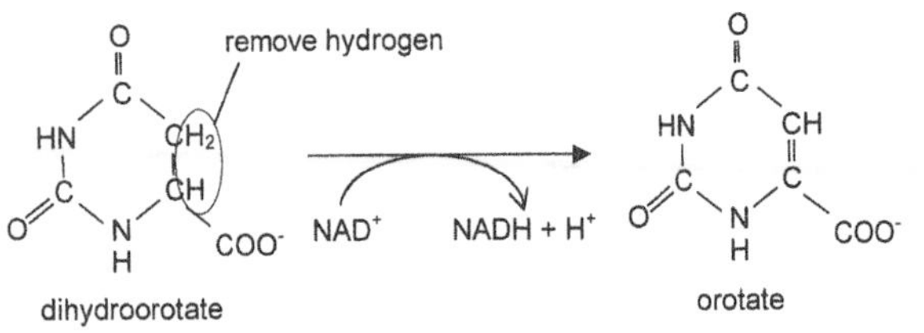

Orotic acid is then linked to 5'-phosphoribosylpyrophosphate to form orotidine-5'-monophosphate (OMP) in a reaction catalysed by orotate phosphoribosyl transferase. It is then decarboxylated to form uridine monophosphate (UMP).

To form uridine-5'-monophosphate, orotidine-5'-monophosphate is decarboxylated. From the diagram above determine where the decarboxylation occurs.

Answer: *The carboxyl group of the purine ring of orotic acid is removed to form uridine monophosphate. Uridine is one of the purine bases found in ribonucleic acid (RNA).*

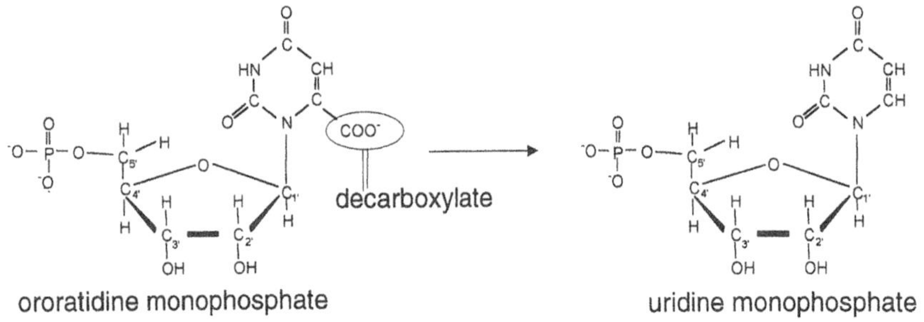

What is the other purine base found in RNA?

Answer: *Cytosine.*

The ribosyl<u>triphosphate</u> form, rather than the monophosphate form of these bases, is needed for the synthesis of RNA. Therefore, uridine monophosphate is not converted directly to cytosine monophosphate, instead it is firstly converted to uridine triphosphate and uridine triphosphate is converted to cytosine triphosphate. The phosphorylation of uridine monophosphate (UMP) occurs in 2 steps. Logically, the phosphates come from ATP because this source is readily regenerated through metabolic pathways. Notice that the enzymes involved are catalysing a phosphorylation and hence are called kinases.

UMP + ATP → UDP + ADP enzyme = uridinylate kinase

UDP + ATP → UTP + ADP enzyme = uridine diphosphate kinase

Once uridine triphosphate has been formed, it is converted into cytosine triphosphate.

What is the difference between uridine triphosphate and cytosine triphosphate?

Answer:

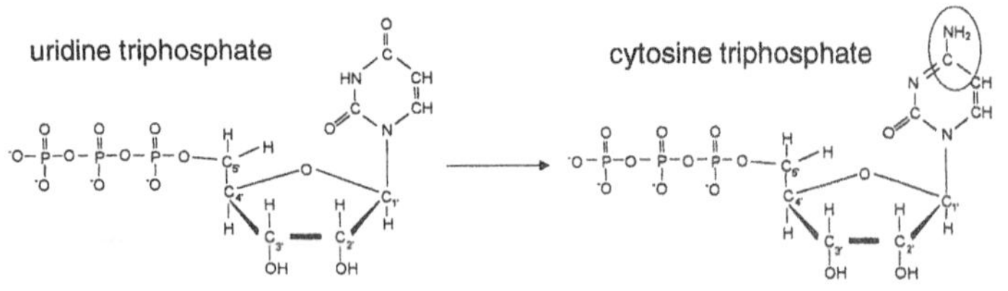

Clearly, the difference is the substitution of an amine group onto the ketone group at C4

How do we transfer amine groups to ketone groups? Hint think about how guanosine and adenosine monophosphates were formed from inosine monophosphate.

Answer: *Amine groups are transferred from either aspartic acid or glutamine. In this case, glutamine is used to form cytosine from uracil.*

> **Q&A 6:** Draw the reaction of the transfer of an amine group from glutamine to C4 of uridine triphosphate.

The control of pyrimidine nucleotide synthesis is mainly product inhibition. Uridine triphosphate (UTP) inhibits cytoplasmic carbamoylphosphate synthase (CPS II). Phosphoribosylpyrophosphate activates it.

In addition to the 2 pyrimidine nucleotides, uridine triphosphate (UTP) and cytosine triphosphate (CTP) whose synthesis is shown above, there is another, but this only occurs in deoxyribonucleic acid. What is the name of the pyrimidine base that only occurs in deoxyribonucleic acid and not ribonucleic acid?

Answer: *Thymidine.*

Thymidine is not synthesised as a ribonucleotide because there is no use for it. It is synthesised as the deoxyribonucleotide. Firstly, uridine diphosphate (UDP) is converted to deoxyuridine diphosphate (dUDP) and from there is converted to deoxythymidine monophosphate (dTMP).

Synthesis of deoxyribonucleotides

What is the difference between a <u>deoxy</u>ribonucleotide and a ribonucleotide?

Answer: *The ribose lacks an oxygen atom on C2.*

uridine diphosphate

deoxyuridine diphosphate

Therefore, for the synthesis of a deoxynucleotide, a ribonucleotide is deoxygenated or reduced. Reduction occurs by loss of an oxygen and in this case the oxygen is combined with 2 hydrogens to form water. Adding hydrogens or losing an oxygen puts energy into a system. The energy in this case comes from the reducing power of NADPH, but how does it get there?

Higher Level

For me, this is one of the most delicious reactions that exist in biochemistry because it illustrates a number of reactions that occur elsewhere.

What would be a suitable name for enzyme that catalyses the reduction of a ribonucleotide?

Answer: *Ribonucleotide reductase.*

This enzyme, ribonucleotide reductase, gives up hydrogen to capture the oxygen being released from ribose group as it forms deoxyribose.

Where can proteins store hydrogen for release?

Answer: *They store it in cysteine groups which form a disulphide bond when they release hydrogen (become oxidised).*

Note: in addition to UDP, ribonucleoside reductase may also use ADP or GDP as substrates, and hence synthesise dADP and dGDP. Once ribonucleotide reductase is oxidised, the sulphur (thio) groups must be reduced again. The hydrogens in this case are donated by another enzyme called thioredoxin (thio = sulphur groups; redoxin = redox reactions). Similar to ribonucleotide reductase, thioredoxin stores its hydrogens in cysteine groups that form a disulphide bridge when they become oxidised.

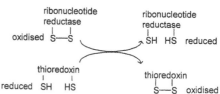

These disulphide bridges of thioredoxin are reduced when hydrogen is transferred from NADPH. The enzyme is called thioredoxin reductase.

This pathway illustrates the reducing power of cysteine groups in proteins, the formation of disulphide bonds in proteins, the reducing power of NADPH and the transfer of energy between different groups.

Deoxyuridine diphosphate (dUDP) is then converted to dUTP with the phosphate coming from any nucleoside triphosphate. The reaction is catalysed by nucleoside diphosphate kinase.

uridine diphosphate

deoxyuridine diphosphate

ribonucleotide reductase

SH   HS

reduced

ribonucleotide reductase

S—S   + H₂O

oxidised

dUTP is then hydrolysed to dUMP and PPi before conversion to deoxythymidine monophosphate (dTMP).

Deoxyuridine monophosphate (dUMP) is then converted to deoxythymidine monophosphate (dTMP) by the transfer of a methyl group ($-CH_3$) to C5. The donor of the methyl group 5,10-methylenetetrahydrofolate, which of course, is a derivative of folic acid.

Another amino acid, serine (**S**), comes into play here. Tetrahydrofolate is re-methylated by serine hydroxymethyl-transferase.

Lecturers' secrets and students' folly

Since deoxythymidine monophosphate dTMP is required for DNA synthesis and not RNA synthesis, rapidly dividing cells e.g. cancer cells, can be targeted by agents which interfere with dTMP synthesis. Some drugs, which work in this manner, are 5-fluorouracil and methotrexate.

The pathways above serve as a beautiful illustration to show the chasm between what the lecturer is thinking and what the student is thinking. After reading through this, use it as a model for all of your studies and for your approach to exam questions.

Students' Folly – I have to memorise these pathways for the exam.

Lecturers' Secret – I cannot memorise these pathways because it is too hard, and too much of my brain is destroyed due to old age or other means. I can only regurgitate these pathways because I understand what is happening. I do not expect the students to memorise these pathways, but I do expect them to have an understanding of the pathways. This understanding will be at different levels and therefore I will set the exam questions accordingly.

Secret – Set a few easy questions which all students should know the answer to and do not require understanding. This will make them all feel comfortable, e.g. name 2 purine bases. Which of the following is a pyrimidine...? Give the names of 2 bases you would find in DNA

Folly – I knew it! The lecturer is giving me memory work. In this case it is true, but these are really basics that are essential in all walks of biology, not just biochemistry.

Secret – slightly harder question, but probably one which the students will expect – draw a purine/pyrimidine ring.

Draw the structure of adenosine monophosphate and number the atoms of both rings. Most will have committed it to memory but the bright ones will answer it because they understand how these structures are synthesised or degraded.

Folly – more memory work. At this stage it is only partly true.

Secret – Now I shall test if they understand anything about the pathways of purine and pyrimidine synthesis. Name 2 differences between purine and pyrimidine synthesis. This is given because there are more than 2 differences and hence it is relatively easy for a student to find 2 differences if they know anything at all about the pathways. If this question is allocated only a couple of marks then the lecturer is not expecting details, just facts (e.g. purine ring is formed on phosphoribosylpyrophosphate whereas the pyrimidine ring is formed and then attached to ribose). Other answers may include different energy requirements, branching of the purine pathway but not the pyrimidine. Another question at this level could be: What substance/compound is most important in regulating the synthesis of guanosine monophosphate from inosine monophosphate?

Folly – more memory work. This is the wrong assumption. It is testing understanding. Be careful of the common student answer – one is synthesised attached to the ribose and the other is not. True, but which is which – not many marks here.

Secret – now for something a little harder, but not too difficult. These are parts of pathways. Describe how adenosine monophosphate is formed from inosine monophosphate and aspartate (**D**). Most lecturers are sympathetic and give you clues. Also the number of marks will dictate the level of expectation. Lots of marks then enzyme names are needed as well. Also, it is a bit devious because telling you that aspartate (**D**) is involved means that there is an expectation that you will say what is produced after it has lost its amine, i.e. fumarate. Leaving this out, even though you knew it, will decrease your mark.

Folly – More memory work. In any case the lecturer is a bit stupid because he has all but given me the answer in the question. Typical answer - an amine group is transferred from aspartate to the ring of inosine to give adenosine monophosphate. In the lecturer's mind (inebriated, twisted and lacking) this is completely wrong. He is expecting as a minimum that it is a 2-step process, that GTP is the energy source, and that the amine is transferred to C6 of the purine ring. Better still would be a drawing of the process and to get full marks name the intermediate and the enzymes involved and all of the products (GDP and fumarate as well as adenosine monophosphate). The real folly is the student who actually knows all these things, but puts in the first part of the answer only.

Secret – something a little harder, but not overwhelming the student with the whole pathway. Draw a purine ring, number the atoms and nominate from which molecules each of the atoms is derived. In this the lecturer is thinking that she is testing whether or not the student understands the synthetic pathway without actually asking for the pathway to be reproduced.

Folly – this is memory work. No! This is testing understanding. To get it absolutely right under the pressure of an exam you must know the synthetic pathway. Also I have seen time and time again students lose marks because they have not underlined the points asked in the question (e.g. <u>Draw</u> a purine ring, <u>number</u> the atoms and <u>nominate</u> from which molecules each of the atoms is derived). Check the answer to see if you have drawn, numbered and nominated. Many lose marks for not numbering the atoms.

Secret – Almost top student finder. These questions usually require detailed knowledge and the students are given a choice e.g. Either describe in detail the synthesis of a inosine or describe in detail the synthesis of deoxythymidine monophosphate.

Folly – memory work. No, you cannot remember this, you must understand it. For the first question students will lose marks for not describing the formation of phosphoribosylpyrophosphate, or small details like the change from the α to the β form of ribose, the controlling points of the pathway, and in which part of the cell it takes place and in which cells it takes place. A really top student will automatically include these things in their answer. More obvious things to include are structures and enzymes and cofactors e.g. ATP where required. The biggest mistake for the second part is that the student just describes the pathway from uridine diphosphate to deoxythymidine monophosphate. The lecturer had in mind (what is left of it) the synthesis of the pyrimidine ring, the conversion of a nucleotide to a deoxynucleotide as well as the

other things mentioned above. The difference between just passing on this question and getting a top mark may simply be that you neglected to express what you knew, rather than that you did not know.

Secret – now a question for the University Medal Holder. This requires piecing together disjoint information over and above knowledge of the pathways, e.g. discuss the role of amino acids in the synthesis of nucleotides.

Folly – this is impossible. This is not impossible, but it does require a little bit of planning. A quick thought will flash through your mind that amino acids are required for the ring synthesis of purines and pyrimidines, and that they are required for the final modifications of these rings to produce specific nucleotides. Use some scrap paper to sketch out these thoughts and pathways, take a few minutes to organise your thoughts, and then go for it. A good starting sentence might be that there are 3 main amino acids involved in nucleotide synthesis and these are .... They are not only used for the initial synthesis of the ring but also for modification of the rings to form specific bases. Do not forget to mention enzymes and structures, and highlight differences between the synthesis of purines and pyrimidines. Comparison with urea synthesis is also a bonus point grabber.

The key is to understand the basics and build on them. It is not a matter of straight memory work. It takes practice.

Secret – We are testing your logic and understanding, not your memory.

Salvage of nucleotides

Clearly, it is more economical in terms of energy for the body to salvage nucleotides when cells die, rather than to synthesise new ones. Nucleic acids can be looked at as polymers of monophosphate nucleosides linked by diphosphoester bonds and hence degradation of nucleic acids would form firstly monophosphate nucleosides (nucleotides).

Their are various enzymes which degrade nucleic acids. Endonucleases (*endo* = within) randomly cut up large pieces of DNA or RNA into smaller fragments. Exonucleases cleave single nucleotides from the end of the DNA or RNA strand or fragment. The nucleotidase used in digestion is produced by the pancreas and give 3'-nucleotides whereas the nucleotidase found in the lysosomes of cells produces 5'-nucleotides.

To better understand the reaction catalysed by the exonucleases, think about how the nucleotides are linked together to form the strands of RNA or DNA. Exonucleases cleave or hydrolyse (add water to) the phosphodiester bond and hence are also called phosphodiesterases. Some add water to the 3'-ester bond to cleave it and others prefer to add water to the 5' ester bond. The products formed are different. Most commonly they hydrolyse the 3'-ester bond resulting in a 5'-phosphonucleoside.

Q&A 7: Draw the reaction and show and name the different products.

Although understanding that the products are different is not exceptionally important in biochemical pathway terms, it serves to illustrate the 'sense' ($5' \rightarrow 3'$, or $3' \rightarrow 5'$) of DNA and RNA. This concept is extremely important in terms of being a practical biochemist because in this day and age all biologists become exposed to molecular biology and its techniques. It is often necessary to choose enzymes that polymerise DNA or RNA in a particular direction, or cleave DNA or RNA in a particular direction.

MAJOR POINT 68: Enzymes which catalyse the degradation or synthesis of DNA or RNA are direction specific ($5' \rightarrow 3'$, or $3' \rightarrow 5'$).

Theoretically, we could just re-use the nucleotides produced by the action of the nucleotidases, but there is one obvious problem: how they are transported from the outside of cells where they are formed to the inside of cells. The enormous negative charge conveyed on these nucleotides by the phosphate (water soluble) makes it very difficult to get them across the cell membrane (lipid).

What would have to happen to these nucleotides in order to get them across cell membranes?

Answer: *They would have to be made more lipid soluble. This is accomplished by removing a phosphate, which converts them to a nucleoside. A nucleotidase (an ecto-5'-nucleotidase) catalyses this reaction. These enzymes are actually found in the plasma membranes of many cell types. The nucleosides formed can be made more lipophilic by removing the base from the ribose group and then the base can be taken up into the cell. A nucleosidase catalyses this reaction.*

nucleotide   nucleoside   ribose-1-phosphate

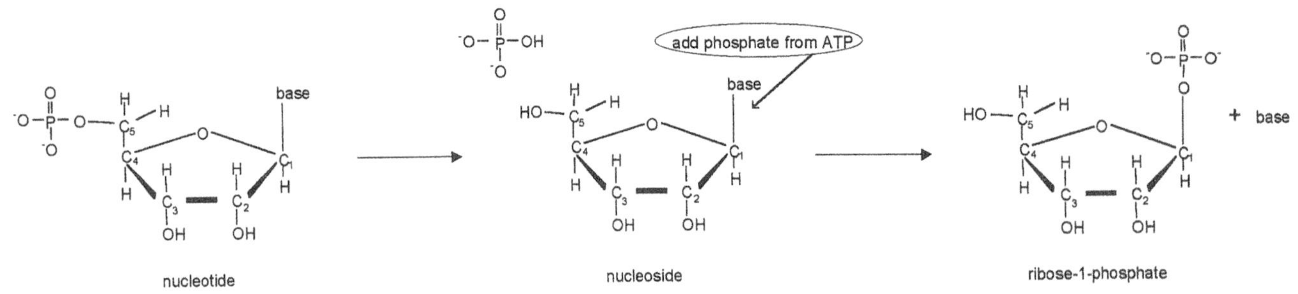

In this form, as a nucleoside or base, they may be transported across the cell membrane by specific transport proteins. This process is very important to tissues that do not readily make bases e.g. blood cells.

Nucleosides are reconverted to nucleotides by adding a phosphate group from ATP to C5 of the ribose. This results in a 5'-phosphonucleoside (nucleotide) and ADP. Remember that the name of enzymes catalyse the transfer of a phosphate are called kinases and hence the enzyme that catalyses this reaction is called a nucleoside kinase.

Bases are reconverted into nucleotides by transferring them to phosphoribosylpyrophosphate to form the corresponding monophosphate nucleoside and pyrophosphate.

Draw the transfer of hypoxanthine and guanine to phosphoribosylpyrophosphate.

Purine salvage

One enzyme catalyses the transfer of adenine to phosphoribosylpyrophosphate to produce adenosine monophosphate (AMP) and pyrophosphate.

Answer:

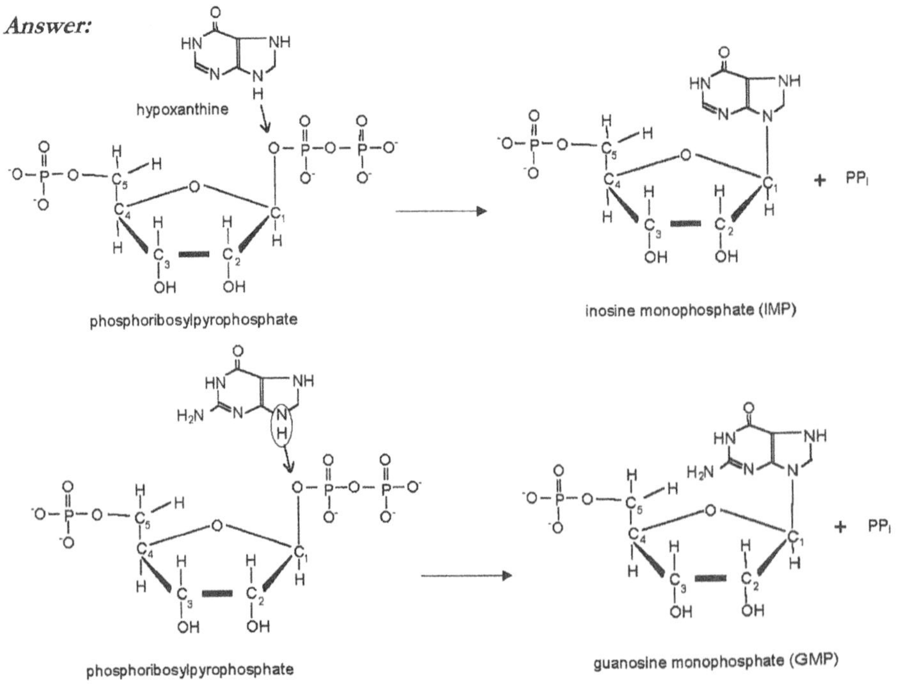

hypoxanthine

phosphoribosylpyrophosphate   inosine monophosphate (IMP)

phosphoribosylpyrophosphate   guanosine monophosphate (GMP)

The PP_i released in the reaction is hydrolysed rapidly into 2 phosphate groups and this renders the reaction irreversible. A similar reaction takes place with the purines, hypoxanthine and guanine. In this case the same enzyme transfers either hypoxanthine or guanine to phosphoribosylpyrophosphate. The transfer of hypoxanthine forms inosine monophosphate and the transfer of guanine forms guanine monophosphate.

What would be an appropriate name for the enzyme which catalyses the transfer of hypoxanthine or guanine to phosphoribosylpyrophosphate?

Answer: *Hypoxanthine-guanine phosphoribosyltransferase.*

The importance of this enzyme hypoxanthine-guanine phosphoribosyltransferase is illustrated in some people who lack it due to a genetic abnormality. One of the strange aspects of this disease, Lesch-Nyhan Syndrome, is a tendency towards self mutilation. The exact reason for this is unclear.

Pyrimidine salvage

Pyrimidines are salvaged in a similar manner to the purines but to a less degree. It differs from purine salvage in that uracil is added to ribose-1-phosphate rather than phosphoribosylpyrophosphate, and thymine is added to deoxyribose-1-phosphate.

Pyrimidine base + (deoxy)ribose-1-phosphate → pyrimidine nucleoside + Pi.

The nucleoside is then converted to a nucleotide by the transfer of phosphate from ATP to form a purine nucleotide and ADP. Of course this reaction is catalysed by nucleoside kinase.

Purine nucleoside + ATP → purine nucleotide + ADP (nucleoside kinase – irreversible).

Catabolism of bases

Degradation of purines to uric acid

The problem with removing purines is that the purine ring cannot be broken down. To excrete the purine ring it must be modified to a form that can be excreted, uric acid. To understand a pathway like this, it is a good idea to firstly examine the differences between the starting products and the end product uric acid. Both adenine and guanine must be deaminated and then oxidised.

adenine   guanine

uric acid

Degradation of adenosine

The enzymes in these reactions describe the process taking place. Adenosine is deaminated (adenosine deaminase) which produces inosine. The ribose is then cleaved from inosine to produce the free base, hypoxanthine and ribose-1-phosphate. This reaction is catalysed by the general enzyme purine nucleoside phosphorylase. Hypoxanthine is then oxidised twice to produce firstly xanthine and then uric acid. The same enzyme, xanthine oxidase, catalyses both reactions. In each of these oxidations, 1 oxygen atom from O_2 is transferred to the ring and the other is combined with NADPH and H^+ to form water and $NADP^+$.

adenosine   inosine   hypoxanthine

deaminate   remove ribose   + ribose-1-phosphate

oxidise

uric acid   xanthine

oxidise

A great success story has been achieved in some individuals deficient in adenosine deaminase. These people have a deficient immune system whereby lymphocytes fail to function properly. Gene replacement therapy has been used successfully to treat this condition.

Degradation of guanosine

The degradation of guanosine follows a very similar pattern to that of adenosine. Ribose is removed (purine nucleoside phosphorylase) to form the free base and ribose-1-phosphate. Guanine is then deaminated (guanase) to produce xanthine. Xanthine is then oxidised by xanthine oxidase as described above.

guanosine   guanine

remove ribose   + ribose-1-phosphate

deaminate

Uric acid   xanthine

oxidise

> **Q&A 9: What cofactors would xanthine oxidase use in the conversion of xanthine to uric acid?**

This process requires 2 organs: hypoxanthine is firstly formed in the **liver** and then oxidised to uric acid in the **kidneys**. Uric acid is very important for birds and many reptiles because this is how they excrete their nitrogen.

Gout is a disease of humans caused by the overproduction or under excretion of uric acid. Due to its poor solubility, excess urate in the blood crystallises and forms deposits in soft tissues such as the kidney and in the toes and joints. Allopurinol, an isomer of hypoxanthine, is used to treat gout. This is converted to oxypurinol, which is an inhibitor of xanthine oxidase. Hypoxanthine and xanthine, which forms due to the lack of production of uric acid, are more soluble than uric acid and are excreted by the kidneys.

Pyrimidine catabolism

Pyrimidines from nucleic acids are acted upon by nucleotidases and pyrimidine nucleoside phosphorylase to yield the free bases. To remove excess pyrimidine bases, the ring is simply cleaved to form soluble products, β-amino acids plus ammonia and carbon dioxide.

The critical step in this pathway is the reduction of the purine at C5 and C6. The addition of an H atom to each of these carbons removes the double bond between C5 and C6. This allows the ring to be broken by hydrolytic cleavage between N3-C4. Ammonia and then carbon dioxide are then removed to form β-amino acids. This pathway for uracil is illustrated below.

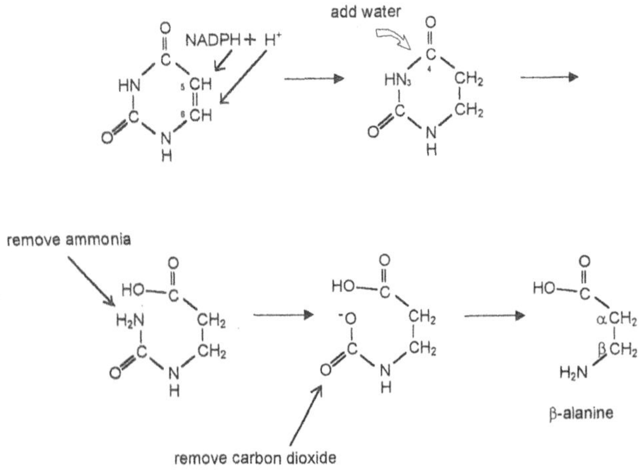

For the catabolism of cytosine there is a simple deamination to form uracil firstly.

> **Q&A 10: Draw the deamination of cytosine to uracil.**

It is hardly a surprise that the enzyme is cytosine deaminase.

> **Q&A 11: Using the above pathway for the degradation of uracil as a model, draw the pathway for the degradation of thymidine.**

> **MAJOR POINT 69:** Purine rings cannot be broken down so the ring is modified to form uric acid, which can be excreted. Pyrimidine rings can be broken down.

Interconversion of nucleotides

The monophosphates are the forms synthesised de novo, although the triphosphates are the most commonly used forms. There are several enzymes classified as nucleoside monophosphate kinases that catalyse the general reaction

$$\text{nucleoside-monophosphate} + \text{ATP} \rightarrow \text{nucleoside-diphosphate} + \text{ADP}$$

e.g. Adenylate kinase: $\text{AMP} + \text{ATP} \rightarrow 2\,\text{ADP}$

The enzymes are relatively specific and hence a different enzyme is used for the conversion of GMP to GDP, and others are used for pyrimidines. There are also enzymes that recognise the deoxy forms.

Similarly, the diphosphates are converted to the triphosphates by nucleoside diphosphate kinase:

$$\text{nucleoside diphosphate} + \text{ATP} \rightarrow \text{nucleoside triphosphate} + \text{ADP}$$

There may be only 1 nucleoside diphosphate kinase with broad specificity.

References and further reading

Alberts, B., Bray, D., Lewis, J., Raff, M., Roberts, K. and Watson, J.D. (1994) *Molecular Biology of the Cell*, 3rd ed.. New York: Garland Publishing.

Arnheim, N. and Ehrlich, H. (1992) Polymerase chain reaction strategy. *Annu. Rev. Biochem.* **61**:131-156.

Blackburn, G.M. and Gait, M.J. (Editors) (1990) Nucleic acids in Chemistry and Biology. New York, Oxford University Press.

Eaton, B.E. and Pieken, W.A. (1995) Ribonucleosides and RNA. *Annu. Rev. Biochem.* **64**:837-863.

Kelman, Z. and O'Donnell, M. (1995) DNA polymerase III holoenzyme: structure and function of a chromosomal replicating machine. *Annu. Rev. Biochem.* **64**:171-200.

Kleinsmith, J.L. and Kish, V.M. (1995) Principles of Cell and Molecular Biology, 2nd ed.. New York: Harper Collins College Publishers.

Wolffe, A.P. (1995) Genetic effects of DNA packaging. *Sci. Amer.* **2(6)**:68-77.

Q&A Answers

1 Adenine, guanine, inosine, xanthine and uric acid are all purine bases, although adenine and guanine are the ones I expected you to name.

2 Folic acid is often recommended as a dietary supplement when children are about 2 years old as there is a rapid growth spurt at this age, and when young women reach puberty. In this case, menstruation increases blood loss and hence a greater demand for new cell synthesis and at the same time, they are entering a growth spurt.

3 The transfer of the amine group of aspartate to the growing purine ring resembles the conversion of citrulline to arginine (**R**) in the urea cycle.

4 In a purine ring N3 and N9 come from glutamine; C4, C5 and N7 come from glycine; C8 and C2 come from N^{10}-formyltetrahydrofolate, N1 comes from asparagine and C6 comes from carbon dioxide.

5 6ATPs are required to convert ribose-5-phosphate to inosine monophosphate. The synthesis of purines is a very energy expensive pathway.

6 The transfer of the amine group of glutamine to uridine triphosphate to produce cytosine triphosphate is illustrated below. This reaction requires ATP which is converted to ADP and the enzyme is cytosine triphosphate synthase.

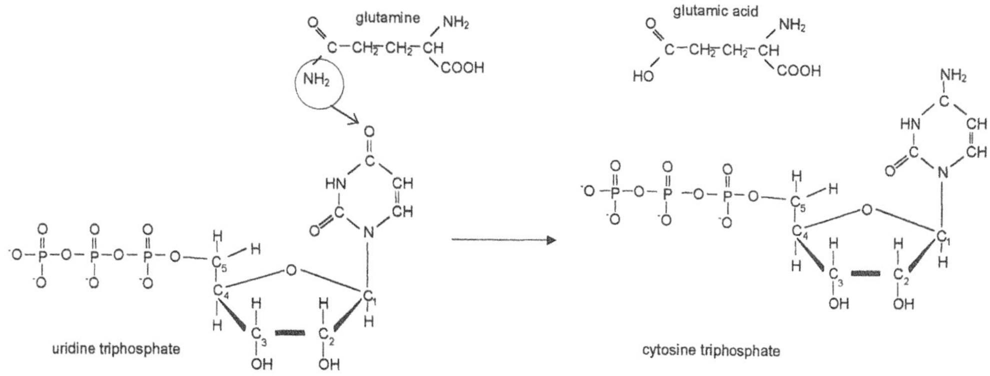

7 Nucleoside-5'-phosphate or nucleoside-3'-phosphate are formed depending upon the cleavage site of the diester bond. Some enzymes prefer the 5' site and others prefer the 3' site.

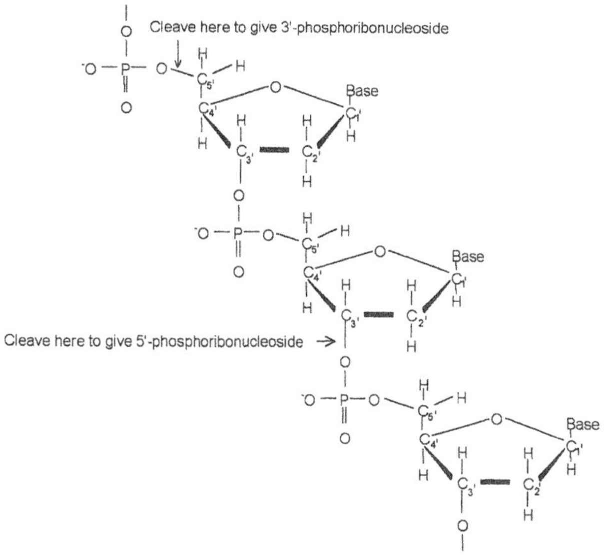

8   The enzyme which catalyses the transfer of adenine to phosphoribosylpyrophosphate is adenine phosphoribosylpyrophosphate transferase. It is part of the salvage pathway as the adenine is salvaged from dying cells.

phosphoribosylpyrophosphate   adenosine monophosphate   (adenylate)   pyrophosphate

9   In converting xanthine to uric acid, xanthine oxidase transfers 1 oxygen atom from O_2 to the purine ring and the other is combined with NADPH and H^+ to form water and $NADP^+$, hence NADPH is the cofactor.

10   For the deamination of cytosine, ammonia is removed from C4.

cytosine   uracil

11   The degradation of thymidine is exactly the same as pathway for the degradation of uracil. The only difference is the persistence of the methyl (CH_3) group at C5.

thymidine   dihydrothymidine   β-aminoisobutyrate

remove carbon dioxide

13 The Structures and Functions of DNA and RNA

In this chapter you will learn:

- the general structure of DNA
- specific forms A-DNA, B-DNA and Z-DNA
- DNA topology
- chromosome structure
- about the breakdown of polynucleotides – the different nucleases
- about restriction enzymes
- the difference between blunt and sticky ends
- the structure of RNA and how it differs from DNA
- the different types of RNA and their functions
- the basic steps in DNA replication
- the basic steps in protein synthesis
- about codons
- the mechanisms of transcription and translation
- the structure of transfer RNAs

The purpose of this section is to give an overview of the structure and the function of nucleic acids. Details about DNA synthesis, protein synthesis, laboratory manipulation of genetic material, differences between bacterial viral and other genetic material has now developed into a subject in its own right, molecular biology, and is beyond the scope of this book. Whole texts are devoted to this subject and these are commended to you.

The structure of DNA

The basis of a DNA molecule is a polymer of deoxynucleotides linked together by phosphodiester bonds.

These occur by forming phosphoester bonds between deoxyribose groups of adjacent molecules. At physiological pH, the chain of phosphate groups is negatively charged and hence very water-soluble. The charge is stabilised with Mg^{2+} or DNA binding proteins.

In actual fact, deoxytriphosphonucleosides are used to build the polymer rather than monophosphonucleosides as illustrated, and pyrophosphate (two phosphates joined together by an acid anhydride bond) is cleaved during polymerisation. Its cleavage provides the energy for the condensation of the phosphonucleosides. The enzymes, which catalyse the polymerisation of DNA, are called DNA polymerases. DNA polymerase requires all 4 deoxyribonucleoside 5' triphosphates and Mg^{2+} to be present for it to be active. The new nucleotides are always added to the 3'C of ribose and hence we always talk about synthesising DNA in the 5'-3' direction. To add more nucleotides to the chain, the 3'C of the ribose must be hydroxylated. This fact can be used to advantage in the laboratory because a DNA strand can be protected by phosphorylating the 3'C of the ribose. This trick is often used to prevent the 2 ends of a DNA molecule (5' and 3') from joining together and thus forming a loop.

> **Q&A 1: Why can't a 5'-2' diester bond form in DNA?**

DNA is a double stranded molecule comprising 2 of these polymers linked together via hydrogen bonds between the bases. This is not a random pairing, but G always pairs with C and A always pairs with T (a purine with a pyrimidine). To remember which pairs with which, G and C are 'curly' letters and A and T are 'angular' letters. In DNA, the hydrogen bonding between the base pairs are constantly being broken and reformed due to changes in conformation. This is known as breathing.

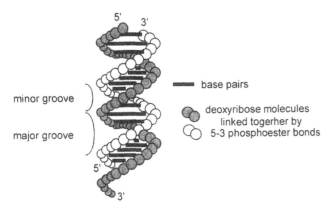

The typical helix that is associated with DNA forms because the hydrophobic bases are pushed together in the interior of the molecule and the charged phosphate groups, which are very water soluble, are pushed towards the outside. Right handed helices are more stable than left handed making this form of DNA more common. This stacking leads to 2 grooves in the molecule – a major and a minor. Within these grooves the bases are exposed to the solvent and hence access is possible for molecules which need to interact with the base pairs.

The 2 strands run in opposite directions, with one being in the 5'-3' and the other in the 3'-5' direction (antiparallel polynucleotide strands). The diameter of the helix is 20Å. The 'width' of an AT pair is exactly the same as the width of a CG pair (11Å). Each pair of complementary bases are like the treads of a spiral stairway being parallel to each other. The distance between the complementary base pairs (the rise) is 3.3Å and is equivalent to the height of a step (in stairs, the piece of wood at the back of each step is called a riser). The distance along the helical axis for one complete turn of the helix is 34Å and known as the pitch. Each base pair is rotated about 36° (34.6°) from the previous base pair. DNA molecules have in the order of 10^6 base pairs and hence we often refer to the size of DNA in terms of number of bases or kilobases. Note: 23 kilobase DNA actually means 23 kilobase pairs.

Q&A 2: How many bases will be in one complete turn of the helix?

MAJOR POINT 70: A double helix is formed from 2 antiparallel polynucleotide strands.

An important feature of DNA laboratory technology is to separate the strands so that copies of the DNA can be made. This can be done by adding acid or alkali to ionise the bases, but much more commonly, the strands are separated by heating the DNA. This occurs at a specified temperature and is called melting. The melting of the DNA is monitored by measuring absorbance of light at wavelength 260 nm and the unstacking leads to increased absorbance. This effect is known as hyperchromism. 260 nm is used because this gives the biggest difference in absorbance between denatured (single stranded DNA) and native DNA (double stranded DNA). The DNA strands spontaneously reform when the temperature is lowered. This process is called annealing.

Q&A 3: Often strands of DNA have more GC bases than AT or vice versa. Would you expect a DNA strand with predominantly more GC bases have a higher melting point than one that has predominantly more AT bases? Explain your answer.

Renaturation, or annealing, is slow because the right complementary base sequences must be found so that the 2 strands of DNA can line up. This occurs more quickly with higher salt concentrations because the negative charges on the single strands are shielded.

Other structural forms of nucleic acids

The form of DNA just described is B-DNA. This structure forms in very aqueous conditions, such as in the cell nucleus, and when the DNA contains a relatively even mixture of bases. Another α-helical form of dinucleotides, A-DNA, can also occur. This forms if water levels are low or if the double stranded structure forms between DNA and RNA or 2 RNA strands rather than 2 DNA strands. The main feature of this structure is that the base pairs almost centre on the axis of the molecule whereas with B-DNA they are offset from the centre.

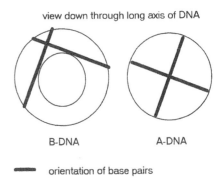

view down through long axis of DNA

B-DNA   A-DNA

▬▬ orientation of base pairs

Another major difference in structure between A-DNA and B-DNA is that in A-DNA the size of the minor and major grooves is almost the same whereas in B-DNA they are very different. A-DNA has about 11 base pairs per turn in the helix and B-DNA has 10. The key to understanding the structural difference lies in the ribose groups. The furanose rings are not flat structures and as a consequence either the C3' or the C2' can either stick up above the plane of the ring on the same side as C5' (3' endo form and 2' endo form). Comparison of these 2 forms shows a dramatic difference in the position of the ester bond to the phosphate group at C3'.

In B-DNA, the ribose is in the 2' endo form (C2' above the ring) and in A-DNA the C3' is above the ring. It is thought that in hydrated DNA (B-DNA), a string of water molecules fit in the minor grove stabilising the structure with hydrogen bonds and forcing the ribose into the 2' endo form, and when dehydrated the water is removed from the minor grove allowing the 3' endo form of ribose to occur. When RNA is involved in the double strand formation, the A form results because there is an OH group on C2' which does not exist in DNA. This extra OH group means that the furan group prefers to be in the 3' endo rather than the 2' endo form resulting in an A-structure.

If DNA is formed from alternating GC residues then a completely different DNA structure forms. Most notable is that a left hand helix forms, there are 12 base pairs per turn and virtually no grooves. The basis for this structure is that for the guanylate groups, the ribose is in the C3' endo form and the base is in the *syn* position. For the cytidylate residue the opposite is true, the ribose is in the C2' endo form and the base is in the *anti* position. This is called Z-DNA.

In nucleic acids, the nucleotides are mainly in the *anti* conformation.

Higher level

Modified nucleotides. The most common form is methylation of the bases, which occurs after the molecules is synthesised. The enzyme is a specific methylase, which recognises a particular base sequence e.g. the Dam methylase is found in *E. coli* and methylates the A in GATC. Methylated units are important as they represent protein recognition sites for recombination and DNA replication. In other cases, they prevent DNA being cleaved by endonucleases.

DNA Topology (study of overall form of DNA). DNA molecules do not exist as nice loops (bacteria) or straight strands (eukaryotes) but bend twist and fold into a variety of shapes. The study of the different shapes of DNA is known as topology. Bacteria and viruses have closed DNA or circular DNA. The loop can be broken and the strands twisted more: overwound or underwound. If this happens and then the ends rejoined then the molecule develops a twist by forming super coils. In life, these are always negative supercoils (underwound not overwound). Relief from supercoiling can be achieved by using zones of Z-DNA because this has a left hand coil. Negative supercoils can also be relived by local unwinding. This tends to be in AT rich regions. This feature is important in DNA replication and transcription.

The coiling can be changed by cutting the DNA, rewinding it and then joining it together. Such strands are called topoisomers and so the enzymes are called topoisomerases. *E. coli* topoisomerase II is also called DNA gyrase and supercoils bacterial DNA.

Chromosomes

One of the problems with the massively long strands of DNA found in eukaryotic cells is how to stop them from becoming hopelessly tangled and how to package them so that they fit into a nucleus. This is accomplished by winding the double stranded DNA around a string of proteins. DNA and protein is called chromatin and one length of DNA and its associated proteins is called a chromosome. Often the term chromosome

sin   anti   sin   anti

is used loosely just to describe the length of DNA without the protein. In humans, we have 46 chromosomes. Bacteria have single chromosomes but it is in a closed loop rather than a string.

Anecdote

This leads to interesting questions regarding chromosome number: which eukaryote has the least number of chromosomes and which has the most? Record holder for the least number of chromosomes <u>was</u> a nematode *Parascaris equorum univalens*, having a single pair of chromosomes. However this animal is unusual because, only the cells of the germ line (those which become eggs or sperm) contain 1 pair of chromosomes. These fragment several times during early development leading to adult cells with up to 70 chromosomes. The new record holder for the least number of chromosomes is to be found at Tidbinbilla Nature Reserve near the capital city of Australia, Canberra. The males of a species of ants, *Myrmecia pilosula* have only 1 chromosome per cell. In Hymenoptera (wasps, ants and bees) reproductive females usually store the sperm and either fertilise or do not fertilise developing eggs. Fertilised eggs become females (including the workers) and unfertilised eggs become males. Hence the males of *Myrmecia pilosula* have only a single chromosome (the females have a single pair of chromosomes), hence are haploid and have no father, although they do have a grandfather. The most chromosomes are found in a species of ferns, *Ophioglossum reticulatum*, which contains 630 pairs of chromosomes, 1260 chromosomes all together. It is only to wonder how they all find a spot on the spindle during mitosis.

The major protein components of chromosomes are called histones. There are 5 histones H1, H2A, H2B, H3 and H4.

> **Q&A 4:** Considering that histones bind to DNA would you expect these histones to contain a predominance of basic amino acids like lysine (K) and arginine (R) or acid amino acids like glutamate (E) and aspartate (D)?

Histones are only found in eukaryotes and are clustered into groups called nucleosomes. These can be seen if chromatin is treated with solutions of low ionic strength. The chromatin unfolds giving rise to a structure that resembles beads on a string; the beads being the nucleosomes. Each nucleosome is composed of 2 molecules of H2A, H2B, H3 and H4 and about 200 base pairs of DNA The DNA is wrapped around the outside about 1.75 turns. The DNA between the core particles is called the linker DNA and is about 54 base pairs long, the fifth histone H1 binds to the linker DNA and to the nucleosome core particle. It is involved in higher order chromatin structures and is depleted in the beads on a string structure. The nucleosomes with their associated DNA are coiled into a solenoid to yield a 30nm fibre. Decondensation which occurs after cell division, is brought about by HMG proteins (high mobility group proteins). These attach to the nucleosomes and prevent binding of histone H1. The 30 nm fibres form large loops that are attached to a protein scaffold of the chromosome.

Breakdown of polynucleotides

Nucleases catalyse the hydrolysis of phosphodiester linkages of nucleic acids. Some are required for the synthesis and repair of DNA and others for the production and degradation of RNA. There are ribonucleases, deoxyribonucleases, exonucleases, and endonucleases. These can form 3' phosphonucleosides or 5' phosphonucleosides depending on which part of the phosphate molecule they cleave. They cleave one or the other but not both.

Exonuclease III is found in *E. coli* and hydrolyses 1 nucleotide at a time from the 3' end of double stranded DNA until only a short region of double-stranded DNA remains. This is used to repair DNA. Restriction endonucleases catalyse hydrolysis of double-stranded DNA at specific sites. These restriction enzymes are commonly used in molecular biology as part of the technique known as gene cloning. Consequently, there are books with lists of different restriction enzymes, which are specific for different DNA sequences. The sites for endonucleases are always palindromes e.g. the site for the restriction enzyme, EcoRI, from *E.coli*.

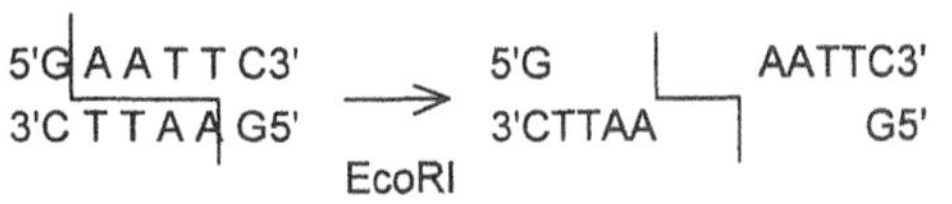

Some endonucleases make blunt ends and others make sticky ends. A sticky end is the result of a staggered cleavage, which produces fragments with a single-stranded extensions at their ends. The advantage of sticky ends, as shown above, is that they can reform without making a mistake. A blunt end can ligate with any other blunt end. Different strains of *E. coli* have different restriction endonucleases Eco K comes from strain K and Eco B from strain B and this prevents exchange of DNA between strains.

Structure of RNA

RNA differs from DNA in 3 major features

* The sugar group is ribose rather than deoxyribose.
  The 2' hydroxy group in RNA makes is more chemically reactive and less stable, e.g. NaOH 0.1M at room temperature will degrade RNA into 2' 3' nucleoside monophosphates. RNases, which break down RNA, are very active and ubiquitous. Their breakdown of RNA is one of the biggest problems in laboratories when making libraries of the genetic information in a cell from the messenger RNA.

* The pyrimidine uracil (U) replaces thymidine (T).

- It occurs usually as single rather than double strands, but complementary regions pair and form hairpin bends. This is important in transcription and is a feature of transfer RNA (tRNA) and ribosomal RNA (rRNA).

There are several different forms of RNA classified according to function.

- Messenger RNA: mRNA - also called polyA RNA. This encodes the sequences of amino acids for the synthesis of proteins. It obtains this message from the nuclear DNA and carries it from the nucleus to the ribosomes in the cytoplasm. 3% total cellular RNA.
- Transfer RNA: tRNA - this carries activated amino acids to the ribosomes for the incorporation into growing peptide chains during protein synthesis. 15% of total cellular RNA.
- Ribosomal RNA: rRNA - this RNA complexes with proteins and provides a docking station for tRNA and mRNA so that proteins can be synthesised. 80% of total cellular RNA.
- Small RNA: sRNA - this catalyses reactions that modify RNA once it has been synthesised.

DNA replication

For a cell to divide the DNA must be duplicated. Essentially to duplicate DNA, the DNA splits down the middle into its 2 complementary strands and then new strands are made from these original templates. Both new strands are made simultaneously.

MAJOR POINT 71: Both new strands of DNA are made simultaneously.

The whole process is much more complicated than it appears. The enzyme that catalyses the synthesis of the new chain is DNA polymerase and it will only synthesise in the 5'-3' direction and will only add a new nucleotide to a 3' hydroxylated deoxyribose or ribose. Therefore, how are bases added to the chain which is increasing from the 5' end, and how does the synthesis start at all since a 3'-hydroxylated deoxyribose is required for DNA polymerase to act upon for chain elongation to begin?

To start the chain, a short piece of RNA is synthesised. This is called an RNA primer. The enzyme that catalyses this is called a primase. Unlike DNA polymerase, it does not need a 3-hydroxylated ribose to begin its action. For the leading strand this happens once, for the lagging strand this happens repetitively. The RNA primer provides the 3-hydroxylated ribose onto which the DNA polymerase can begin to add nucleotides.

DNA polymerase III (in *E. coli*) synthesises both the leading and the lagging strands. When the lagging strand reaches the downstream RNA primer there is a problem: the RNA primer must be replaced with DNA. This occurs 1 base at a time and therefore 1 RNA nucleotide is nicked out and replaced by a deoxyribose nucleotide. This is called nick translation (the alternative would be to remove RNA as a block and then replace it. This would be gap translation). DNA polymerase I carries out this function. Once

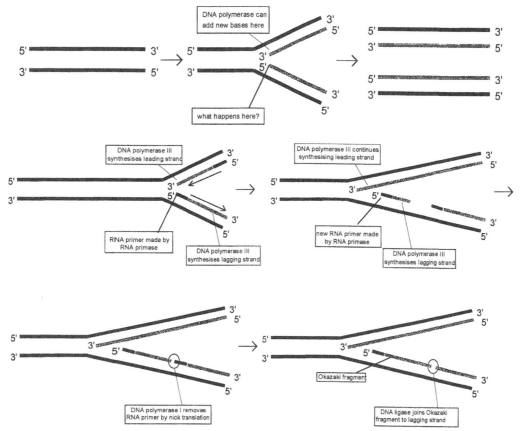

this is accomplished the lagging fragment, also known as the Okazaki fragment, must be ligated to the rest of the lagging strand. Another enzyme, DNA ligase in conjunction with DNA polymerase I is responsible for this. DNA polymerase I also repairs damaged DNA. It has a proof reading function and if it detects an incorrect base then it cleaves this out (exonuclease activity) and replaces it with the correct base.

A problem occurs with the lagging strand when synthesising linear DNA (i.e. in eukaryotes rather than in bacteria, which have closed loop DNA) and this problem is how to reproduce the terminal end. If no special step is taken then the chromosome would become shorter each time it replicates. This occurs because the RNA primer needs to be replaced by a lagging strand using nick translation and DNA ligase. Right at the very terminal there is no lagging strand to promote this function and therefore the RNA primer has to be dealt with in a different way. This problem is combated in 2 ways. The average chromosome terminates in a stretch of up to 10 000 base pairs consisting of short repetitive sequences with the template strand rich in T and Gs. These terminal regions are known as telomeres. Some shortening of this region does not affect the genome and may be the reason for the limitation on numbers of cell divisions. Shortening is also minimised by enzymes called telomerases. These contain a RNA molecule whose sequence is complementary to the DNA ends and hence it can bind to the terminus and allow DNA replication back from the fork. Histones, the proteins about which DNA is folded in eukaryotes, are reproduced at the same time as DNA is being replicated.

Three other proteins are needed for this replication to take place. A protein is needed to unwind the DNA strand. This is known as a helicase and there is 1 for each strand. The one associated with the lagging strand is bound with the primase and together they are called a promosome. For the helicases to work, they need energy input and this comes from hydrolysis of ATP. A protein is also needed to bind to the unbound bases so that the original strands do not fuse together again before replication. This is called a binding protein. As the DNA strand unwinds, the remaining parental DNA duplex strand becomes supercoiled, which places enormous stress on this part of the DNA molecule. Unless this were relieved, the molecule would become impossibly tangled and break. A topoisomerase is working hard in this region breaking and rejoining the DNA so that this supercoiling is relieved. Once the DNA has been duplicated the cell can divide.

DNA replication in eukaryotes is similar to DNA replicaton in prokaryotes. There are some obvious differences. Eukaryotes have substantially more DNA than prokaryotes and the DNA is linear (several chromosomes) rather than in a loop. If, like bacterial DNA, it only had 2 origins of replication on each chromosome, replication would be extremely slow due to the large size of the chromosomes. To overcome this each chromosome has a large number of independent origins of replication and replication proceeds bidirectionally from each origin and the forks from different origins move towards one another. This enables replication to take about the same amount of time as prokaryotic genomes. The origins of replication are generally AT rich sites (autonomously replicating sequence or ARS consensus sequences) and this is where initiator proteins bind.

> **Q&A 5:** Why would it be expected that origins of replication sites of DNA be rich in AT rather than GC bases?

There are several different DNA polymerases in eukaryotes. DNA polymerases α, δ, and ε are found in the nucleus γ in mitochondria and another in chloroplasts. DNA polymerase α is responsible for replication at the lagging strand. Similarly, there are many different helicases, topoisomerases and ligases in eukaryotic cells. The protein that protects the separated strands of DNA from fusion is called replication protein A.

Protein synthesis

Genes are regions of DNA that are transcribed. Some of these regions are transcribed into rRNA and others are transcribed into mRNA. The regions transcribed into mRNA code for proteins. Note that this means that not all genes code for proteins.

The information to produce proteins is contained in the sequence of bases in the DNA. This is called the genetic code. The information is first transcribed to a similar material, mRNA, so that the information can be transported out of the nucleus to the cytoplasm. In the cytoplasm, the code is translated into a different language, amino acids, which are linked together via peptide bonds to form a protein.

Only some of the DNA in a cell contains information for building proteins, and the rest is used for housekeeping. Within a gene, a sequence of 3 bases read in the direction of the 5' to the 3' ribose linkage, codes for 1 amino acid. Such a sequence is called a codon. Since there are 4 different bases, there are 4^3 different ways of grouping them into 3s. There are only 20 amino acids used in proteins and hence some of the amino acids are represented by more than 1 of the 64 codons. This is called the degenerate codon system. The codons and the amino acids they represent is given below for RNA. In RNA the pyrimidine uracil replaces thymidine. Some of the sequences do not code for amino acids but are stop or start codons. This informs where the gene starts and where it stops.

UUU	Phe	UCU	Ser	UAU	Tyr	UGU	Cys
UUC	Phe	UCC	Ser	UAC	Tyr	UGC	Cys
UUA	Leu	UCA	Ser	UAA	Stop		
UUG	Leu	UCG	Ser	UAG	Stop	UGG	Trp
CUU	Leu	CCU	Pro	CAU	His	CGU	Arg
CUC	Leu	CCC	Pro	CAC	His	CGC	Arg
CUA	Leu	CCA	Pro	CAA	Gln	CGA	Arg
CUG	Leu	CCG	Pro	CAG	Gln	CGG	Arg
AUU	Ile	ACU	Thr	AAU	Asn	AGU	Ser
AUC	Ile	ACC	Thr	AAC	Asn	AGC	Ser
AUA	Ile	ACA	Thr	AAA	Lys	AGA	Arg
AUG	Start Met	ACG	Thr	AAG	Lys	AGG	Arg
GUU	Val	GCU	Ala	GAU	Asp	GGU	Gly
GUC	Val	GCC	Ala	GAC	Asp	GGC	Gly
GUA	Val	GCA	Ala	GAA	Glu	GGA	Gly
GUG	Val	GCG	Ala	GAG	Glu	GGG	Gly

Transcription and translation

The message on the DNA is transcribed to mRNA in much the same way as the DNA strands are duplicated. The DNA is partially uncoiled and RNA polymerase synthesises a strand of RNA. The message is transcribed from the complementary DNA strand so the mRNA strand has the same message as the sense DNA strand. Note that there is not a sense and an antisense DNA strand as such. The genes for some proteins are contained on 1 strand and the genes for other proteins on the other. The mRNA moves into the cytoplasm where it docks with a ribosome. The ribosome is a mixture of proteins and RNA called ribosomal RNA (rRNA). Besides a docking station for mRNA, it is also a docking station for transfer RNA (tRNA). tRNAs reside in the cytoplasm. On one end they have an antisense codon (3 antisense bases) representing an amino acid and the other end a binding site for a specific amino acid. Hence there are about 61 of them. These bind to specific amino acids and transfer them to the ribosome where their antisense codons dock with the mRNA according to the order prescribed by the mRNA template. In this way, amino acids are brought together in the ribosome in a prescribed sequence and they are condensed to form peptide bonds.

RNA transcription from DNA is begun at promoters. These are equivalent to, but not the same as, the origins of duplication used for DNA replication. The promoter regions are located upstream from the DNA sequence (gene) that codes for the protein to be produced. Typically there is a TATAAT in the -10 region and a TTGACA in the -35 region in E. coli, and in eukaryotes there is a TATA rich section in the -35 region. This is called the TATA box. It is not uncommon for 2 genes, which code for proteins that are part of the same metabolic pathway, to be coded for next to each other. Both genes then have the same promoter and hence both proteins are produced simultaneously. Such a region of DNA is called an operon.

RNA polymerase also has a helicase and primase function. These functions are carried out by different enzymes in DNA replication. At termination of the RNA synthesis, the DNA template comprises a series of interrupted complementary base pairs (complementary base pairs with a region of bases between them, which are not complementary). When these sequences are copied to the RNA strand, the RNA folds back onto itself with the complementary bases aligning with each other and the non-complementary region in the middle provides the hinge region to allow it to fold back. This puts a hairpin bend in the RNA which dislodges it from the DNA. At the very terminus there is dA, rU pairing which is relatively weak and further encourages the RNA to break away from the DNA template. Once activated, hundreds of copies of mRNA are made from the gene. RNA polymerase does not have exonuclease proofreading activity and hence there are more mistakes in RNA than occur in DNA. This is not very important because many of the mistakes will lead to the same protein being produced anyway due to amino acids having more than 1 codon (see above), and if an altered protein were formed it would only be a small fraction of the total protein produced from that particular gene.

RNases are very active and newly synthesised RNA is degraded quickly. It has an average half life of 30min in eukaryotes (3 min in prokaryotes). mRNA is protected to some degree by the addition of a polyA tail. Poly A polymerase adds up to 250 As to the 3' end. Such modifications are called post translational modification. mRNA is often referred to as polyA RNA because of this modification. The poly A tail protects the reading sequence from RNA exonuclease activity i.e. it takes longer for the RNA exonuclease to reach the reading sequence. The 5' ends of messenger RNA are capped with GTP to create a GTP 5'5' triphosphate linkage. This protects this end from degradation.

There are 3 different RNA polymerases in eukaryotes. RNA polymerase type I transcribes ribosomal genes, RNA polymerase type II transcribes the protein genes and type III transcribes small RNA molecules.

In eukaryotes different cells have different functions and yet all contain the same genetic information. Therefore, which genes are transcribed is different in different cells and must be controlled. Transcription can be negatively regulated by a repressor protein or positively regulated by an activator protein or both. A repressor protein binds in the box region of the DNA and prevents replication. Signals to the cell can activate other proteins, which bind to the repressor and inactivate it. These proteins are called inducers. Activator proteins bind upstream of the TATA box and make binding of RNA polymerase to this region easier. Some genes have dual promoters and hence will be activated in one cell type with one signal and in another cell type by a different signal.

Transfer RNA

tRNAs are made from large primary transcripts and then trimmed to their mature lengths by ribonucleases. Many of the bases in tRNA undergo post translational modification, particularly methylation. Complementary regions means that it folds into a T shape. Three dimensionally, it actually has helical structure as well.

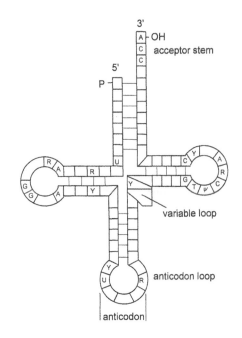

The process of translation is illustrated below. Note that several ribosomes will be active on the one messenger RNA strand and hence multiple copies of the protein are being made at once.

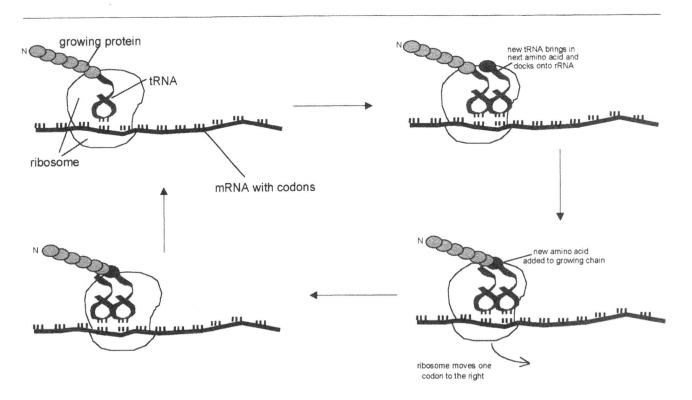

tRNA is loaded with a specific amino acid by firstly transfer of the amino acid to ATP and then from the aminoacyladenylate to the 3' or 2' C of the 3' terminal ribose of the tRNA. The enzyme that catalyses the transfer to the 2'C is called aminoacyl-tRNA synthetase class I and to the 3'C, aminoacyl-tRNA synthetase class II.

action of aminoacyl transferase class II

References and further reading

Alberts, B., Bray, D., Lewis, J., Raff, M., Roberts, K. and Watson, J.D. (1994) *Molecular Biology of the Cell,* 3rd ed.. New York: Garland Publishing.

Aldhous, P. (1993) Managing the genome data deluge. *Science* 262:502-503.

Arnheim, N. and Ehrlich, H. (1992) Polymerase chain reaction strategy. *Annu. Rev. Biochem.* **61**:131-156.

Blackburn, G.M. and Gait, M.J. (Editors) (1990) *Nucleic acids in Chemistry and Biology.* New York: Oxford University Press.

Chan, L. and Seeburg, P.H. (1995) RNA editing. *Sci. Amer. Science and Medicine* **2(2)**:68-77.

Clayton, D.A. (1995) Transcription of the mammalian mitochondrial genome. *Annu. Rev. Biochem.* **53**:573-594.

Conaway, R.C. and Conaway, J.W. (1993) General initiation factors for RNA polymerase II. *Annu. Rev. Biochem.* **62**:161-190.

Echols, H. and Goodman, M.F. (1991) Fidelity mechanisms in DNA replication. *Annu. Rev. Biochem.* **60**:477-511.

Ellis, R.J. and vand der Vies, S.M. (1991) Molecular chaperones. *Annu. Rev. Biochem.* **60**:321-347.

Friedberg, L.C. (1996) Relationships between DNA repair and transcription. *Annu. Rev. Biochem.* **65**:15-42.

Kleinsmith, J.L. and Kish, V.M. (1995) *Principles of Cell and Molecular Biology,* 2nd ed.. New York: Harper Collins College Publishers.

Kornberg, A and Baker, T.A. (1992) *DNA replication.* 2nd Ed. San Francisco: W.H. Freeman and Co.

MacDougald, O.A. and Lane, M.D. (1995) Transcriptional regulation of gene expression during adipocyte differentiation. *Annu. Rev. Biochem.* **64**:345-373.

Noller, H.F. (1991) Ribosomal RNA and translation. *Annu. Rev. Biochem.* **60**:191-227.

Noller, H.F. (1993) tRNA-mRNA interactions and peptidyl transferase. *FASEB J.* 7:87-89.

Wood, R.D. (1996) DNA repair in eukaryotes. *Annu. Rev. Biochem.* **65**:135-168.

Zimmermann, R.A. (1995) Ins and outs of the ribosome. *Nature* **376**:391-392

Q&A Answers

1   A 5'-2' diester bond cannot form in DNA because <u>deoxy</u>nucleotides are strung together. These do not have an hydroxyl group at C2' of the ribose (deoxy).

2   One complete turn of the helix is 360° and there is about a 36° turn between each base pair, hence there are 10 bases in each complete turn of the helix.

3   A strand with predominantly GC bases would have a higher melting point than a strand with predominantly AT bases because each GC pair has 3 H bonds whereas each AT pair has only 2.

4   Histones have a predominance of basic amino acids, lysine (**K**) and arginine (**R**), as they are binding to an acid (deoxyribonucleic acid). These are positively charged at neutral pH and hence bind to the negatively charged phosphate (phosphoric acid) backbone.

5   An AT rich region has only 2 hydrogen bonds between the base pairs and the GC regions have 3. Hence, it is very much weaker and needs less energy to separate the strands to begin replication.

14 Metabolism and its Control

In this chapter you will learn:

- the definitions of metabolism, catabolism and anabolism
- the concept of oxidation as an energy source
- about energy units
- energy requirements and basal metabolic rate
- macronutrients and micronutrients
- carbohydrates, fats and proteins as energy sources
- dietary fibre
- essential and non-essential amino acids
- essential fatty acids
- vitamins and minerals
- fat-soluble and water soluble vitamins
- vitamin A and its role in vision
- the effects of vitamin deficiencies and vitamin excess.
- about digestion
- about the control of metabolism in muscle, liver and fat cells
- how metabolism in muscle differs in anaerobic and aerobic conditions
- how glycolysis is regulated in muscle and liver
- the roles of insulin glucagon, adrenaline and corticosteroids in controlling metabolism
- regulation of glycogen metabolism in muscle and liver
- about the control of the TCA cycle and choice between substrates for the TCA cycle
- about the control of lipid oxidation in muscle
- how the liver as a supplier of glucose for the rest of the body and how gluconeogenesis is controlled
- about glycogen synthesis and storage in the liver
- about metabolism of proteins by the liver
- about the function of the TCA cycle in the liver
- about the formation of ketones from fatty acids and amino acids in the liver
- about synthesis and storage of fats by adipose tissue

Metabolism and energy

Nomenclature: Metabolism comprises catabolism and anabolism. Catabolism is all of the processes where molecules are broken down by the body. This usually generates energy. Anabolism is all of the biosynthetic pathways where new chemicals are made by the body. These pathways require energy.

Food as an energy source

We need energy to warm our houses, to generate electricity and to move motor cars. To achieve this we burn fuel. What does burn actually mean? Another word for burning is <u>oxidation</u>. We need a hydrocarbon, such as wood or oil, and oxygen. A hydrocarbon is made up of hydrogen and carbon and hence oxidation of this gives oxides of carbon and hydrogen - carbon dioxide (CO_2) and dihydrogen oxide (H_2O) and of course energy in the form of heat and light.

To obtain energy in our bodies, we oxidise fuel (food) to produce carbon dioxide, water and energy. However, we cannot have little fires all over our body and therefore the burning process in our bodies is very controlled and is called respiration. Like fire, the oxidation releases some heat, but other energy is captured in chemical bonds.

Besides being fuel, food is also a source of complex chemical structures that can be made by certain plants and bacteria, but not by us e.g. vitamins, essential fatty acids, essential amino acids, and minerals.

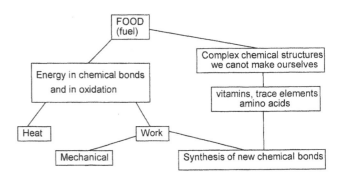

The primary source of energy to carry out these processes is light. In plants, the energy from light is captured and used to form chemical bonds in a process called photosynthesis. Animals then eat plants (or eat animals which have eaten plants) to obtain energy by breaking the chemical bonds and by oxidising some of the atoms to form CO_2 and H_2O. The energy gained is used to carry out anabolic processes and to provide heat.

Energy in food is measured in Calories or kilojoules. In Australia, the SI unit kilojoules (kJ) has been adopted, but it is still very common for people to refer to Calories in food. In the United States of America, Calorie is still the accepted measure. One calorie is the amount of energy required (heat) to increase the temperature of 1 gram of water by 1°C. One

Calorie (with a capital C) is 1000 calories (lower case c) or 1 kilocalorie (kcal). One kilojoule (1000 joules) is 0.239 Calories (Kcal) or 1 Kcal is 4.184 kJ.

We need a minimum amount of energy to survive. This energy is used for essential processes: cell metabolism, heart and lung function, digestive and kidney function and the maintenance of body temperature. This is related to the basal metabolic rate (BMR) which is generally regarded as the amount of energy consumed during early morning sleep.

We all require about the same amount of energy to make our hearts beat and to breathe. The biggest factor affecting basal metabolic rate is the requirement for heat. The amount of heat we lose is dependent on our surface area and therefore, BMR is given in terms of surface area. For females this is about 155 kJ/m²/h (37 kcal/m²/h) and for males 170 kJ/m²/h (40 kcal/m²/h). Charts called nomograms, which relate height and weight to surface area, are available for determining surface area. BMR is affected by hormones. It is increased by adrenaline, thyroid hormones and androgens. The amount of energy required to just stay alive for 24 h is called the minimum energy requirement or MER and is 24 x BMR.

Nutritional value of food

Food contains the energy and the atoms and molecules we need for building new molecules. It is commonly divided into 2 major groups: macronutrients and micronutrients. The macronutrients are carbohydrates (polysaccharides), proteins and fats. The micronutrients are vitamins and minerals.

Macronutrients

All of the macronutrients are sources of energy, and carbon and hydrogen atoms. The proteins are also a source of nitrogen and sulphur atoms. In addition, we cannot synthesise some of the amino acids and fatty acids we need, and hence these essential amino acids and essential fatty acids must be obtained in the proteins and fats we eat. The energy yields for the different major macronutrients and alcohol are given below.

Food type	Recommended % of dietary intake.	kJ/g oxidised
Protein	14	17
Fat	16	37
Carbohydrate (mainly complex)	67	17
Alcohol	3	29

Carbohydrates (polysaccharides)

For most of the world's people, carbohydrates, such as rice, beans, millet, cassava, pasta or bread, are the staples in their diets. In Australia and other developed countries, carbohydrates

are spurned for foods that are rich in protein and fat. The primary use for carbohydrates is to produce energy. It is recommended that the total energy intake derived from complex carbohydrates should be 58% and that no more than 10% be derived from simple sugars.

Unlike protein and fat, carbohydrates can be converted to energy by every cell in the body. Red blood cells and some cells in the brain and kidney must have a **constant** supply of carbohydrate for their energy. Under normal circumstances, the brain uses roughly 125 g of carbohydrate (over 2000 kJ, a bit less than the energy contained within 2 Mars bars) per day. Hence, carbohydrates are brain food! Carbohydrate intake is very important to diets (especially a weight reduction diet) because it ensures that protein can be used for growth and repair rather than energy. This is called the 'protein sparing' function of carbohydrate.

Many processed foods have refined simple sugars added e.g. sucrose. A 200 g can of cola drink has about 41 g of sugar. Such foods, typical in Western diets, have high sugar content but few nutrients. Hence these foods are called 'empty calories'. It is much better to have fruit, vegetables, grains and other nutrient dense foods.

There is no direct link between obesity and excessive sugar consumption. However, sugar consumption certainly contributes to obesity. Foods that are high in sugar tend to be high in fat and in general, fat intake increases in direct proportion to sugar intake. Cutting down on sugar is probably paralleled by a reduction in fat intake.

Dietary fibre

Another form of carbohydrate in the diet is as dietary fibre. This form of carbohydrate is not digested. There are 2 types: insoluble and soluble. Soluble fibre forms a gel-like solution when combined with water. Soluble fibre slows down the passage of food through the digestive tract. Oat bran, dried beans, vegetables and the pulp of fruits are rich sources of soluble fibre. Some types of soluble fibre appear to decrease cholesterol absorption which has lead to the popularity of oat bran. Insoluble fibre facilitates the movement of food through the digestive tract, but it tends to bind with certain minerals during digestion and decrease their absorption. The bran that covers wheat, rice, corn and other plant whole grains is rich in insoluble fibre.

Deficiency of dietary fibre intake appears to increase the likelihood of having haemorrhoids or cancer of the colon. Americans eat about 13 g of fibre each day and an intake of 25–35 g/day is considered desirable. Make sure that if you decide to increase the fibre content of your diet that you do it slowly because it takes the intestines and their bacterial colonies time to adjust to the larger supply of fibre. If it is increased too quickly, diarrhoea or constipation could occur.

Proteins

Proteins are essential structural components of all living matter, are involved in almost all biological steps (enzymes) and constitute more than half the dry weight of cells (but not adipocytes of course).

> **Q&A 1:** What element(s) is contained in proteins and not in lipids or carbohydrates.

In some third-world countries, protein deficiency is a major nutritional problem. In Australia, the typical diet is high in protein, but this is not necessarily associated with benefits to health. This is because high-protein diets are usually rich in fat. For example, although a hamburger is a good source of protein, almost half of the energy in a hamburger (47.5%) comes from fat.

Essential and non-essential amino acids. Proteins are made up of about 20 amino acids. Nine are essential and eleven are non-essential. The body can manufacture the non-essential amino acids from carbohydrates and nitrogen but it cannot synthesise the essential amino acids, which it must obtain from the diet.

Make a list of essential and non-essential amino acids.

Answer:

Essential (9)	**Non-essential** (11)
leucine	alanine
isoleucine	arginine
lysine	asparagine
phenylalanine	aspartic acid
threonine	cysteine
tryptophan	glutamic acid
valine	glutamine
methionine	glycine
histidine	proline
	serine
	tyrosine

Learning strategy: <u>No</u> amino acid with a name starting with A or G (think of agriculture) is an essential amino acid. Amino acids with branched R groups or complex ring structures are all essential except for tyrosine (**Y**), which can be made from phenylalanine (**F**).

Most proteins found in animal products contain all of the amino acids and hence are complete proteins; whereas most of those found in plant products are incomplete proteins. Vegetarians need to choose complementary foods and eat them at the same meal. Corn lacks lysine (**K**), threonine (**T**) and tryptophan (**W**) (for corn, these are 'limiting' amino acids); beans lack cysteine (**C**) and methionine (**M**) - but the Indians in South America combined both in a meal to make a dish called succotash and therefore remained healthy.

> **MAJOR POINT 72:** The lack of just 1 essential amino acid halts protein production.

> **Q&A 2:** Each day we need about 0.75 g of protein/kg body weight. If you weigh 80kg, how much protein would you require per day?

Fats

Like carbohydrates, fats consist of carbon, oxygen and hydrogen. However, fats contain more hydrogen and less oxygen than carbohydrates. Simply put, fats are less oxidised that sugars and hence are capable of delivering about twice as much energy/g as carbohydrates. Synthesis of fats requires a large amount of energy and hence we tend to store fats from our diet rather than use them for energy. We only use them as energy sources if other sources of energy are insufficient. This is why it is so easy to store fat and so difficult to lose fat when dieting.

Linoleic, linolenic and arachidonic acids are essential fatty acids. Although linolenic and arachidonic acids are synthesised by the body from linoleic acid, they are synthesised in insufficient quantities for our needs. The essential fatty acids are needed for proper cell membrane formation and for synthesis of a group of chemicals called prostaglandins.

Fats obtained from animals are generally saturated and those from plants are commonly polyunsaturated. However, there are some exceptions: coconut, palm and cocoa oils are highly saturated and oils from mackerel, salmon and herring are polyunsaturated.

It is recommended that only about 30% of our daily energy comes from fat and that equal amounts of polyunsaturated, monounsaturated and saturated fats are consumed. Cholesterol is also a fat and is needed for synthesising cell membranes and steroid hormones. It is only found in animals and we can make all that we require. Therefore, we do not need cholesterol in our diets and a high intake of cholesterol contributes to high blood cholesterol levels, which is a major factor in cardiovascular disease.

Micronutrients

Micronutrients are vitamins and minerals. The word vitamin is derived from amines vital for life. Vitamins A, B [Thiamine (B_1), Riboflavin (B_2) and Niacin (B_3)] and C are commonly listed in food content tables. Vitamin A was discovered first, followed by vitamin B. It was later discovered that vitamin B was actually several different compounds and so it is also called vitamin B complex. Other compounds were thought to be vitamins, but then found not to be. Thus vitamins B_4, B_5, B_7, B_{11}, F, G, H, I and J are not vitamins. The following vitamins are now recognised: A, B_1, B_2, B_3, B_6, B_{12}, C, D, E, K and other unlettered vitamins. Vitamins act as coenzymes and antioxidants.

> **MAJOR POINT 73:** Some vitamins are lipid soluble and others water-soluble.

Lipid soluble vitamins are stored in our fat deposits, and water-soluble vitamins are constantly flushed from our bodies. Therefore, we can do without lipid soluble vitamins for a reasonable amount of time, but we must keep replacing the water-soluble vitamins.

Water soluble vitamins		Fat soluble vitamins	
B_1	thiamine	A	trans retinal
B_2	riboflavin	D	antirachitic factor (calciferol)
B_3	niacin	E	α-tocopherol
B_6	pyridoxine	K	coagulation factor
B_{12}	cyanocobalamin		
C	ascorbic acid		

Vitamin A

Vitamin A, or all-*trans*-retinol, is lipid soluble and is found in meat, particularly the liver of some animals. It has an alcohol group and therefore can form ester bonds with fatty acids. This occurs in the liver where it is stored as esters. It can be derived from the pigment ß-carotene (a provitamin). The conversion of β-carotene to retinol is about 6:1 (6µg of ß-carotene = 1µg of retinol).

all-*trans*-retinol
vitamin A

retinoic acid

can form esters with fatty acids for storage

β-carotene

in vision all-*trans*-retinol is converted to 11 *cis*-retinal

all-*trans*-retinol

11 *cis*-retinal

Q&A 3: Which vegetable would be a rich source of the orange pigment β-carotene?

Vitamin A has an important role in vision. In photoreceptors called rods, all-*trans*-retinol is converted to 11-*cis*-retinal (a dehydrogenation and isomerisation) in the dark. The aldehyde terminus is then linked to a protein called opsin to form rhodopsin. Rhodopsin is a photopigment and can capture the energy in photons. When light strikes it, the 11-*cis*-retinal component of rhodopsin is converted to all-*trans*-retinal. The pigment, now called bathorhodopsin, then triggers a series of events that leads to neuronal <u>inhibition</u>. What a surprise! Our photoreceptors are actually inhibited by light! One effect of vitamin A deficiency is decreased vision in dim light, known as night blindness.

Both retinol and retinoic acid bind to intracellular receptor proteins and then bind to chromosomes to regulate gene expression and cell differentiation. Therefore, vitamin A deficiency causes cessation of growth and dryness of epithelium,

particularly of the eye. **RDI** (recommended daily intake) of retinol = 750µg.

Anecdote – an Antarctic tale

In 1913, the Australian geologist Douglas Mawson was exploring near the South Magnetic Pole with 2 companions, Mertz and Ninnis. They were over 480 km from home base when Ninnis and 1 of the 2 dog sledges fell into a huge crevasse. In bad weather, Mawson and Mertz pressed on with the remaining sledge and few supplies. With food supplies low they were forced to eat their dogs. Following this, both Mawson and Mertz suffered from blurred vision, pain in the bones and joints, headaches, dry skin and poor appetite. After 320 km, Mertz died. Mawson continued and when his boots gave way, he walked on the ice in feet wrapped in rags. It took him 1 month in appalling weather to cover the last 160 km to home-base. It is believed that the symptoms of both Mawson and Mertz and the death of Mertz was caused by vitamin A excess. The dogs, which they ate, were fed seal meat. The seals ate deep sea fish which are extremely rich in vitamin A. Hence the vitamin A accumulated down the food chain and so when Mawson and Mertz ate the dogs (particularly liver) they took in toxic doses of vitamin A.

Vitamin B complex (thiamine, riboflavin and niacin)

These vitamins (B_1, B_2 and B_3) play an essential role in the metabolism of carbohydrate, fat and protein and thus their RDIs are very dependent upon energy requirements.

Thiamine (B_1) occurs only in small amounts in animals and plants. The richest source is yeasts, but this is not an excuse for drinking beer. Refining of cereal products removes much of the thiamine content and therefore, in many of these products, vitamins are re-added to ensure that daily requirements are met. Chronic alcoholics tend to rely on alcohol for their calories and hence eat little or no food for extended periods. This frequently leads to thiamine deficiency. Severe deficiency leads to beriberi. This is characterised by peripheral neuropathy, which leads to hypersensitivity of the skin. Touching the skin, especially the feet, is extremely unpleasant and eventually so bad that walking becomes impossible. In some cases oedema, breathlessness and tachycardia (racing heart) occur, indicating the heart muscle is involved. **RDI** = 1.1 mg.

Thiamine is converted to thiamine pyrophosphate, which is a coenzyme for the decarboxylationa of α-keto acids.

thiamine

thiamine pyrophosphate

Acting as a coenzyme, the thiazole ring of thiamine pyrophosphate provides a mechanism for drawing electrons away from α-keto acids and hence releasing carbon dioxide. This makes vitamin B_1 an essential element of energy production and carbon skeleton cycling via the TCA cycle.

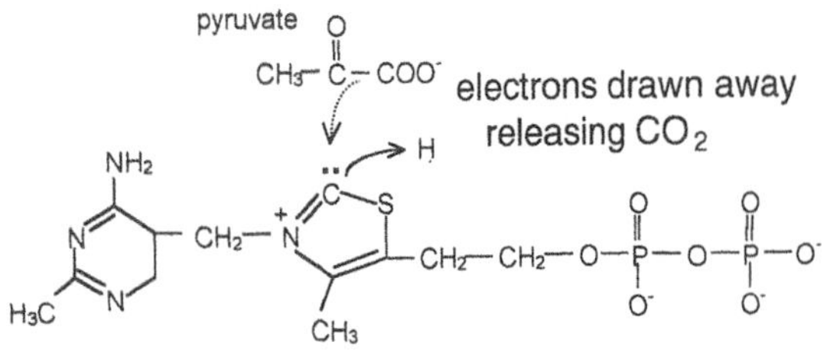

pyruvate

electrons drawn away releasing CO_2

Riboflavin (B₂) occurs in most plant and animal tissues, with the main sources in the Australian diet being milk and dairy products (also nuts, cereals and organ meats). It is the precursor for the co-enzymes flavin adenine dinucleotide (FAD) and flavin mononucleotide (FMN).

FMN and FAD take part in oxidation-reduction reactions – degradation of pyruvate, fatty acids and amino acids – and the transfer of electrons in the electron transport chain. Also used in the metabolism of other vitamins such as in the formation of pyridoxal phosphate and folinic acid. These are necessary for nucleic acid production and hence necessary for cell growth. Riboflavin deficiency is rarely seen, but is characterised by changes to the mucous membranes and skin, glossitis, and sore throat. **RDI** = 1.6 mg.

Niacin (B₃) is the generic term for 2 closely related compounds - nicotinamide (niacinamide) and nicotinic acid. About half the required niacin is obtained from offal, meat and legumes. The other half comes from the conversion of dietary tryptophan (**W**) by the liver. It is used to synthesise nicotinamide adenine dinucleotide (NAD⁺) and nicotinamide adenine dinucleotide phosphate (NADP⁺). In general NAD⁺ is a co-enzyme for dehydrogenases and NADPH is a co-enzyme for reductases.

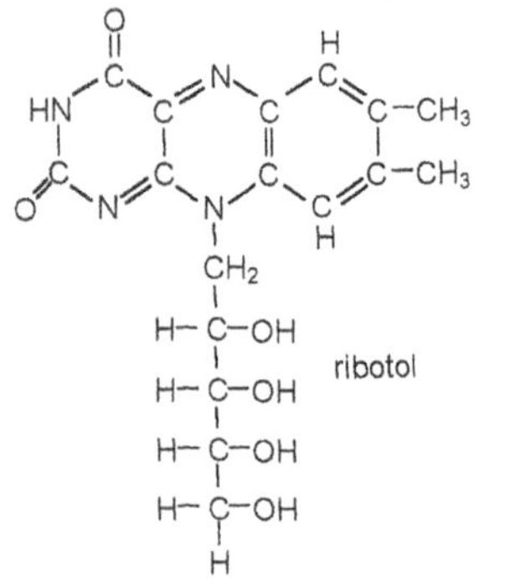

isoalloxazine ring system

ribotol

riboflavin

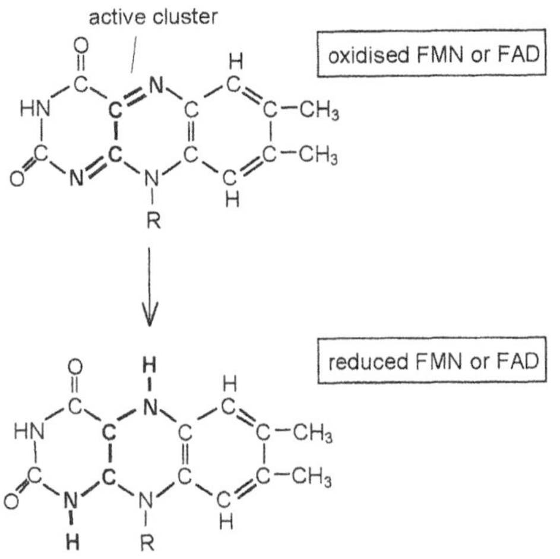

flavin mononucleotide (FMN)

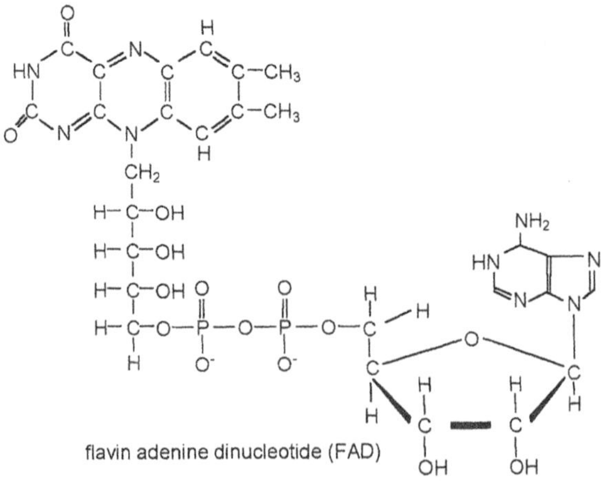

active cluster

oxidised FMN or FAD

reduced FMN or FAD

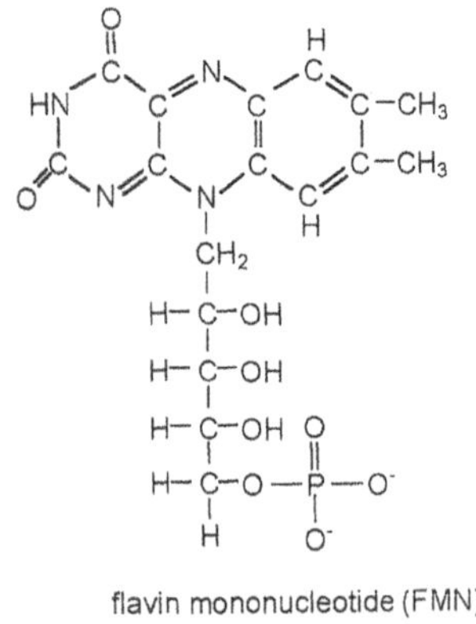

flavin adenine dinucleotide (FAD)

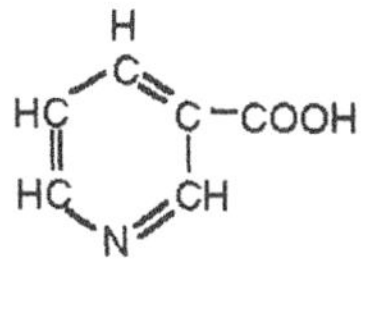

nicotinic acid

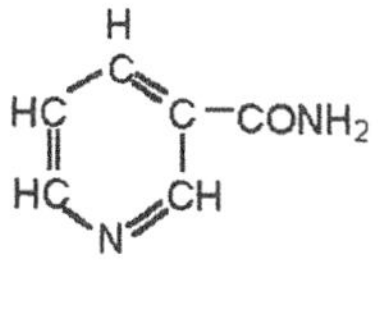

nicotinamide

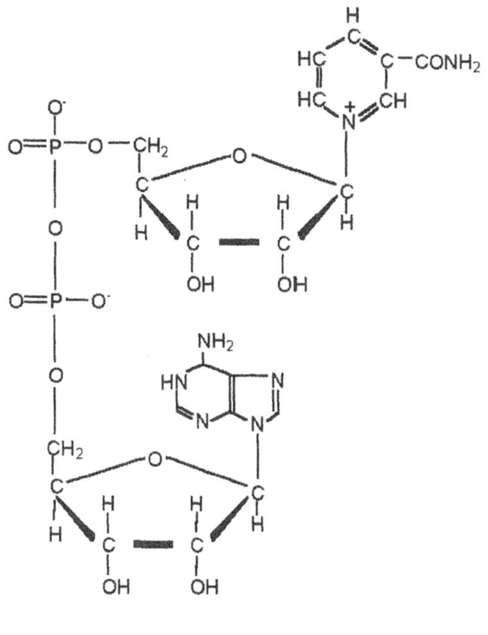

nicotinamide adenine dinucleotide (NAD⁺)

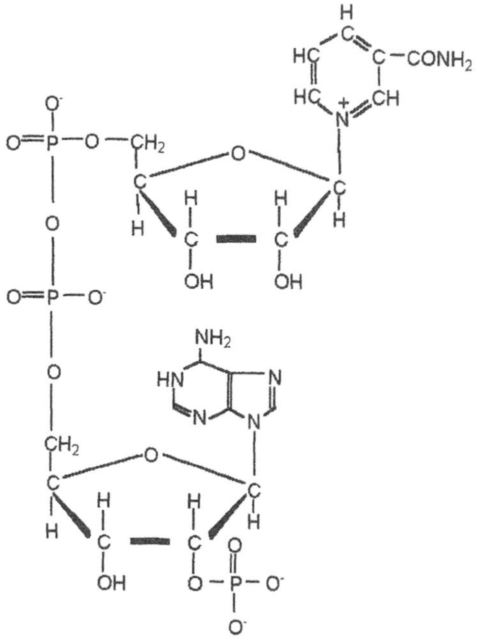

nicotinamide adenine dinucleotide monophosphate (NADP⁺)

A deficiency of niacin is associated with pellagra. Victims of pellagra suffer from 'the 4 Ds' – dermatitis, diarrhoea, dementia and death. Crusting and scaling of exposed skin is also evident. **RDI** (niacin equivalents – NEs) = 10 mg.

Anecdote – another Antarctic tale

Raoul Amundsen reached the South Pole on 15 December 1911; Robert Scott reached the South Pole on 17 January 1912. Amundsen used dogs to haul his sledges; Scott manhauled his. Amundsen's basic sledging ration was (per man per day): biscuits – 400 g; dried milk – 75 g; chocolate – 125 g; pemmican – 375 g (lean dried ground meat mixed with melted fat. A time honoured Polar sledging ration, pemmican was the most concentrated nourishment available before the advent of dehydrated food). This is a total of 975 g which contained 19 079 kJ (4560 kcal) and **thiamine – 2.09 mg; riboflavin – 2.87 mg; niacin – 25.85 mg**. Scott's sledging ration consisted of (per man per day): tea – 0.7 oz. (20 g); biscuits – 1 lb (454 g); cocoa – 0.86 oz (24 g); pemmican – 12 oz (340 g); butter – 2 oz (56.75 g); sugar – 3 oz (85.13 g). This is a total of 2 lb 3 oz (980 g) containing 18 535 kJ (4430 kcal) and **thiamine – 1.26 mg; riboflavin – 1.65 mg; niacin – 18.18 mg**.

Amundsen needed 18 828 kJ (4500 kcal) daily for the work he was doing (skiing beside dogs in sub-zero temperatures); Scott probably needed more than 23 012 kJ (5500 kcal) daily for the grotesque labour he undertook (manhauling sledges in worse than sub-zero temperatures). Scott and his companions suffered from fatigue, depression and susceptibility to the cold. They died and, at the time of death, it was clear that they were suffering from scurvy. However, scurvy does not cause the other symptoms they suffered. These are more typical of vitamin B deficiency.

> **Q&A 4:** Calculate the thiamine, riboflavin and niacin requirements of the 2 explorers; Amundsen (19 079 kJ); Scott (18 535 kJ) and determine if Scott's party was deficient in vitamin B.

Note: The difference in B complex content in the above diets was partly the result of Amundsen taking more chocolate and dried milk (remember that sugar is empty energy), but mostly it was due to the biscuits used by each expedition. Both were especially produced for concentrated nourishment. However, Scott's biscuits contained white flour with sodium bicarbonate for leavening; Amundsen's biscuits were based on wholemeal flour and crude rolled oats with yeast as the main leavening. Yeast and whole grains are potent sources of vitamin B.

Vitamin B₆ Vitamin B$_6$ (folic acid) exists in food as pyridoxine, pyridoxal and pyridoxamine. The phosphates, pyridoxal phosphate and pyridoxamine phosphate, are co-enzymes needed in transamination reactions.

pyridoxine (vitamin B$_6$)   pyridoxal   pyridoxal phosphate

pyridoxamine   pyridoxamine phosphate

Deficiency leads to a megaloblastic anaemia but does not cause polyneuritis. Folate is found in green leafy vegetables, nuts, legumes, whole grains, and citrus fruits. **RDI = 2 mg.**

Vitamin B₁₂ (cyanocobalamin). Structurally, it is very complex comprising a corrin ring (similar to a porphyrin ring with cobalt in its centre). It has an important role as a cofactor in reactions involving the transfer of a methyl group. We have enough stored to last about 5 to 7 years. Deficiency is not unusual though, because to absorb vitamin B$_{12}$ from the diet, we need a substance in the intestines called intrinsic factor. Some people lack this factor and hence are unable to absorb vitamin B$_{12}$. Vitamin B$_{12}$ is essential for generating tetrahydrofolate which is required for synthesis of the bases in DNA. Rapidly dividing cells (red blood cell precursors) need large amounts of new DNA and are thus affected by a deficiency. The first symptoms of a deficiency are a megablastic anaemia characterised by a decrease in red blood cells and larger than normal red blood cells. There is also a neurological disorder associated with vitamin B$_{12}$ deficiency. This is thought to be due to the lack of synthesis of methionine (**M**) from homocysteine. Vitamin B$_{12}$ is only found in animal foods: all meat and dairy products. **RDI = 1.0 mg.**

Vitamin C

Vitamin C is the generic descriptor for compounds having antiscorbutic (anti-scurvy) activity, chiefly ascorbic acid. Most mammals can synthesise ascorbic acid from glucose, but primates, fruit-eating bats and guinea-pigs cannot. Structurally, it is a lactone of a sugar acid.

It is a cofactor in a number of oxidative reactions. It participates in the degradation of phenylalanine (**F**) and tyrosine (**Y**), and it is necessary for the hydroxylation of proline (**P**) and lysine (**K**). Hydroxyproline and hydroxylysine are components of collagen and are necessary for the cross linking of the triple helical polypeptide chains. Without them, collagen breaks down. Hence the symptoms of vitamin C deficiency, scurvy, are skin lesions, fragile blood vessels, bleeding gums and loose teeth. These symptoms are usually not manifest until after 3 months on a diet deficient in vitamin C, because deficiencies in normal collagen are not apparent until then. Megadoses (1000 mg/day) are used prophylactically to prevent influenza. Its antioxidant properties do seem to help in this regard, but any true advantages are still controversial. Citrus fruits, broccoli, cauliflower, strawberries, tomatoes and potatoes are rich sources of vitamin C. **RDI = 40 mg.**

Historical note: In 1740 (30 years before Cook's discovery of Australia), George Anson of the British navy set sail around the world with 6 ships and 1955 men. When he returned with only the lead ship, scurvy had claimed the lives of more than half of the seamen. James Lind found that oranges and lemons were most effective in curing scurvy and thereafter, the Royal Navy forced British sailors (limeys) to eat limes on long sea voyages. In his 3 long voyages of discovery, James Cook did not lose a man to scurvy.

As stated above, Robert Scott and his party perished on the ice after failing in their attempt to be the first to reach the South Pole. Unwilling to declare a monumental error of dietary planning by a British naval officer, scurvy was dismissed as a possible cause of death in the official report. However, it is now believed that Scott's party was severely weakened by vitamin deficiency. Although the diets of both Scott and Amundsen were almost devoid of vitamin C, Amundsen was back to sources of vitamin C (seal meat) after only 2.5 months. Scott began on 1 November 1911 and died on 31 March 1912 (est.) – 5 months.

Vitamin D

This family of sterols structurally resemble cholesterol and are lipid soluble. They promote absorption of calcium and phosphorus from the intestine and the modeling of bone. 7-dehydrocalciferol, an inactive from of vitamin D, is converted to cholecalciferol by ultraviolet irradiation exposure through the skin. Cholecalciferol receives 2 hydroxyl units, one in the liver at C25 (25-hydroxycholecalciferol) and the second in the kidney to become 1,25-dihydroxycholecalciferol which is the biological active form of vitamin D.

cholecalciferol (vitamin D$_3$)

Q&A 5: Use the above diagram to indicate where the liver hydroxylates cholecalciferol and where the kidney hydroxylates 25-hydroxycholecalciferol.

Lack of sunshine or dietary intake of vitamin D can lead to a deficiency. Vitamin D is used in conjunction with parathyroid hormone to increase levels of calcium in the bloodstream. It causes resorption from bone, which is important in bone remodeling, and absorption of calcium from the gastrointestinal tract, which is important in providing calcium for the body. Children with insufficient vitamin D or not enough sunshine, develop bowed leg bones. This is called rickets. Vitamin D is found in small amounts only in liver, milk, butter and fatty fish, and none is found in plant foods. **RDI** = 2.5 mg.

Vitamin E

Unfortunately, the rumour that it is an aphrodisiac is untrue. However, lack of vitamin E in rats does cause sterility. Vitamin E is a lipid-soluble vitamin and its main role appears to be as an antioxidant and in particular the prevention of oxidation of polyunsaturated fatty acids. This leads to a protection of cell membranes.

Vitamin E deficiency leads to muscular weakness in man, and the impotence observed with rats has not been evident. It is in high concentrations in vegetable oils, particularly wheat germ oil, margarine, whole grains, dark green leafy vegetables, nuts and legumes. **RDI**=9 mg.

Vitamin K

Vitamin K (potassium) is similar structurally to vitamin E and is lipid soluble. It is particularly abundant in green leafy vegetables such as spinach and cabbage, and fruit such as bananas.

It is the nutritional factor necessary for blood clotting. It is a cofactor for the carboxylation of the γ-carbon of glutamic acid (**E**) in some of the blood clotting proteins. The carboxylation allows these proteins to bind Ca^{2+}, which in turn activates them.

Some medications, the coumarins, deliberately cause a deficiency in vitamin K so that blood clotting is inhibited. People who have had cardiac valve replacements or who have suffered a stroke, are often given the coumarin, warfarin, to prevent blood clots. Warfarin is also used to lace wheat as a poison for rats. The rats eat the wheat and die somewhere else from internal haemorrhages. This is quite useful because there is no carcass to dispose of and they decompose out of 'noseshot'.

Digestion

The first process in obtaining energy and nutrition from food is called digestion. This is the process of breaking down food into smaller particles so that it may be absorbed by the body. Firstly, it is broken down mechanically by the teeth and then by enzymes. Some of the enzymes break down complex carbohydrates into simple sugars e.g. amylases. Some enzymes break down proteins into amino acids e.g. trypsin and chymotrypsin. Other enzymes breakdown fats into fatty acids and glycerol, e.g. lipases. These simpler components are then either broken down further by the body (catabolism) or used to synthesise new molecules (anabolism).

MAJOR POINT 74: Food is broken down to structures that can be used in common pathways.

Common elements in pathways are pyruvic acid, acetyl-CoA, elements of the TCA cycle and oxidative phosphorylation. If food were not broken down to common pathways e.g. glucose and fatty acids both to acetyl-CoA, we would have a special storage and processing pathway for tomatoes, another for potatoes and still another for eggs.

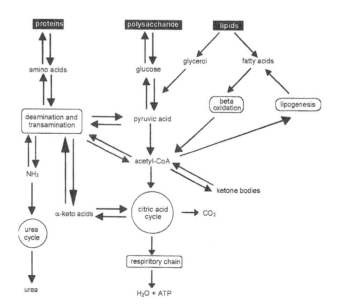

Control of metabolism

Once food is digested, the nutrients are metabolised depending upon the body's needs. Metabolism is controlled by activation or inhibition of particular enzymes in the metabolic pathways. They are controlled intrinsically by levels of metabolites within a cell, or extrinsically by hormones. All rate-controlling reactions are far from equilibrium, but this is not to say that a reaction far from equilibrium is a rate controlling reaction. All enzymes are affected by the availability of substrate.

When you eat your food what are you most commonly doing?

Answer: *Sitting — you are not running, walking, or doing activities that need the energy gained by eating*

In addition, the pattern of our eating is bizarre. We often overeat or undereat, and the frequency of meals and the type and quantity of food varies considerably. We do not use all of the energy contained in a meal immediately but save most of it for later use. In terms of usage, different organs have quite different nutritional and energy requirements, e.g. the requirements of skeletal muscle, the brain and the liver are all quite different. Therefore, there is a major problem for our body. We eat food when we do not need it and do not eat food when we need it most. The food we eat is inconsistent in both quality and quantity and different organs have different nutritional requirements.

The study of control of metabolism is examining how the body buffers the wild disparities between nutritional intake and nutritional needs. To understand this requires a knowledge of the functions of 3 major organs/tissues in the body: liver, muscle, and adipose (fat) tissue and the special needs of the brain for glucose. The only way to unravel the apparent complexity of the following information is to really think (probably harder than ever before) as you go through it. If this is done, then most of the metabolic control becomes obvious and expected.

Metabolism in muscle

What is the main function of muscle?
a   *to produce heat (70%) and movement (30%)*
b   *to produce heat (30%) and movement (70%)*
c   *fuel storage*
d   *metabolism (catabolism and anabolism)*
e   *catabolism only*

Answer: *Surprisingly **a** is right. Seventy per cent of the energy used by muscle is used in generating heat. We need this heat for our enzymes and other proteins to function correctly.*

> **MAJOR POINT 75:** Metabolism in muscle is to provide heat, and ATP for contraction.

Muscle is a consumer. It preferentially burns glucose and fatty acids, which are readily obtained from the bloodstream.

An excess of glucose is converted to glycogen and stored in the muscle, but the capacity for muscle to store glycogen is very limited.

If muscle is a consumer, then what are the most active metabolic pathways?

Answer: *The most active pathways are glycolysis, β-oxidation of fatty acids, the TCA cycle and oxidative phosphorylation. Glycogen synthesis will also occur when there is an excess of glucose and glycogenolysis will occur in conditions when glucose is lacking. Gluconeogenesis does not occur in muscle, and fatty acid synthesis is minimal.*

In **anaerobic** conditions, oxidative phosphorylation and β-oxidation cannot occur and hence energy can only be obtained from glycolysis with glucose as the substrate and lactic acid as the end-product. Energy from fatty acids (β-oxidation) can only be obtained in **aerobic** conditions. In fact, with plenty of oxygen, red muscle fibres prefer to oxidise fatty acids rather than glucose. This is the reason why aerobic exercise is recommended for weight reduction. Cardiac muscle almost exclusively uses fatty acids and hence cannot be deprived of oxygen at all. Small reductions in oxygen (blood flow) to the heart may lead to a severe pain called angina pectoris and if reduction of oxygen becomes severe, the cardiac muscle dies. This is called cardiac infarct or a heart attack.

Regulation of glycolysis

Glycolysis is regulated at 3 points of which 2 are obvious: the transport of glucose into the cell and phosphofructokinase (which is the enzyme used for glycolysis and not for gluconeogenesis). The less obvious one is hexokinase, which phosphorylates glucose at C6 when it enters the cell.

Glucose transport into muscle cells

When would you expect glucose to be transported into muscle cells?

Answer: *Glucose transport into muscle cells would be expected when glucose levels in the bloodstream are high (after eating), when oxygen levels are low (the only source of energy in anaerobic conditions), when there are low levels of ATP (muscle is hard at work), and when alternative fuels such as fatty acids are not available.*

Insulin is a hormone secreted by the pancreas. Its levels rise after a meal in response to an increase of glucose in the blood, and it promotes the uptake of glucose by cells. In muscles, insulin causes a marked increase in glucose uptake and hence intracellular glucose. The mechanisms in muscle for how insulin activates the glucose transporter are currently unknown, but in fat cells, insulin directly increases the number of transporters in the cell membrane. High levels of free fatty acids (alternative fuel source) inhibit the glucose transporter in muscle cells.

The control of glycolysis by phosphofructokinase

Phosphofructokinase is a critical enzyme in glycolysis because it converts fructose 6-phosphate to fructose 1,6-bisphosphate, but the reverse reaction requires a different enzyme, fructose 1,6-bisphosphatase (gluconeogenesis). The ATP/AMP ratio is the most significant factor for controlling phosphofructokinase in muscle. It would be expected that if there were high levels of ATP, then glycolysis would not be required and therefore ATP would inhibit phosphofructokinase. This is the case. Phosphofructokinase has 2 ATP binding sites and ATP inhibits the enzymes activity. A fall in ATP relieves inhibition and glycolysis proceeds. Inhibition of phosphofructokinase by ATP is opposed by AMP. Therefore, a rise in AMP or a decrease in ATP, both signal that the muscle is working hard and that it is in a low energy state. This removes the inhibition of phosphofructokinase and increases glycolysis. Inorganic phosphate also increases the activity of phosphofructokinase. Inorganic phosphate increases when ATP is converted to ADP.

ATP is interesting because it is a substrate for phosphofructokinase and as such should increase its activity, but at the same time, it is an end-product of glycolysis and so should decrease its activity. This anomaly is overcome by the ATP/AMP ratio. AMP is only 2% of the levels of ATP in resting muscle and therefore a small percentage fall in ATP gives a relatively large increase in AMP. This occurs because of the following equilibrium equation

$$\text{ATP + AMP} \rightleftharpoons \text{2ADP}$$

adenylate kinase

A fall in ATP during metabolism leads to an increase in ADP. This decrease in ATP and increase in ADP will push the above equilibrium equation to the left increasing AMP, and therefore increase the AMP/ATP ratio. The levels of AMP in absolute amounts are relatively small and the levels of ATP are relatively high. Because of this, a 15% decrease in ATP gives 300% increase in AMP. The AMP binds to phosphofructokinase, inhibiting binding of ATP and increasing the activity of phosphofructokinase.

The end product of glycolysis is pyruvate (or lactic acid) and energy. This can be carried out under aerobic conditions, but under these conditions, other fuels, particularly fats, can be used instead for energy. Therefore, it would be a waste to use glucose for energy if other fuels could be used. It would make better sense to preserve glucose for use when there was not enough oxygen available. Clearly, glycolysis needs to be inhibited if there are other sources of fuel available. Catabolism of these sources, fatty acids or proteins, is via the TCA cycle and oxidative phosphorylation. When these fuels are being used, the TCA cycle becomes 'overloaded', particularly with acetyl-CoA. To re-

lieve this, citrate is formed by combining the acetate part of acetyl-CoA with oxaloacetate. It is then shuttled across the mitochondrial membrane, leading to increased levels of citrate in the cytoplasm. Therefore, high citrate levels in the cytoplasm is a signal that the TCA cycle is overloaded. Citrate enhances the effect of ATP blockage on phosphofructokinase and hence also inhibits glycolysis. In starvation, there is also an increase in citrate due to alternative fuel usage and hence this preserves glucose usage for brain.

Futile cycle Although muscle has a clear need for phosphofructokinase (glycolysis), muscle has no obvious need for fructose 1,6-bisphosphatase which converts fructose 1,6-bisphosphate to fructose 6-phosphate. This is the pathway to glucose synthesis and muscle cannot synthesise glucose. However, muscle has fructose 1,6-bisphosphatase. This enzyme together with phosphofructokinase allow a futile substrate cycle between fructose 1,6-bisphosphate and fructose 6-phosphate. The purpose of such a cycle is twofold. The first is to generate heat. Excess ATP formed from glucose taken up after a meal is used in the futile cycle is used to generate heat. We know this happens because we feel warm after a meal. The second is to amplify the activation of glycolysis caused by a fall in ATP and an increase in AMP. This is possible because AMP not only activates phosphofructokinase but also inhibits fructose 1,6 bisphosphatase.

To understand how this futile cycle rapidly increases glycolysis in response to a fall in energy state, a concept of flux is necessary. Suppose at rest, the rate of conversion of fructose 6-phosphate to fructose 1,6-bisphosphate is 2 molecules per minute and the reverse reaction is 1 molecule per minute, then the net flux through the cycle in the direction of glycolysis will be 1 molecule per minute. If AMP increases, and as a result causes the activity of phosphofructokinase to double and the activity of fructose 1,6-bisphosphatase to be halved, then the flux will increase from 1 to 3.5 molecules per minute in the direction of glycolysis. So a twofold change in activity of each enzyme results in a 3.5 fold change in flux. If the flow difference at rest were not 2 to 1 but 100 to 99, then the net flux would still be 1 in the direction of glycolysis.

Double the activity of phosphofructokinase and halve the activity of fructose 1,6-bisphosphatase. What will the new flow in the direction of glycolysis be and what will the flow in the direction gluconeogenesis be? From this work out the increase in flux.

Answer: *The flow in the direction of glycolysis will double to 200 and the flow in the reverse direction will halve to 50, a difference of 150. Therefore the twofold change in enzyme activity causes a 150 fold change in flux. This example is close to the real situation where it has been found that a fall in ATP levels of 10% causes a 500% increase in glycolytic flux.*

Regulation of glycolysis by hexokinase

Under conditions of excess glucose, hexokinase, which phosphorylates glucose to prevent it diffusing back out of the cell, becomes the rate limiting enzyme for glycolysis. Glucose is able to enter the muscle cell at a faster rate than it can be converted to glucose 6-phosphate by hexokinase and hexokinase is inhibited by glucose 6-phosphate.

Glucose 6-phosphate is then either converted to glucose 1-phosphate for glycogen synthesis or to fructose 6-phosphate for glycolysis. It is important to realise that both of these pathways are nearly at equilibrium in the cell and hence an increase in glucose 1-phosphate, or fructose 6-phosphate, will cause them to be readily converted to glucose 6-phosphate. This will of course also cause a decrease in hexokinase activity.

This means that glucose 6-phosphate is a pivotal step between glycolysis and glycogen synthesis. If the energy state of the muscle cell is high, or other fuels are available then there is an increase in fructose 6-phosphate, which is converted to glucose 6-phosphate. Glucose 6-phosphate cannot be converted to glucose in muscles and therefore it is converted to glucose 1-phosphate for glycogen synthesis. If glycogen stores are full then the levels of glucose 1-phosphate increase and no more glucose 6-phosphate will be converted. The levels of glucose 6-phosphate will increase, hexokinase will be inhibited and glucose will not enter the cell.

breakdown glycogen. You would expect that activation of one would inactivate the other. In a nutshell, glycogen synthetase is inhibited by phosphorylation, and glycogen phosphorylase is activated by phosphorylation.

Glycogenolysis Muscle glycogen can rapidly supply the muscle with glucose 6-phosphate for glycolysis. It never acts as a source of circulating glucose because muscle lacks glucose 6-phosphatase, which catalyses the breakdown glucose 6-phosphate to glucose. The amount of glycogen available in muscle lasts about 2 minutes under heavy exercise and hence, storage is limited.

Glycogen breakdown in muscle is dependent on the activation of glycogen phosphorylase. Glycogen phosphorylase cleaves the terminal glucose units from glycogen to form glucose 1-phosphate. Adrenaline indirectly activates glycogen phosphorylase activity. Adrenaline is a hormone released from the adrenal medulla under sudden stress and prepares the body for a fight or for flight. In both cases, skeletal muscles are about to be used and instant activation of glycogen phosphorylase leads to the provision of fuel from glycogen.

Adrenaline is a water-soluble hormone and cannot cross the cell membrane and hence binds to a receptor embed-

Regulation of glycogen metabolism

When would you expect glycogen to be synthesised in muscle and when do you think it would be used by muscle as a fuel source?

Answer: *It is quite obvious that muscle will synthesise glycogen when there is an excess of glucose available and the muscle is at rest (a high energy state, high ATP/AMP ratio). It will break down glycogen when the muscle is in an energy deficient state and there is little circulating glucose or other fuels. It also occurs more in oxygen depleted states when the only source of ATP is from glycolysis.*

Critical to the control of glycogen metabolism are the activities glycogen synthetase, which catalyses the synthesis of glycogen, and glycogen phosphorylase, which catalyses the

ded in the cell membrane (an adrenergic receptor). The activated receptor activates an enzyme called adenylate cyclase which converts ATP to cAMP. Many enzymes in cells are sensitive to cAMP or cGMP and when they bind these substances they change their activity.

cAMP is bound by a protein kinase, which becomes active. Remember that the function of kinases is to catalyse phosphorylations. Glucose kinase phosphorylates glucose and protein kinase phosphorylates proteins. In this case, the inactive protein kinase is called protein kinase b and in its active form i.e. bound to cAMP, it is called protein kinase a. The protein which protein kinase a phosphorylates is a phosphorylase kinase. The inactive phosphorylase kinase is called phosphorylase kinase b and the active form phosphorylase kinase a.

Phosphorylase kinase will catalyse the phosphorylation of which enzyme? Hint: it tells you in its name.

Answer: *Phosphorylase kinase catalyses the phosphorylation of glycogen phosphorylase b which becomes the active form, glycogen phosphorylase a. Glycogen phosphorylase a catalyses glycogenolysis.*

Glycogen phosphorylase b is a dimer with the subunits having molecular weights of about 100 000. Phosphorylase kinase adds a single phosphate to each subunit which causes it to form a tetramer by the dimerisation of 2 glycogen phosphorylase b molecules. This tetramer is glycogen phosphorylase a.

Such pathways are called cascades. They amplify the initial signal e.g. each protein kinase may activate 10 phosphorylase kinases and each of these 10 phosphorylase kinases might activate 10 glycogen phosphorylases. This would give a 100-fold amplification of the initial signal.

Glycogen synthesis Glycogen synthesis is dependent upon the activity of glycogen synthetase (UDP-glucoglycosyltransferase), which catalyses the incorporation of glucose into glycogen from UDP glucose.

It is controlled both positively and negatively by phosphorylation and the increase or decrease in its activity depends upon the site on the enzyme which is phosphorylated. It is phosphorylated by at least 5 different protein kinases

1. Cyclic AMP-dependent protein kinase phosphorylates sites 1a, 1b and 2
2. Phosphorylase kinase phosphorylates site 2
3. Glycogen synthetase kinase 3 phosphorylates sites 3a, 3b and 3c.
4. Glycogen synthetase kinase 4 phosphorylates site 2
5. Glycogen synthetase kinase 5 phosphorylates site 5

From the list above, and knowing that activation of glycogen synthetase causes increased glycogen synthesis, phosphorylation of which sites would decrease its activity?

Answer: *The activity is reduced by phosphorylation at sites 1a, 2, 3a, 3b and 3c. These sites are phosphorylated by the enzymes which stimulate glycogenolysis. As more sites are phosphorylated, the more the inhibition. Phosphorylation of sites 1b and 5 do not affect the activity. It means that the activity of the enzyme is progressive.*

The phosphorylation of site 2 (inhibition), caused by the cAMP activation of phosphorylase kinase, makes a great deal of sense because it stops wasteful glycogenesis when glycogenolysis is being promoted.

Glycogen synthesis should respond to the levels of glucose in the circulation. Insulin in-

Higher level

Phosphorylase kinase is actually a tetramer made from 4 different subunits α, β, γ and δ. Phosphorylation occurs on both the α and β subunits and the γ subunit is the active site. The δ subunit is calmodulin and binds calcium which activates the enzyme. When muscle is activated to contract, Ca^{2+} is released from the sarcoplasmic reticulum into the cytoplasm. This not only causes muscle contraction, but also stimulates the breakdown of glycogen to provide energy for the contraction. Furthermore, there are major differences in the adrenergic receptor in cardiac and skeletal muscle and the receptor in smooth mus-

cle. In cardiac and skeletal muscle, it is a β receptor and activates adenylate cyclase, but in smooth muscle it is an a receptor, and activates Ca^{2+} entry into the cell.

The phosphorylation (activation) of glycogen phosphorylase b is made easier by AMP (increasing AMP decreases the Km for Pi of glycogen phosphorylase b).

To summarise: glycogen phosphorylase is activated 3 factors: adrenaline through the cAMP-kinase cascade, muscle contraction through Ca^{2+} activation and signals which indicate a low energy state and lack of glucose, a rise in AMP and Pi, a fall in ATP and glucose 6-phosphate

creases when glucose levels are high and hence there would be an expectation that insulin may affect glycogen synthetase and this indeed is the case. Exactly how this is accomplished is not known. Insulin does not increase the levels of cAMP or the extent of phosphorylation of sites 1 or 2. It does cause a reduction in the level of phosphorylation of site 3. This is probably via inhibition of glycogen synthetase kinase 3 since there are no known phosphorylases of site 3. Another possibility is the regulation of glycogen synthetase kinase 5 whose only function appears to be control over the sensitivity of glycogen synthetase to phosphorylation by glycogen synthetase kinase 3.

Therefore, if the levels of circulating glucose are high (high insulin) and the muscle is at rest (high energy state), glycogen synthesis will be promoted. Insulin decreases the phosphorylation of glycogen synthetase at site 3 which causes its activation. The high energy state decreases the activity of phosphofructokinase which increases the levels of glucose 6-phosphate. This is converted to glucose 1-phosphate which can then be used for glycogen synthesis.

Higher level

In addition to the kinases, there are protein phosphatases which remove phosphates from the various enzymes. The activity of protein phosphatase 1 is responsible for the dephosphorylation of most phosphoproteins in the glycogen metabolism. Its activity is not well regulated and provides a background activity. However, it is known that a rise of cAMP leads to inhibition of protein phosphatase 1. Remember from above that a rise in cAMP is the signal that leads to the activation of protein kinases, which phosphorylate proteins, and therefore there would be an expectation that this signal would also inhibit phosphatase activity. This inhibition occurs indirectly via activation of phosphatase inhibitor protein, an enzyme whose activity inhibits protein phosphatase 1. Phosphatase inhibitor protein is activated by its phosphorylation catalysed by a cAMP dependent protein kinase. However, this creates a problem. If protein phosphatase 1 is inactivated because of the phosphorylation of phosphatase inhibitor protein, then how is the phosphorylase inhibitor dephosphorylated to reactivate it to relieve its inhibition of phosphorylase? The answer is that phosphatase inhibitor protein is phosphorylated on the threonine residue instead of serine residues like the other proteins involved in glycogen synthesis

and hence its dephosphorylation is catalysed by a different enzyme, protein phosphatase 2.

Glycogen itself inhibits glycogen synthetase activity (substrate inhibition). The mechanism is that glycogen synthetase binds to glycogen and this inhibits its dephosphorylation and hence, as the levels of glycogen increase there is less dephosphorylation of glycogen synthetase and glycogen synthesis decreases.

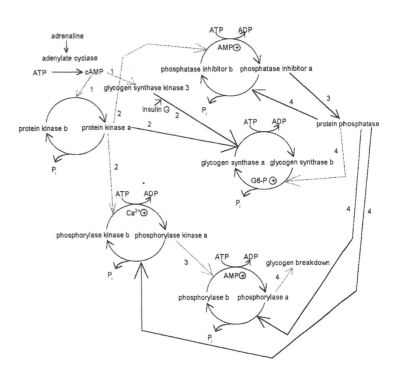

There is nothing like a simple diagram to explain what is happening. It is a pity about the one above. The dotted lines indicate that the effect is to activate the pathway pointed to, and the solid line indicates that the effect is inhibiting the pathway pointed to. For example, begin at adrenaline and follow the numbers that lead to glycogen synthesis: cAMP activates protein kinase a (1), which phosphorylates and activates phosphatase inhibitor (2). Phosphatase inhibitor inhibits protein phosphatase (3) and therefore protein phosphatase does not dephosphorylate (activate) glycogen synthetase. For glycogen synthesis the effects are reversed.

Control of the TCA cycle

Choosing between glucose and fatty acids For muscle, if fatty acids were available for energy production, it would be a waste of glucose to channel it into the TCA cycle. It would be much better to use the glucose for glycogen synthesis. Similarly, if the cell were in a high energy state it would be unnecessary to channel glucose through the TCA cycle. Therefore, there must be signals indicating that the TCA cycle is full and to stop glycolysis.

For glucose to enter the TCA cycle, it is converted to pyruvate (glycolysis) and then to acetyl-CoA with pyruvate dehydrogenase catalysing the conversion of pyruvate to acetyl-CoA + NADH. This is not a single enzyme, but a multienzyme complex and it is controlled by product inhibition with NADH and acetyl-CoA both being inhibitors. When fatty acids are being used as a fuel, β-oxidation loads the mitochondria with acetyl-CoA and NADH and therefore pyruvate dehydrogenase is inhibited.

Phosphorylation of the first enzyme in the pyruvate dehydrogenase complex also causes inhibition. The enzyme that catalyses its phosphorylation is a mitochondrial protein kinase. This enzyme is inhibited by pyruvate and ADP, and activated by acetyl-CoA and NADH.

The enzyme that catalyses the dephosphorylation of pyruvate dehydrogenase and hence reactivates it, is pyruvate dehydrogenase phosphatase. It is activated by low levels of Ca^{2+}. During muscle contraction, levels of Ca^{2+} rise and activate pyruvate dehydrogenase phosphatase and hence pyruvate dehydrogenase, which allows pyruvate to enter the TCA cycle as acetyl-CoA.

The TCA cycle is, to a large extent, controlled by the levels of citrate. The important enzymes here of course are citrate synthetase and isocitrate dehydrogenase. Citrate synthetase, which catalyses the condensation of acetyl-CoA and oxaloacetate to form citrate, is inhibited by citrate. In the TCA cycle, citrate is rapidly converted to isocitrate which is then converted to α-ketoglutarate. This second reaction is catalysed by isocitrate dehydrogenase, which is inhibited by NADH and activated by Ca^{2+} and ADP. Therefore, it will be inhibited in relaxed muscle which is in a high energy state and low in Ca^{2+}, and activated in active muscle. Its inhibition leads to a build-up of citrate in the mitochondria. Excess citrate is transported out of the mitochondria into the cytoplasm where it inhibits phosphofructokinase, stopping glycolysis (see diagram below).

Lipid oxidation in muscle

In skeletal muscle (particularly red fibres) and cardiac muscle, fatty acids are used in preference to glucose. The availability of substrate controls lipid oxidation. It is mobilised under conditions of exercise and starvation. In starvation there is also a mobilisation of ketones. Mobilisation of fatty acids from their stores occurs when glucose levels are low. Glucagon and adrenaline are released during exercise and starvation and these promote lipid oxidation – insulin inhibits it.

MAJOR POINT 77: Adrenaline stimulates and insulin inhibits the release of free fatty acids from fatty tissue.

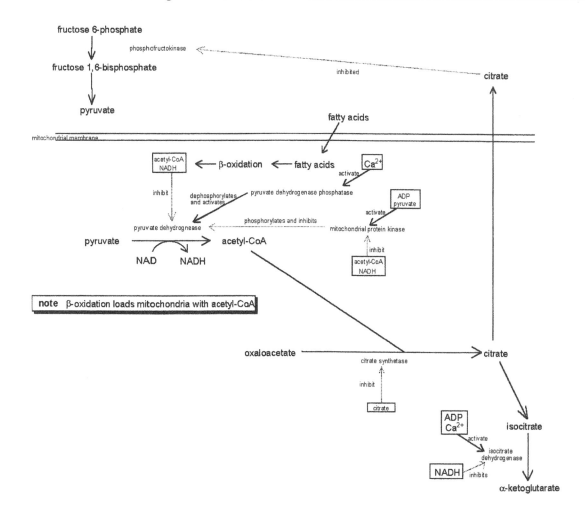

The ultimate limitation for fatty acid oxidation in muscle appears to be the availability of CoA-SH of which there is a limited in quantity in mitochondria. Thus, as it is acetylated to acetyl-CoA, there is less available for lipid oxidation.

Liver metabolism

What is the function of the liver in metabolism?

Answer: *The function of the liver is to provide a constant supply of dietary nutrients to the body even though there is enormous variation of dietary intake.*

The venous blood from the gastrointestinal tract passes to the liver where it once again enters a capillary bed. This means that the liver is the first organ that has the opportunity to take up nutrients which have been absorbed from the gastrointestinal tract. It does this very efficiently and then processes these for release into the bloodstream. It performs all interconversions between lipids, carbohydrate and amino acids and is the only organ capable of forming ketone bodies from fatty acids. It synthesises urea and lipoproteins. Excess blood glucose leads to glycogen and then fatty acid synthesis. Here we shall concentrate on its major role as a supplier of glucose for the rest of the body.

> **MAJOR POINT 78:** The liver buffers glucose levels in the body, particularly supplying the body with glucose when levels are low.

Supplier of glucose

The liver supplies glucose for the rest of the body when circulating glucose is low, such as during fasting, starvation or heavy exercise. This is especially important for the brain because it has an absolute demand for some of its energy in the form of glucose. It is not surprising that the liver itself has little need for glucose other than to store it. Glucose from the liver is supplied from breakdown of stored glycogen or from synthesis of glucose from lactate, pyruvate and amino acids. The liver and the kidney cortex are the only tissues which can actually make glucose because they have glucose 6-phosphatase, which catalyses the removal of phosphate from glucose 6-phosphate to form glucose. Therefore, in liver metabolism, the focus is on the signals and controls that promote gluconeogensis over glycolysis.

Glycolysis and gluconeogenesis

As for muscle, there are 3 points in glycolysis where there is a large change in energy: the conversion of glucose to glucose 6-phosphate, fructose 6-phosphate to fructose 1,6-bisphosphate and phosphoenol pyruvate to pyruvate. For gluconeogenesis, different enzymes control the reverse reactions at these points and hence they form futile cycles which are likely regulation sites in the pathway.

The glucose transporter in the liver, unlike the glucose transporter in muscle, is relatively insensitive to insulin and therefore substantial uptake does not take place until the levels of glucose are above 12mmol/L in the blood. This is about twice the normal levels of glucose. This means that glycolysis is a minor pathway in the liver and only takes place to a substantial degree if the blood levels of glucose are very high, such as after a meal.

In the liver, glucose transport into the cells is not rate limiting (an equilibrium constant of near 1) and hence glucose is nearly the same concentration inside and outside the cell. For this to happen, the phosphorylation of glucose must be relatively slow, otherwise the levels of glucose would be lower inside the cell than outside, which is not the case.

The liver uses glucokinase to catalyse the phosphorylation of glucose rather than the hexokinase used by muscle and fat. Glucokinase, which is only found in the liver, has a relatively high K_m and low V_{max} (does not bind the substrate easily and does not convert it quickly to product) which makes it rate limiting in the liver. In muscle and fat, hexokinase has low K_m and high V_{max}, which means that glucose entering these cells is rapidly converted to glucose 6-phosphate. The only regulator of glucokinase appears to be the concentration of glucose and substantial conversion of glucose to glucose 6-phosphate only occurs if the concentration of glucose is above 10mmol/L, which means that the concentration in the blood must be above 10mmol/L since the concentration inside the cell is about the same as in the blood.

At normal levels of glucose in the blood, 5 mmol/L, there is virtually no glycolysis taking place, but glucose kinase is still operating at about 25% of its maximal rate. This means that the synthesis of glucose, catalysed by glucose 6-phosphatase must be operating at the same rate. This cycle between glucose and glucose 6-phosphate provides a buffer system for the blood glucose levels. Since neither enzyme is operating near its maximal capacity, a fall in blood glucose levels will cause more flux from glucose 6-phosphate to glucose and an increase in blood glucose levels will cause an increase in flux from glucose to glucose 6-phosphate.

During prolonged starvation, glucocorticoids are released from the adrenal cortex. These cause the synthesis of glucose 6-phosphatase, hence favouring gluconeogenesis by the liver, making more glucose available for the rest of the body.

The next controlling step in glycolysis is the conversion of fructose 6-phosphate to fructose 1,6-bisphosphate. This reaction is controlled by phosphofructokinase and the reverse reaction, in the direction of gluconeogenesis, is controlled by fructose 1,6-bisphosphatase. The V_{max} for these enzymes is substantially different in the liver compared with muscle. These differences favour gluconeogenesis. V_{max} for fructose 1,6-bisphosphatase is about 4 times the V_{max} of phosphofructokinase whereas in muscle V_{max} for fructose 1,6-bisphosphatase is about 10% of the V_{max} of phosphofructokinase. Like muscle, phosphofructokinase is inhibited by ATP and activated by AMP, and fructose 1,6-bisphosphatase is activated by citrate and inhibited by AMP.

Higher level

Sometimes the liver might need energy, but the rest of the body needs it more. We can lose about 50% of our liver without disastrous consequences (useful information for alcoholics). Therefore, a signal is needed to override a falling ATP/AMP ratio to promote gluconeogenesis over glycolysis. This is the fructose 2,6-phosphate system and it is controlled by glucagon which is released into the bloodstream by the pancreas during fasting.

The liver has another phosphofructokinase and this catalyses the conversion of fructose 6-phosphate to fructose 2,6-bisphosphate. Fructose 2,6-bisphosphate activates phosphofructokinase 1 and inhibits fructose bisphosphatase and hence stimulates glycolysis and inhibits gluconeogenesis. The enzyme that catalyses the conversion of fructose 6-phosphate to fructose 2,6-bisphosphate is known as phosphofructokinase 2 to distinguish it from the enzyme that catalyses the conversion of fructose 6-phosphate to fructose 1,6-bisphosphate which is known as phosphofructokinase 1. Similarly, there is a fructose bisphosphatase 2 which catalyses the dephosphorylation of fructose 2,6-bisphosphate. Phosphofructokinase 2 is indirectly inhibited by glucagon and fructose bisphosphatase 2 activated by it.

During fasting, glucagon is released into the bloodstream and binds to a receptor in the cell membrane of the hepatocytes. This leads to the activation of adenylate cyclase which catalyses the synthesis of cAMP from ATP. Protein kinase A is activated by the high levels of cAMP and phosphorylates phosphofructokinase 2 and fructose bisphosphatase 2. Phosphorylation causes an inhibition of phosphofructokinase 2 activity and stimulates fructose bisphosphatase 2 activity. This causes a decrease in fructose 2,6-bisphosphate, which decreases the activity of phosphofructokinase 1 and glycolysis. At the same time, it increases gluconeogenesis.

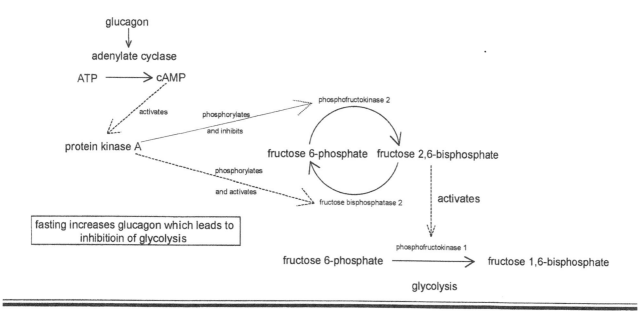

Interconversion between pyruvate and phosphoenolpyruvate

The liver, unlike muscle, has need to convert pyruvate to phosphoenolpyruvate which causes pyruvate to flow in the direction of gluconeogenesis. The pyruvate may originate from glycolysis, lactic acid or particular amino acids.

This reaction occurs in 2 steps and requires the equivalent of 2 high-energy phosphate bonds. Pyruvate carboxylase (mitochondria) catalyses the conversion of pyruvate to oxaloacetate and then phosphoenolpyruvate carboxykinase (mitochondrial, cytoplasmic or both depending upon the species) catalyses the conversion of oxaloacetate to phosphoenolpyruvate. Pyruvate carboxylase is activated by acetyl-CoA, and acetyl-Co and NADH decrease the activity of pyruvate dehydrogenase (gluconeogenic conditions). This inhibits the oxidation of pyruvate obtained from lactate or alanine. Acetyl-CoA cannot be used as a source of glucose, so any pyruvate converted to acetyl-CoA is lost as a source of glucose.

If phosphoenolpyruvate carboxykinase is exclusively cytoplasmic, then oxaloacetate must be transferred across the mitochondrial membrane for it to be converted to phosphoenol pyruvate. This is done by converting it to malate which has the advantage in that it also transfers the reducing power of NADH from the mitochondria to the cytoplasm. This drives the glyceraldehyde 3-phosphate in the direction of gluconeogenesis.

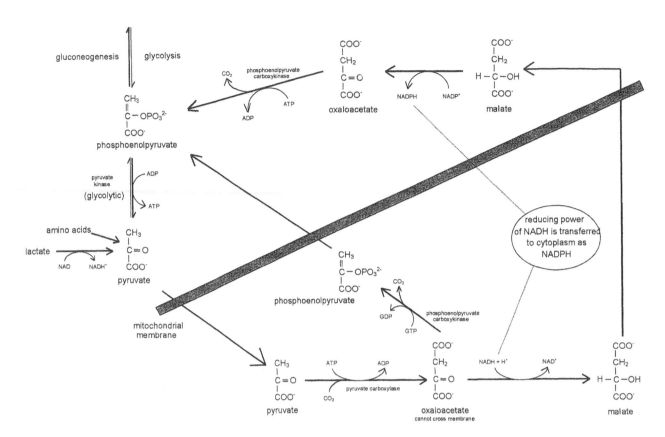

If lactate is the substrate, the oxidation of lactate to pyruvate provides a source of cytoplasmic NADH. In this case, oxaloacetate can be transferred without transferring the reducing power. It is converted to aspartate, glutamate and α-ketoglutarate.

Pyruvate kinase, which catalyses the conversion of phosphoenolpyruvate to pyruvate, is highly regulated. Pyruvate kinase is inhibited by alanine (**A**), phenylalanine (**F**) and ATP, and activated by fructose 1,6-bisphosphate. Alanine (**A**) and phenylalanine (**F**) are gluconeogenic substances. It is also phosphorylated by protein kinase A which inhibits its activity. During prolonged starvation the levels of pyruvate kinase decrease in the liver due to the action of either glucagon or corticosteroids.

Glycogen storage

Is glucose a good substrate for glycogen synthesis by the liver in the well fed state?

Answer: *Surprisingly, no. Lactate and fructose are the most important substrates quantitatively for the synthesis of glycogen in the liver. This makes sense since the function of the liver is to supply glucose to the rest of the body, not to use it.*

However, glycogenesis is activated by high levels of blood glucose (10–12 mmol/L). This leads to an increase of glucose 6-phosphate which is probably the cause of activation of glycogenesis. Insulin also activates glycogen synthesis and inhib-

its glycogen breakdown. Insulin probably reduces the phosphorylation of glycogen synthetase.

Glycogenolysis

Only enough glycogen is stored in the liver to cope with short-term fluctuation of glucose in diet. There is not enough to maintain glucose through extended starvation. Under passive conditions, there is enough to last 24 hours, but with severe exercise it lasts about 90 minutes.

Unlike muscle, in the liver the energy state of the cells is unimportant in terms of glycogenolysis and in these cells ATP/AMP ratios change little in the first 24h of fasting. Glucagon, which is secreted when glucose levels are low, controls the activation of glycogenolysis and is a signal to the liver to supply the rest of the body with glucose.

Glucagon binds with a receptor on the liver cell membrane which activates glycogenolysis through the, cAMP cascade. It activates adenylate cyclase which increases cAMP. cAMP activates a protein kinase which catalyses the phosphorylation (activates) of phosphorylase kinase. This in turn catalyses the phosphorylation and activation of glycogen phosphorylase. Glycogen phosphorylase catalyses the breakdown of glycogen. This is exactly the same cascade as seen in skeletal and cardiac muscle, but these tissues are more sensitive to adrenaline than glucagon. Some other hormones, vasopressin, angiotensin and adrenaline also activate glycolysis. These cause an increase in cytoplasmic Ca^{2+}. This high level of Ca^{2+} allows easier phosphorylation of phosphorylase kinase.

It would be expected that during glycogenolysis, glycogen synthesis would be inhibited. What would be phosphorylated by protein kinase A and by phosphorylase kinase to cause its inactivation?

Answer: *Glycogen synthetase. Therefore glucagon inhibits glycogen synthetase activity.*

MAJOR POINT 79: Starvation is signalled by a rise in glucagon, adrenaline and corticosteroids and a fall in insulin.

Proteins and the liver

The liver is the most important organ in the body for protein and amino acid metabolism. In particular, it synthesises and degrades amino acids. There is a shuttle of amino acids from the liver to the blood stream to the body tissues and from the body tissues to the liver.

Under what conditions would there be an overall transfer of amino acids to the body tissues and under what conditions would the reverse would be true?

Answer: *Following eating, there would be a transfer of amino acids from the liver to the body tissues and when fasting, the reverse occurs with some amino acids being used to form glucose and others to form ketones.*

During fasting, amino acids are transferred to the liver where they are deaminated and have their carbon skeletons shuttled into the TCA cycle for subsequent gluconeogenesis or the formation of ketones. The ketones and glucose are then shipped off to the brain and muscles as energy sources. When there is an abundance of amino acids, growth hormone, thyroid hormone and insulin increase amino acid uptake. Amino acids can enter the TCA cycle at α-ketoglutarate and oxaloacetate (glucogenic) and at succinyl-CoA (ketogenic). Others are converted directly to pyruvate and then enter the cycle (glucogenic).

TCA cycle

The oxidation of acetyl-CoA through the TCA cycle in the liver is relatively low compared with muscle. Therefore, the main function of the TCA cycle in the liver is as a source of carbon skeletons for gluconeogenesis or amino acid synthesis. The part of the cycle most used for this is the part from α-ketoglutarate to oxaloacetate.

Abundance of α-ketoglutarate leads to an abundance of oxaloacetate. Oxaloacetate will either be converted to malate and transferred to the cytoplasm for gluconeogenesis as indicated above, or will be combined with acetyl-CoA using the enzyme citrate synthetase to form citrate. Excess citrate is transferred into the cytoplasm and inhibits phosphofructokinase.

Fatty acids could also be used in the TCA cycle providing that there is an abundance of oxaloacetate. Fatty acids are converted to acetyl-CoA by β-oxidation and the acetyl-CoA generated is combined with oxaloacetate to form citrate in a reaction catalysed by citrate synthetase. However, a large amount of NADH is generated during β-oxidation which increases the $NADH/NAD^+$ ratio (the liver consumes relatively little energy). This favours malate formation from oxaloacetate (reversing the normal cycle) rather than the conversion of malate to oxaloacetate. This means that the levels of oxaloacetate become relatively low. It also tends to cause fatty acids to be converted to ketones and amino acids to glucose, which are both fuel sources for the rest of the body.

Lipid metabolism

Once again, keep in mind that the purpose of the liver is to supply fuel to the rest of the body, and therefore a major purpose of fatty acid oxidation in the liver is to use the carbon skeletons as fuel, ketones, for other tissues in the body. It is not for the liver to obtain large amounts of energy, although it does gain some energy from this process.

Recall that after fatty acids enter the cell, they are esterified to CoA-SH and then to carnitine for transport across the mitochondrial membrane where they undergo β-oxidation. The transferase in the cytoplasm that catalyses the transfer of the acyl group from acyl-CoA to carnitine is inhibited by malonyl-CoA, which is an intermediate in fatty acid synthesis. Malonyl-CoA is formed during fatty acid synthesis by the carboxylation of acetyl-CoA in the cytoplasm. This reaction is catalysed by acetyl-CoA carboxylase. Therefore, as fatty acid synthesis is increased, malonyl-CoA is increased and inhibits β-oxidation.

The main control of fatty acid oxidation in the liver appears to be the availability of fatty acids in the bloodstream. This increases after eating fatty meals and also after fat cells have been stimulated to release fats. Adrenaline causes release of fatty acids from lipid stores during exercise, and glucagon causes their release during fasting or starvation. Insulin inhibits lipolysis and a fall in insulin when fasting increases lipolysis.

Abundant carbohydrate leads to fatty acid synthesis. The normal substrates are glucose, fructose and lactate. The conversion of glucose or fructose into fatty acids leads to a net increase in ATP. Since the liver does not require much ATP, fatty acid synthesis gives the liver most of its energy without the necessity of using the TCA cycle and oxidative phosphorylation. Lactate, or the carbon skeletons of some amino acids, can also be used for fatty acid synthesis.

Ketone body synthesis

Acetoacetate and β-hydroxybutyrate are synthesised in liver mitochondria and hydroxymethylglutaryl-CoA synthetase

(HMG-CoA synthetase) is thought to be the rate limiting enzyme in this process. During starvation, mitochondrial acetyl-CoA increases due to β-oxidation which stimulates HMG-CoA synthetase. Acetyl-CoA cannot pass across the mitochondrial membrane and hence must be changed into another form to enable the fuel supply to pass out of the mitochondria. Ketone bodies are formed and these can be transported across the mitochondrial and cell membranes into the bloodstream and hence become available to other tissues (cardiac muscle and brain) as fuel. A second advantage of ketone synthesis is that it restores the supply of CoA-SH in mitochondria. 2 acetyl-CoA → acetoacetate + 2CoA-SH. This allows fatty acid oxidation to continue and not be restricted by the supply of free CoA-SH in mitochondria which is limited.

During starvation, the heart and skeletal muscle have large capacity for ketone usage and during prolonged starvation, the brain adapts to obtain 70% of energy form ketone bodies. (Chapters 8 and 11.)

Control of metabolism in adipose tissue

The main purpose of fat cells is to store large quantities of fuel for use during starvation. Fats are ideal because they do not require water to store them (compare this with glycogen and glucose) which means it takes up little space. I know that this is hard to believe when most people see themselves naked, but it is true. Therefore, the main metabolic processes in fat cells are to convert glucose into fatty acids when there is excess, and to release fatty acids into the circulation during exercise or fasting. Mammals have 2 types of fats, white and brown. Brown fat occurs in human new-borns and animals which hibernate. Its main function is to generate heat. The cells are packed with mitochondria and they oxidise fat. The oxidation can be uncoupled from oxidative phosphorylation and hence the energy generated forms heat rather than ATP. White fat is used to store and release lipid and hence it is a fuel store.

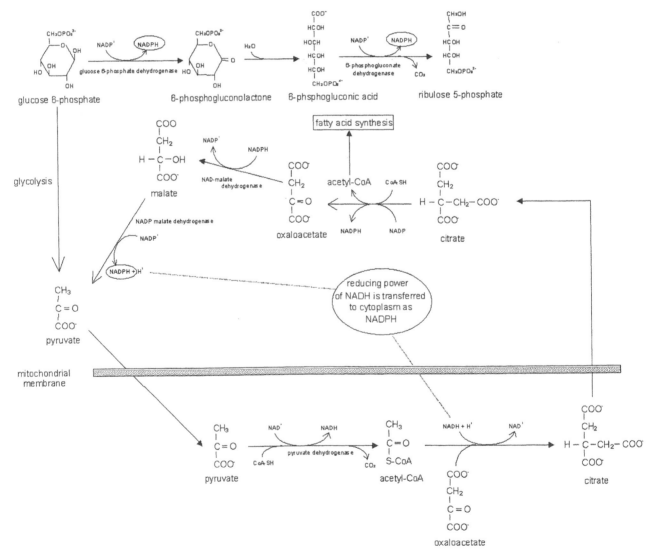

White fat synthesises fatty acids from glucose and takes up fatty acids from circulating lipoproteins and esterifies them with glycerol to form triacylglycerols. In humans, uptake and storage of fats from the diet is by far the main mechanism for fat storage and **little is synthesised from glucose**. Fasting or exercise leads to hydrolysis of triacylglycerols to glycerol and fatty acids which are released into the bloodstream.

Fatty acid biosynthesis

The main concern with fatty acid synthesis is to supply enough reducing power in the cytoplasm for this to occur. Therefore, the key to understanding this process is to know how the reducing power is formed in the cytoplasm.

Glycolysis in the fat cells converts glucose to pyruvate and then pyruvate dehydrogenase in the mitochondria converts pyruvate to acetyl-CoA. The acetate group of acetyl-CoA is transferred to the cytoplasm via the citrate shuttle and is converted to fatty acids. Other substrates for fatty acid synthesis are pyruvate from other sources such as amino acids, and acetate. Acetate is converted to acetyl-CoA in the cytoplasm by acetyl-CoA synthetase

Acetyl-CoA is activated by carboxylation to malonyl-CoA and this reaction is catalysed by acetyl-CoA carboxylase. Remember from earlier that CO_2 is essential for this part of fatty acid synthesis but not actually incorporated into the fatty acids (Chapter 11).

About half of the NADPH required for fatty acid synthesis comes from the citrate shuttle. In the cytoplasm, citrate is converted to oxaloacetate, to malate and then pyruvate which has the effect of transferring NADH in the mitochondria to NADPH in the cytoplasm (see diagram previous page). The other NADPH required for fatty acid synthesis comes from the pentose phosphate pathway. This is an alternative pathway to glycolysis for the oxidation of glucose. Glucose 6-phosphate dehydrogenase catalyses the dehydrogenation of glucose 6-phosphate to 6-phosphogluconolactone and hydration breaks the ester bond and 6-phosphogluconic acid is formed. This is then converted to ribulose 5-phosphate by a decarboxylation and a dehydrogenation. This reaction is catalysed by 6-phosphogluconate dehydrogenase. Ribulose 5-phosphate is converted back to the glycolytic pathway by a series of interconversions. At each of the steps for the formation of ribulose 5-phosphate from glucose 6-phosphate, NADP is converted to NADPH. This appears to give 3 molecules for synthesis of fatty acids. However, each molecule of glucose forms 2 molecules of acetyl-CoA and hence 2 molecules of NADPH from the pentose phosphate pathway and 2 molecules of NADPH from the transfer of acetyl-CoA from the mitochondria to the cytoplasm results in 4 molecules of NADPH for each glucose molecule incorporated into fatty acids or 2 molecules for each acetyl-CoA.

Nomenclature: A lactone forms when a sugar acid forms an ester bond with itself.

gluconic acid   gluconolactone

Conversion of glucose to fatty acids leads to a net production of ATP. Therefore, there is no requirement for the TCA cycle to operate to produce energy in fat cells. The activity of aconitase is also very low and hence the citrate formed is not converted to isocitrate.

Short-term regulation of fatty acid synthesis is activated by insulin, and inhibited by glucagon and adrenaline. Insulin stimulates glucose transport into the cells, similar to muscle cells. Fat cells contain significant amounts of glycogen, which is used as a substrate for fatty acid biosynthesis in the absence of extracellular glucose. Insulin activates glycogen breakdown in the absence of glucose. Pyruvate dehydrogenase is activated by insulin, but the mechanism for this is not known although Ca^{2+} concentrations have been proposed. The levels of AMP are always high enough so as not to inhibit phosphofructokinase and hence glycolysis is never rate limiting in fatty acid synthesis.

Acetyl-CoA carboxylase, which catalyses the conversion of acetyl-CoA to malonyl-CoA in the cytoplasm, is also regulated by insulin. Activation causes polymerisation of 10 or more inactive monomers. This process is inactivated by the presence of acyl-CoA (fatty acid oxidation). Conversely, adrenaline and glucagon, which promote fatty acid oxidation, would increase acyl-CoA and hence inhibit acetyl-CoA decarboxylase.

References and further reading

(1992) Special edition *Trends Biochem. Sci.* **17**:367-443.

Alborn, H.T., Turlings, T.C.J., Jones, T.H., Stenhagen, G., Loughrin, J.H. and Tumlinson, J.H. (1997) An elicitor of plant volatiles from beet armyworm oral secretion. *Science* **267**:945-949.

Alonso, M.D., Lomako, J., Lomako, W.M. and Whelan, W.J. (1995) A new look at the biogenesis of glycogen. *FASEB J.* **9**:1126-1137.

Bessman, S.P. and Capenter, C.L. (1985) The creatine-creatine phosphate energy shuttle. *Annu. Rev. Biochem.* **54**:831-862.

De Luca, L.M., Darwiche, N., Jones, C.S. and Scita, G. (1995) Retinoids in differentiation and neoplasia. *Sci. Amer. Science and Medicine* **2(4)**:28-37.

Derewenda, Z.S. (1994) Structure and function of lipases. *Adv. Prot. Chem.* **45**:1-52.

Felig, P. (1975) Amino acid metabolism in man. *Annu. Rev. Biochem.* **44**:933-955.

Gelb, M.H., Jain, M.K., Hanel, A.M. and Berg, O.G. (1995) Interfacial enzymology of glycerolipid hydrolases. Lessons from secreted phospholipase A2. *Annu. Rev. Biochem.* **64**:653-688.

Greenstein, B. and Greenstein, A. (1996) *Medical Biochemistry at a Glance.* Oxford, Blackwell Science.

Greenstein, B. (1994) *Endocrinology at a Glance.* Oxford: Blackwell Science.

Hayashi, H. Wada, H., Yoshimura, T., Esaki, N. and Soda, K. (1990) Recent topics in pyridoxal 5'-phosphate enzyme studies. *Annu. Rev. Biochem.* **59**:87-110.

Johnson, L.N. (1992) Glycogen phosphorylase:control by phosphorylation and allosteric effectors. *FASEB J.* **6**:2274-2282.

Lienhard, G.E., Slot, J.W., James, D.E. and Mueckler, M.M. (1992) How cells absorb glucose. *Sci. Amer.* **266(1)**:34-39.

Mattevi, A., Obmolova, G., Schulze, E., Kalk, K.H., Westphal, A.H., De Kok, A. and Hol, W.G.J. (1992) Atomic structure of the cubic core of pyruvate dehydrogenase. *Science* **255**:1544-1550.

McGarry, J.D. and Foster, D.W. (1980) Regulation of hepatic fatty acid oxidation and ketone body production. *Annu. Rev. Biochem.* **49**:395-420.

Newsholme, E.A., Challiss, R.A.J. and Crabtree, B. (1984) Substrate cycles: their role in improving sensitivity in metabolic control. *Trends Biochem. Sci.* **9**:277-280.

Nilsson-Ehle, P., Garfinkel, A.S. and Schotz, M.C. (1980) Lipolytic enzymes and plasma lipoprotein metabolism. *Annu. Rev. Biochem.* **49**:667-693.

Nilsson-Ehle, P., Garfinkel, A.S. and Schotz, M.C. (1980) Lipolytic enzymes and plasma lipoprotein metabolism. *Annu. Rev. Biochem.* **49**:667-693.

Packer, L. (1994) Vitamin E is nature's master antioxidant. *Sci. Amer. Science and Medicine* **1(1)**:54-63.

Pilkis, S.J., Claus, T.H., Kurland, I.J. and Lange, A.J. (1995) 6-Phosphofructo-2-kinase/fructose-2,6,bisphosphatase: A metabolic signalling enzyme. *Annu. Rev. Biochem.* **64**:799-835.

Pilkis, S.J., Elmaghrabi, M.I. and Claus, T.H. (1988) Hormonal regulation of hepatic gluconeogenesis. *Annu. Rev. Biochem.* **57**:755-784.

Roehrig, K.L. (1984) *Carbohydrate Biochemistry and Metabolism.* Westport, Conneticut: AVI Publishing Company.

Silverman, M. (1991) Structure and function of glucose transporters. *Annu. Rev. Biochem.* **60**:757-794.

Snell, K. (1979) Alanine as a gluconeogenic carrier. *Trends Biochem. Sci.* **4**:124-128.

Snyder, F. (1995) Platelet-activating factor and its analogs: Metabolic pathways and related intracellular processes. *Biochem. Biophys. Acta* **1254**:231-249.

Tall, A. (1995) Plasma lipid transfer proteins. *Annu. Rev. Biochem.* **64**:235-257.

Welch, W.J. (1993) How cells respond to stress. *Sci. Amer.* **268(5)**:34-41.

Willett, W.C. (1994) Diet and Health: What should we eat? *Science* **264**:532-537.

Q&A Answers

1   Like carbohydrates and fats, proteins contain carbon, oxygen and hydrogen. However, proteins also contain nitrogen and some sulphur.

2   If you weigh 80 kg, you require a mere 60 g of protein per day.

3   The vegetable that is a rich source of β-carotene is carrots

4   The amount of vitamin B needed is dependent upon energy consumption.
Amundsun (19079 kJ) Thiamine 0.1 x 19079/100 = 1.9 mg, riboflavin 0.15 x 19079/1000 = 2.86 mg, niacin 1.6 x 19079/1000 = 30.5 mg
Scott (18535 kJ) Thiamine 0.1 x 18535/1000 = 1.85 mg, riboflavin 0.15 x 18535/1000 =2.78 mg, niacin 1.6 x 18535/1000 = 29.66 mg
Therefore Scott's diet is deficient in all vitamin B groups and Amundsen's diet is actually about 15% deficient in niacin. However, the RDI for niacin, 1.1 mgNE/1000 kJ, includes a safety factor of 50%.

5

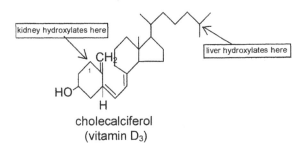

cholecalciferol
(vitamin D₃)

15 Cellular Communication

In this chapter you will learn about:

- the differences between hormones and neurotransmitters
- the main types of cellular receptors; ligand gated ion channels, transducer type receptors, and steroid receptors
- the importance of ion distribution across cell membranes and ion channels
- the main second messenger pathways in cells and how they are activated
- G proteins, kinases, calcium, cAMP, diacylglycerol, inositoltrisphosphate
- how prostaglandins are formed and some of their functions
- the location and mechanism of action of steroid receptors

Overview of intercellular messengers and receptors

With few exceptions, cells use chemicals such as hormones, neurotransmitters and cytokines to communicate with each other. This concept can be extended further to include communication between intracellular compartments, particularly the cytosol with the nucleus.

> **Q&A 1: What is: a hormone; neurotransmitter; and cytokine?**

Study tip: Questions like the definitions above are commonly set as exam questions and seem simple, but if you enter an exam without having practised answering definitions, then expect to do poorly. The problem is that when you are studying you think that you know these and therefore do not bother to practise them. In an exam, when you actually have to write the definition you struggle to find the right group of words that define, e.g. a hormone. Your thoughts will be something like this: 'I know this! It's a ummm it's a...you know, it's released and well it ummm..., I know but I just cannot explain it.' Not only have you wasted time, you have not answered a simple question, which in effect, you do know the answer to. This is a further disaster because you begin to lose confidence in your ability to answer other questions.

To solve this problem, make sure that you can answer such questions by practising them at home. Do not assume that you

know information when you are studying, but actually test yourself by writing or saying aloud the answer. Do not use a textbook definition, but put together your own group of words. Textbook definitions are useful until you have to regurgitate them in an exam and then you tend to get them a bit muddled because they do not use the style or vocabulary that you would normally use.

For the above questions, it is not enough to say that a hormone is a chemical messenger, which sends signals to a distant organ. This does not distinguish it from a neurotransmitter. It is also advisable to give an example in such questions because it can overcome some of the difficulties associated with the explanation and it indicates to the marker a greater depth of knowledge. Make sure the example is right!

Consider the hormone adrenaline being released from the adrenal medulla and its target is the heart. How does it get there?

Answer: *Via the bloodstream.*

This question is not as simplistic as it seems. For something to travel in the bloodstream easily it has to be water soluble. Indeed, most chemical messengers are water soluble. Assume the adrenaline has reached the heart and needs to get is message to the <u>inside</u> of a heart cell. The message must cross the cell membrane, but now there is a problem - how to get the message across the cell membrane. The cell membrane is primarily lipid and therefore adrenaline, which is water soluble, cannot cross the membrane. Instead, adrenaline binds to a protein, called a receptor, embedded in the cell membrane. This protein spans the cell membrane so that parts of it are on the outside of the cell, and parts on the inside. The parts that are on the outside have specific binding sites for adrenaline. When adrenaline binds to this protein, the protein changes its shape and the message is conveyed to the inside of the cell. Structurally, receptors have regions of non-charged amino acids, which are lipid soluble and embedded in the cell membrane, and regions of charged amino acids, which are water soluble and lie either outside the cell membrane or in the cell cytosol. This means that the receptor proteins span the cell membrane, or weave in and out of the membrane with some parts on the inside and other parts on the outside.

The general name for a substance that binds to a receptor is called a ligand. A particular feature of a receptor is that it is relatively specific for a particular ligand. This property is similar to enzymes, which are relatively specific for a particular substrate. For example, some receptors are called adrenergic receptors and these will bind adrenaline but will not bind the neurotransmitter acetylcholine. However, adrenergic receptors are not absolutely specific for adrenaline; they will also bind noradrenaline, but the amount of noradrenaline required to activate the receptor differs from that of adrenaline. This feature is very important because only cells with a receptor for a particular substance will respond to that substance. For example, this means that only cells with adrenergic receptors will respond to adrenaline.

In addition to the water-soluble chemical messengers, there are also some fat-soluble messengers. These are the steroid hormones: the sex hormones, the corticosteroids and the mineralocorticoids.

Would you expect the cell membrane to be a barrier to these hormones?

Answer: *No, the cell membrane is not a barrier to these hormones and they pass into the cytosol of their target cells and bind with proteins inside the cell and activate them.*

Receptors

Receptors are 1 of 3 general types. Some are ligand gated ion channels, others are transducer proteins, and the third type is the intracellular receptor, which is activated by steroid hormones as mentioned above.

Distribution of ions across a cell membrane

Before investigating the nature of ion channels and how they are activated, we have to examine the ion distribution in a cell and how this is controlled. In our bodies, Na^+, K^+, and Ca^{2+} are the major cations, and Cl^- is the major anion associated with cell function with respect to channels. These ions are distributed unevenly between the inside and the outside of a cell.

Nomenclature: An anion is an ion that moves towards the anode or positive electrode: therefore these are negatively charged ions.
A cation is an ion that moves towards the cathode or negative electrode: therefore these are positively charged ions.

Draw a cell and show the relative distribution of the major ions Na^+, K^+, Ca^{2+} and Cl^-, comparing the inside with the outside of the cell.

Answer: *The major cation inside a cell is K^+, and the major cation outside a cell is Na^+. Calcium has an interesting distribution. It is for all intents and purposes in relatively high concentrations outside the cell. The relatively high concentration of Ca^{2+} within the endoplasmic reticulum (or sarcoplasmic reticulum of muscle) is packaged away from the inside of the cell and hence is really outside the cell. This is analogous to food in our intestines, which is really outside of our bodies. You only have to vomit to realise this is true, but I do not*

recommend it, despite the nature of this topic. The major anion, Cl^-, is distributed relatively evenly inside and outside the cell, but there is slightly more on the outside.

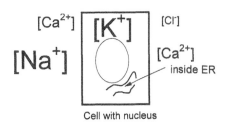

Cell with nucleus

When a person dies, the ions slowly redistribute to be even across the cell membrane. Therefore there is something keeping them separated when you are alive. This something is the Na^+K^+-ATPase pump, which is a transporter located in the cell membrane that uses the energy of ATP to drive Na^+ out of the cell and K^+ into the cell. For Ca^{2+} things are a little more complicated. Across the cell membrane, Ca^{2+} on the inside of the cell is exchanged for Na^+ (Na^+/Ca^{2+} exchanger). The Na^+ is then pumped back out of the cell via the Na^+K^+-ATPase pump. The endoplasmic reticulum has a Ca^{2+} uptake protein, which transports Ca^{2+} from the cytosol back into the endoplasmic reticulum. With this uneven distribution of ions, the cell has a potential difference across the cell membrane. The potential difference across a cell membrane, comparing the inside of a cell with the outside, is between -70mV and -40mV depending upon the cell type.

MAJOR POINT 80: The potential difference across a cell membrane is such that the inside of a cell is negative compared with the outside of a cell.

If we opened a channel in the cell membrane that was selective for Na^+, then Na^+ would flow through it. Channels are not directional, that is, when they open it, ions can travel in either direction, but the net direction of flow depends upon the concentration of ions and the membrane potential of the cell at the time.

Given that Na^+ is in high concentrations on the outside of a cell and that the inside of a cell is relatively negative, what would happen to the membrane potential if a Na^+ selective channel were opened? Explain your answer.

Answer: *Clearly, Na^+ would move into the cell down its concentration gradient and the negative charge inside the cell would also draw Na^+ into the cell. As a result the membrane potential of the cell would become less negative as positive ions (Na^+) were added to the inside of the cell. This would cause the potential difference across the cell membrane to move towards 0mV and hence the cell would become less polar or <u>depolarise</u>.*

This change in membrane potential is a signal for the cell to do something different. More difficult is to determine what would happen to the membrane potential if a K^+ channel were opened. On one hand the high concentration of K^+ inside the cell would tend to drive K^+ out of the cell, but the negative charge on the inside of the cell would tend to hold K^+ inside the cell. The net effect of these opposing forces means that K^+ actually moves out of the cell making the cell more negatively charged or more polarised. This is called hyperpolarisation of the cell. If the membrane potential of the cell were about -90mV instead of -40mV then the negative charge tending to hold K^+ in the cell would balance the force caused by the concentration difference trying to drive K^+ out of the cell. Under these conditions, when a K^+ channel is opened, the net movement of K^+ would be 0. This is actually called the reversal potential for K^+ because if the membrane potential were less than -90mV then K^+ would move into the cell instead of out of the cell.

> **Q&A 2:** When a Cl^- channel opens Cl^- moves into the cell. What would happen to the membrane potential of the cell?

The signal for a cell to do something is a depolarisation. In particular, there are voltage-gated Ca^{2+} channels which open when a cell reaches a certain potential, e.g. it could be -10mV. The entry of Ca^{2+} into a cell is very important because it is the entry of Ca^{2+} that causes the cell to do something (see below). A hyperpolarisation, such as when a K^+ channel opens, quietens a cell down making it necessary for a cell to receive more depolarising signals before it will do something.

Ligand-gated ion channel receptors

Ligand-gated ion channels are one class of receptor. They are special proteins in the cell membrane. These proteins have quaternary structure. That is, they are made up of several protein subunits, some of which may be the same and others are different. One or more of these subunits has a binding site for a ligand. When a ligand binds to the subunit, it leads to changes in the shape and arrangement of the other subunits (allosteric or distant interactions), which causes a hole to form in the membrane. This hole is actually a channel, and lets particular ions flow across the membrane.

Any one neuron in the brain has a mixture of ligand-gated ion channels. Some of these depolarise the cell and others hyperpolarise the cell when opened. Therefore, at any one time, a neuron is receiving signals to keep it quiet (hyperpolarising signals) and other signals to activate it (depolarising signals). The balance of all of these inputs determines the final response of the neuron. Nearly all neurones in the brain have GABA-ergic receptors. These are ligand-gated ion channels, which are sensitive to GABA (see Chapter 4). When activated, they open a Cl^- channel which hyperpolarises the cell and makes it difficult to depolarise or activate the cell. Therefore, GABA keeps the brain quiet. The anxiolytic medications known as benzodiazapines (e.g. Valium) work by making it easier for the Cl^- channels to be opened by GABA, and therefore the neurones become hyperpolarised and much harder to depolarise. The brain (and the person) becomes quieter and less anxious. A common excitatory amino acid is the amino acid glutamate. It tends to open non-specific cation channels. These let all cations through, but in a physiological situation Na^+ is the ion which mainly passes through these channels and it causes depolarisation of neurones. Some of these glutamate sensitive channels are associated with both short and long term memory. As you repeat something over and over again, the neurones become more and more depolarised until a Ca^{2+} channel opens. The Ca^{2+} binds to proteins inside the neuron. The additional positive charge on these proteins causes the proteins to change shape and you learn whatever it was you were trying to learn.

Higher level

Ion channels are opening and closing all of the time, not just when a ligand binds to the receptor. When a ligand binds to the receptor, it increases the probability of the ion channel being open rather than closed. Therefore, when dealing with ion channels, we talk about the probability of the ion channel being open. This can refer to the frequency with which it opens or the length of time for which the channel is open. Both increase the conductance through the channel.

The ligand-gated ion channel on skeletal muscle is known as a nicotinic receptor. It is opened when acetylcholine released from a nerve terminal innervating the muscle binds to the receptor. It opens a non-specific cation channel that, as mentioned before, is capable of letting any cation through, but under physiological conditions, Na^+ is the major ion that is conducted through the channel. It therefore causes the muscle to depolarise, which in turn, activates a voltage-sensitive Ca^{2+} channel located in the sarcoplasmic reticulum. The Ca^{2+} stored in the sarcoplasmic reticulum is released into the cytoplasm of the cell where it binds to proteins involved with contraction. These proteins are myosin and the tropomysin-troponin complex. The change in the shape of the tropomysin-troponin complex when it binds Ca^{2+} causes it to expose binding sites for myosin on the actin filaments. Actin and myosin can then interact to cause contraction of the muscle.

In this case, there are 2 messages that tell the cell to contract. The first message is acetylcholine, which binds to the receptor in the cell membrane, and the second message is Ca^{2+}, which is released into the cytoplasm where it binds with specific proteins that directly lead to contraction. The first message acts on the outside of the cell and its signal activates the second message, which acts inside the cell. Therefore, Ca^{2+} is known as a second messenger.

In other cell types, release of Ca^{2+} into the cytoplasm activates different calcium binding proteins causing these cells to do something, e.g. it may lead to release of a vesicle.

> **MAJOR POINT 81:** Entry of Ca^{2+} into the cytoplasm of a cell binds to calcium binding proteins, such as calmodulin, causing the protein to have a massive positive charge where it previously did not. This causes the protein to change its shape and do something different.

Transducer-type receptors

Binding of a ligand to these receptors causes a conformational change in their structure that leads to activation of enzyme activity inside the cell. This is possible because these receptors span the cell membrane.

Receptors that bind to G proteins

The full name for a G protein is a GTP-binding transducer protein. The receptor is coupled to a G protein, and ligand binding to the receptor causes an exchange of GTP for GDP on the α subunit (G_α) of the G protein trimer, followed by subunit dissociation. It dissociates into a G_α subunit and a $G_{\beta\gamma}$ subunit. The G_α then modulates the activity of an effector molecule such as adenylate cyclase, phosphodiesterase, phospholipase C or an ion channel. The activity of the effector molecule is terminated when GTP is hydrolysed to GDP and inorganic phosphate is released. The subunits of the G protein re-associate so that the reaction may begin again.

In the case illustrated on the right, the βγ subunit also has a role. A receptor kinase binds to it and is activated. This then phosphorylates the receptor which causes the receptor to change shape and the ligand dissociates from the receptor and the receptor is uncoupled form the G protein. This provides a form of tolerance and the receptor cannot be activated again until it is dephosphorylated.

> **MAJOR POINT 82:** Transducer-type receptors are used to amplify the initial signal.

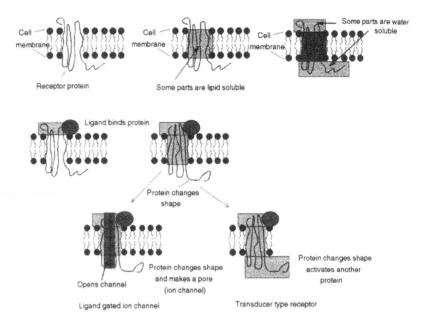

The most important concept about transducer type receptors is that they amplify the initial signal. For those coupled to G proteins, once it has activated a G protein, it is then available to activate a second G protein and then a third and so on. This continues until the ligand is dissociated from the receptor. The enzyme activated by the G_α subunit will continue to be active until the GTP is hydrolysed to GDP. During this time it will produce a large amount of substrate.

To illustrate this, if an activated receptor activates 10 G proteins before the ligand is dissociated from the receptor, and each G protein activates a adenylate cyclase, and each adenylate cyclase converts 10 ATPs to 10 cAMPs before the G_a subunit is inactivated, what would be the amplification?

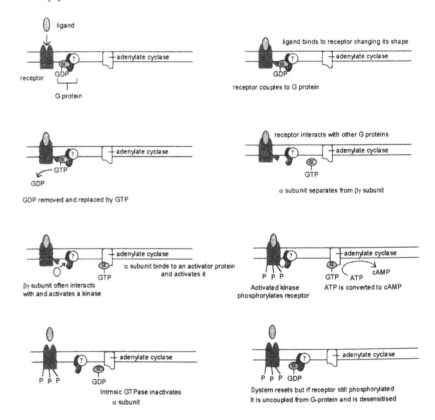

Answer: *Clearly, the initial signal is amplified 100 times - 1 receptor activated leading to 10 activated G proteins and each of these is responsible (indirectly) for the production of 10 molecules of cAMP. Therefore, this system is exquisitely sensitive with a large response occurring for a minimal stimulus.*

The term G protein refers to a family of proteins, not just one, and different cell types will have different G proteins. There are at least 10 G proteins, e.g. G_t (transducin) activates phosphodiesterase; G_o closes calcium channels and perhaps opens potassium channels; G_i controls potassium channels and may regulate adenylate cyclase activity; G_s also modulates adenylate cyclase activity.

Proteins activated or inhibited by G proteins

Keep in mind that, in some specific cases, instead of activating an enzyme, some of the G proteins actually inhibit an enzyme. Some main enzymes activated by G proteins are adenylate cyclase, phosphodiesterase, phospholipase C and phospholipase A. Some are coupled to Ca^{2+} channels. The effects of activating these proteins on further processes in the cell will be discussed in the next section.

Adenylate cyclase and phosphodiesterase

Adenylate cyclase catalyses the conversion of ATP to cAMP (Chapter 3) and therefore its activation leads to an increase in cAMP. cAMP is a second messenger because several enzymes in cells are activated by cAMP. When cAMP binds to these enzymes, it puts a negative charge (the phosphate group) on the protein causing the enzyme to change shape and hence its activity. One particular enzyme activated by cAMP is protein kinase A. Protein kinase A catalyses the serine phosphorylation of particular proteins; that is, it catalyses the attachment of phosphate via an ester bond to the hydroxyl group (-OH) of serine. This places a massive negative charge in the protein where there was a small positive charge ($\delta+$) previously. This change causes the protein to change its shape and do something it was not doing before. Therefore, activation of the receptor leads to an increase in cAMP which activates protein kinase A and the phosphorylation of particular proteins. Note that which proteins are finally phosphorylated will depend upon the cell type. A very important aspect of this is amplification. Protein kinase A can phosphorylate many proteins before its activity is switched off.

Phosphodiesterase switches off the pathway adenylate cyclase switches on. It catalyses the breakdown (hydrolysis) of cAMP to AMP. Given the information above, this would decrease the levels of the second messenger cAMP, which would decrease the activity of protein kinase A and in turn inhibit the phosphorylation of proteins. Once again these proteins would have a different shape and therefore a different function.

One of the mechanisms of action of methylxanthines (the active ingredient in chocolate) and the closely related caffeine is to inhibit phosphodiesterase activity. This then increases the levels of cAMP in a cell and activates it. This is why drinks of cocoa and coffee are stimulants.

Phospholipase A and Phospholipase C

Some of the phospholipids in the cell membrane are phosphatidylinositols. Phospholipase C hydrolyses phosphatidylinositol, releasing inositol 1,4,5-trisphosphate and diacylglycerol (see chapter 11).

> **Q&A 3:** Draw this reaction, illustrating the structures of phosphatidylinositol, inositol 1,4,5-trisphosphate and diacylglycerol, and clearly indicate where the cleavage occurs.

Inositol 1,4,5-trisphosphate is released into the cytosol. It is a second messenger. The endoplasmic reticulum has receptors for inositol 1,4,5-trisphosphate and when these receptors are activated, they open Ca^{2+} channels in the endoplasmic reticulum which causes Ca^{2+} to be released into the cytosol. This can then bind to calmodulin which can then activate other proteins. Among proteins which are calmodulin dependent is a multiprotein kinase called CaM kinase. In other cells, the Ca^{2+} released binds to protein kinase C which catalyses the serine phosphorylation of specific proteins.

> **Nomenclature:** Proteins which are calmodulin dependent often have an abbreviated name which includes the 'Ca' signature, such as CaM kinase.

The diacylglycerol, which is the other product formed from the hydrolysis of phosphatidylinositol, is also active (another second messenger) and increases the activity of protein kinase C. Therefore, activation of phospholipase C leads to the activation of the second messengers inositol 1,4,5-trisphosphate, Ca^{2+} and diacylglycerol (commonly abbreviated to DAG).

Higher level

Muscarinic receptors are receptors for acetylcholine and they receive innervation from the parasympathetic nervous system. These receptors are not all the same and the different receptors have different numbers, *m1, m2, m3, m4, m5*. This means that stimulation of an organ with acetylcholine, which activates a muscarinic receptor, will give quite different responses depending upon the organ and second messenger system activated. The receptors *m1, m3* and *m5* are associated with stimulation of phospholipase C, and *m2* and *m4* are associated with inhibition of adenylate cyclase. All of the receptors except *m5* modulate K^+ channels.

Activation of phospholipase A leads to hydrolytic cleavage of fatty acids from the glycerol backbone of phospholipids in the cell membrane (see chapter 11). A particular phospholipase A, phospholipase A_2, cleaves fatty acids from C2 of the glycerol backbone.

> **Q&A 4:** Given the following phospholipid, show the hydrolytic cleavage which would be catalysed by phospholipase A_2.

Quite commonly, the fatty acid at C2 is arachidonic acid (20C and 4 double bonds) and is released into the cytoplasm by the action of phospholipase A_2. It is then cyclysed to form the basis of eicosanoids (bioactive 20C structures) among which are the prostaglandins. The eicosonoids are involved with the inflammatory response, the production of pain and fever, the regulation of blood pressure, the induction of platelet clotting, the sleep wake cycle, and the induction of labour. Their actions are local and they are relatively unstable.

The enzyme which catalyses the cyclysation of arachidonic acid is prostaglandin endoperoxidase. It has 2 activities: a cyclooxygenase which catalyses the addition of 2 molecules of O_2 and cyclysation, and the second which leads to the formation of the hydroxyl group. The PGH_2 formed, is the precursor of many other active derivatives.

> **Nomenclature:** In naming prostaglandins, the PG stands for prostaglandin, the next letter (E,F,etc) is for a specific prostaglandin reflecting changes around the ring, and the subscript is the number of double bonds in the molecule, usually 2. The α or β reflects the whether particular groups are below or above the ring, e.g. $PGF_{2\alpha}$.

In platelets, PGH_2 is converted to thromboxanes which cause the platelets to become sticky and promote coagulation, and vasoconstriction. Endothelial cells make PGI_2 which has the opposite effect. It dilates blood vessels and prevents platelets from aggregating.

> **Q&A 5:** Non-steroidal anti-inflammatories (NSAID), which are typified by aspirin, inhibit the cyclooxygenase activity of prostaglandin endoperoxidase. How might this lead to the anticoagulant property of NSAIDs in some cases and to the anti-inflammatory effects in other cases?

Steroids inhibit the synthesis of phospholipase A, and hence there is less arachidonic acid cleaved and therefore less inflammatory agents such as the prostaglandins and leukotrienes (modified linear forms of arachidonic acid).

Fish oils, which are taken as anticoagulants and anti-inflammatories, have fatty acids that are similar to arachidonic acid, but have 5 double bonds instead of 4. This means that the leukotrienes and thromboxanes formed have an extra double bond and are biologically less active.

Tyrosine kinase receptors

A closely related group of receptors have enzyme activity themselves. Within this group are the insulin receptor and various growth factor receptors such as epidermal growth

Prostaglandins

factor (EGF), platelet-derived growth factor (PDGF), colony-stimulating factor 1 (CSF-1), and fibroblast growth factor (FGF). The receptor protein has a binding region for the ligand on the outside of the cell, a single membrane spanning region and intracellular region, which has enzymatic activity. These receptors do not have second messengers associated with them.

There are 3 main subclasses which have common characteristics.

TYROSINE KINASE RECEPTORS

Epidermal growth factor receptor is a typical class I receptor, insulin receptor typical of class II and platelet derived growth factor receptor class III. The class II receptors are made from several subunits, e.g. the insulin receptor is a tetramer made from 2 α and 2 β subunits, which are stabilised by disulphide bonds. The α chain is the binding site for insulin at its C terminus, and the β chain spans the membrane and carries the tyrosine kinase activity on its intracellular C terminus.

One common feature of all of these receptors appears to be autophosphorylation. It is now believed that binding of the ligand to the receptor causes clustering of the 2 or more receptors and the tyrosine kinase activity of one phosphorylates the other receptor and vice versa – a bit like mutual masturbation really. These receptors do not function without autophosphorylation. A very good example of this is some forms of insulin-independent diabetes in which the autophosphorylation of the insulin receptor does not occur.

The tyrosine kinase activity causes the phosphorylation of tyrosine in particular proteins (depending upon the cell).

This places a large negative charge onto the phenol group of tyrosine changing it from a hydrophobic group to a hydrophilic group. This changes the shape of the protein and its activity.

Steroids and their receptors

If you have been through puberty or a sex change, you would be aware that steroids affect the growth of cells. Therefore, it would be expected that they affect the nucleus of a cell. This is the case. However, steroids have to firstly get from their site of production to their site of action, which is difficult because they are lipid soluble. To overcome this problem they are transported in the bloodstream by a specific carrier protein. Once they reach their target organ, being lipid soluble they simply pass through the membrane of the target cell by diffusion. They then bind to specific receptor proteins in the nucleus or the cytoplasm. This hormone receptor complex then binds to regions of special DNA sequences called hormone response elements and alter gene expression. Depending upon the cell, the hormone receptor can either enhance or inhibit expression of genes. The glucocorticoid receptors are found in the cytoplasm, oestrogen and progesterone receptors are predominantly nuclear and thyroid receptors is exclusively nuclear. It appears that the receptors are protected from activity by another protein in close association. This protein is a heat shock protein, e.g. hsp90. When the steroid binds to the receptor, 2 receptor steroid complexes form a dimer which relieves them from protection by the heat shock protein and also causes them to form the right conformation to interact with the DNA. Thyroid hormones exert their action in a similar manner.

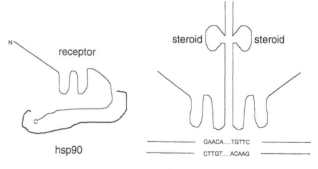

hormone response element

References and further reading

Berridge, M.J. (1985) The molecular basis of communication within the cell. *Sci. Amer.* **253**(4):142-152.

Bootman, M.D. and Berridge, M.J. (1995) The elemental principles of calcium signalling. *Cell* **83**:675-678.

Bray, D. (1992) *Cell Movements*. Garland publishing.

Catterall, W.A. (1995) Structure and function of Na^+, Ca^{2+}, and K^+ channels. *Annu. Rev. Biochem.* **64**:493-531.

Chao, W. and Olson, M.S. (1993) Platelet activating factor: Receptors and signal transduction. *Biochem. J.* **292**:617-629.

Chinkers, M., and Garbers, D.L. (1991) Signal transduction by guanylyl cyclases. *Annu. Rev. Biochem.* **60**:553-575.

Cohen, P. (1982) The role of protein phosphorylation in neural and hormonal control of cellular activity. *Nature* **296**:613-620.

Cooper, D.M.F., Mons, N. and Karpen, J.W. (1995) Adenylyl cyclases and the interaction between calcium and cAMP signalling. *Nature* **374**:421-424.

Crivici, A and Ikura, M. Molecular and structural basis of target recognition by calmodulin *Annu. Rev. Biophys. Biomol. Struct* 1995 24:85-116

Hanson, P.I. and Schulman, H. (1992) Neuronal Ca^{2+}/calmodulin-dependent protein kinases. *Annu. Rev. Biochem.* **61**:559-601

Klarlund, J.K., Guilherme, A., Holik, J.J., Virbasius, J.V., Chawla, A. and Czech, M.P. (1997) Signalling by phosphotide-3,4,5-trisphosphate through proteins containing pleckstrin and Sec7 homology domains. *Science* **275**:1927-1930.

Linder, M.E. and Gilman, A.G. (1992) G proteins. *Sci. Amer.* **267**(1):36-43.

Linder, M.E. and Gilman, A.G. (1992) G Proteins. *Sci. Amer.* **267**(2):56-65.

Luna, E.J. and Hitt, A.L. (1992) Cytoskeletal-plasma membrane interactions. *Science* **258**:955-964.

Needleman, P., Turk, J., Jakschik, B.A., Morrison, A.R. and Lefkowith, J.B. (1986) Arachidonic acid metabolism. *Annu. Rev. Biochem.* **55**:69-102.

Shimiy, T. and Wolfe, L.S. (1990) Arachidonic acid cascade and signal transduction. *J. Neurochem* **55**:1-15.

Simon, M.I., Stratmann, M.P. and Gautam, N. (1991) Diversity of G proteins in signal transduction. *Science* **252**:802-808.

Swope, S.L., Moss, S.J., Blackstone, C.D. and Huganir, R.L. (1992) Phosphorylation of ligand gated ion channels: a possible mode of synaptic plasticity. *FASEB J.* **6**:2514-2523.

Yuen, PST and Garbers, D.L. (1992) Guanylyl cyclase-linked receptors *Annu. Rev. Neurosc.i* **15**: 193-225.

Q&A Answers

1. A hormone is a chemical messenger released by a gland (endocrine) into the <u>bloodstream</u> where it is transported to other distant organs where it exerts a physiological effect, e.g. insulin.

   A neurotransmitter is a chemical messenger released by a neuron from an axon terminal at a synapse. It acts locally in this synaptic region, e.g. acetylcholine.

   A cytokine is a chemical messenger which is released by cells, generally in response to an immune insult. These chemicals have relatively local effects, e.g. the interleukins. Although nearly all cell types may produce cytokines, it is one of the most important roles of cells of the immune system.

2. When a Cl⁻ channel is opened in the cell membrane, Cl⁻ moves into the cell making the cell more negatively charged or hyperpolarised.

3,4 See next page for answers to these questions.

5. Non-steroidal anti-inflammatories, e.g. aspirin, inhibit the cyclisation of arachidonic acid, which is a crucial step in the synthesis of prostaglandins and thromboxanes. Inhibition of thromboxane synthesis in platelets prevents them from becoming sticky and they do not clot, hence the anticoagulant activity of NSAIDS. Prostaglandins are produced by many cells, but particularly those of the immune system. These are mediators of the inflammatory response and so their inhibition decreases inflammation.

3

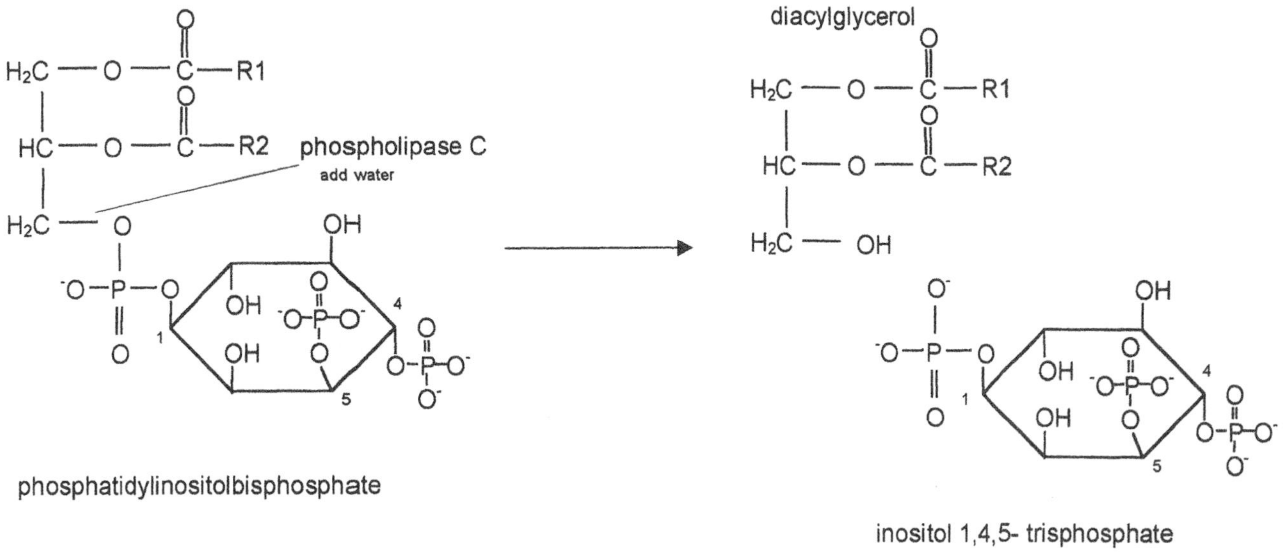

phospholipase C
add water

diacylglycerol

phosphatidylinositolbisphosphate

inositol 1,4,5- trisphosphate

4

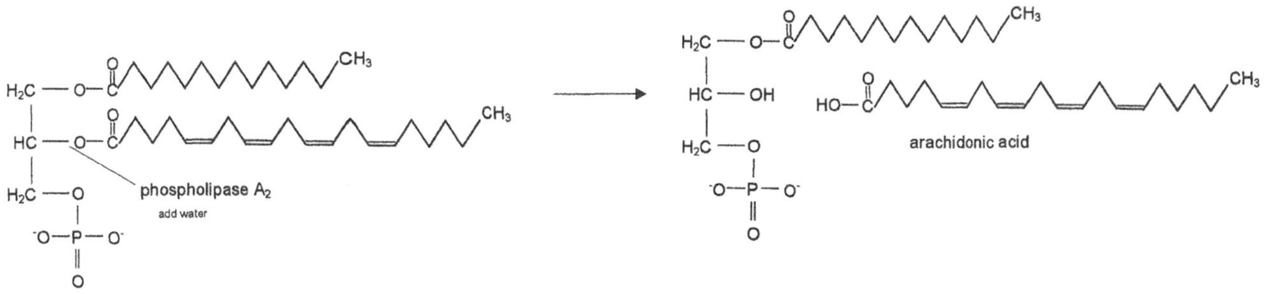

phospholipase A₂
add water

arachidonic acid

Index

Printed and bound by CPI Group (UK) Ltd, Croydon, CR0 4YY

01/11/2024

01782610-0017